OPERATOR TECHNIQUES IN ATOMIC SPECTROSCOPY

PRINCETON LANDMARKS
IN MATHEMATICS AND PHYSICS

Non-standard Analysis,
by Abraham Robinson

General Theory of Relativity,
by P.A.M. Dirac

Angular Momentum in Quantum Mechanics,
by A. R. Edmonds

Mathematical Foundations of Quantum Mechanics,
by John von Neumann

Introduction to Mathematical Logic,
by Alonzo Church

Convex Analysis,
by R. Tyrrell Rockafellar

Riemannian Geometry,
by Luther Pfahler Eisenhart

The Classical Groups,
by Hermann Weyl

Topology from the Differentiable Viewpoint,
by John W. Milnor

Algebraic Theory of Numbers,
by Hermann Weyl

Continuous Geometry,
by John von Neumann

Linear Programming and Extensions,
by George B. Dantzig

Algebraic Theory of Numbers,
by Hermann Weyl

Operator Techniques in Atomic Spectroscopy,
by Brian R. Judd

OPERATOR TECHNIQUES IN ATOMIC SPECTROSCOPY

BRIAN R. JUDD

With a new preface by the author

PRINCETON UNIVERSITY PRESS
PRINCETON, NEW JERSEY

Published by Princeton University Press, 41 William Street,
Princeton, New Jersey 08540
In the United Kingdom: Princeton University Press,
Chichester, West Sussex

First published in 1963 by McGraw-Hill Book Company, Inc.
Reprinted by arrangement with the author

Library of Congress Cataloging-in-Publication Data

Judd, Brian R.
Operator techniques in atomic spectroscopy / Brian R. Judd.
p. cm.—(Princeton landmarks in mathematics and physics)
Originally published: New York : McGraw-Hill, 1963.
Includes bibliographical references and index.
ISBN 0-691-05901-2 (pbk. : alk. paper)
1. Atomic spectroscopy—Mathematics. 2. Calculus, Operational.
3. Continuous groups. I. Title. II. Series.
QC454.A8J82 1998
539.7'028'7—dc21 98-9450

Princeton University Press books are printed on acid-free paper and meet the guidelines for permanence and durability of the Committee on Production Guidelines for Book Longevity of the Council on Library Resources

First printing, in the Princeton Landmarks in Mathematics and Physics series, 1998

http://pup.princeton.edu

Printed in the United States of America

1 2 3 4 5 6 7 8 9 10

CONTENTS

PREFACE TO THE PAPERBACK EDITION

The original publication of this book took place at a time when rare-earth elements were becoming more readily available and when computing methods were beginning to make it possible to study *f*-electron configurations in detail. The acceleration of these developments has enabled relatively small effects, such as those coming from the magnetic interactions between the *f* electrons, to be examined. We have now reached a point where large matrices can be diagonalized without too much trouble, and the effects of configuration interaction in the analyses of rare-earth and actinide spectra are now routinely represented by multi-electron operators. However, the basic theory of angular momentum, as developed by Racah and others, remains intact, and it is probably still best exemplified for the *f* shell by the topics that were originally chosen for that purpose.

The rotation group in three dimensions lies at the heart of angular-momentum theory. In describing its generalizations, the text can still play a useful role. Modern computing techniques easily cope with multi-electron operators, but it is advantageous to define such operators in terms of the irreducible representations of suitably chosen Lie groups. This aids in the construction of orthogonal operators whose strengths, as measured by a set of adjustable parameters, are minimally correlated. The matrix elements of such operators can be evaluated by the generalized form of the Wigner-Eckart theorem. It is frequently necessary to calculate the coupling coefficients for the Lie groups, and this calls for a familiarity with the topics covered in the later chapters of this book. Particle physicists, whose interests lie in groups other than the ones introduced for the *f* shell, may well find something to interest them here.

Of course no subject stands still. Diagrammatic methods now go far beyond the merely picturesque representation of coupled angular momenta. An advance in another direction stems from second quantization, in which operators acting on n electrons at a time are represented by coupled products of n creation and n annihilation operators. A development of these ideas leads to the notion of quasispin, which enables all the properties of the seniority number v to be expressed in the language of angular-momentum. However, such elaborations of the theory, no matter how elegant, might well be considered unnecessary distractions by the many chemists who, perhaps surprisingly, have formed a major component of the readership. Their interest in efficient methods of understanding spectra often take precedence over mathematical style.

Some readers appear to have found the problems at the ends of the chapters difficult. My intent was to indicate the range of the theory to a casual reader as well as to provide an obstacle course for the serious student. Some of the problems hint at the directions taken in the last thirty-five years in forming what is now a vast literature on the theory of rare-earth and actinide spectroscopy. The book was written at a time when only a few of the excited states of the rare-earth and actinide ions had been identified, but at least the direction to take in future research seemed clear. For the reprint, my thanks go to Princeton University Press for their interest and for the opportunity to correct some typographical and other minor errors. Some were brought to my attention by penciled corrections in library copies. However, I can assure all those who have put modulus signs around some of the symbols m in equations (1-3) and (1-4) that those equations are correct as they stand. Finally, I thank my many colleagues and friends who have made my life with the f shell such a happy one.

Brian R. Judd

PREFACE

Between 1942 and 1949, Racah published four papers in the *Physical Review* under the general title *Theory of Complex Spectra*. Before the appearance of these papers, the standard method for finding expressions for the energies of the terms of a complex atom was to follow a set of rules formulated in 1929 by Slater. In attempting to improve on this prescriptive approach, Racah introduced a number of concepts that have since proved to have widespread applications. In the second of his four papers, he defined a tensor operator; in the third, he developed the relation between a spectroscopic term and its parents and defined the coefficients of fractional parentage; in the fourth, perhaps the most remarkable of the series, he applied the theory of continuous groups to the problem of finding the term energies of configurations of equivalent f electrons. Although immediately relevant to atomic spectroscopy, these papers made a greater impact on nuclear physics. It is in this field that further applications of the theory of continuous groups have been made, principally by Jahn, Flowers, and Elliott. The comparative indifference of atomic spectroscopists is not difficult to understand, however. Most of the theoretical problems connected with their work could be handled by the classical techniques described by Condon and Shortley in their book *The Theory of Atomic Spectra*, a volume seemingly insusceptible of improvement. In addition, the frequent occurrence in the new algebra of what are now known as 6-j symbols, often tedious to evaluate, served as an effective barrier to all but the most determined experimentalists. The use of continuous groups, viewed from behind the added hurdle of fractional parentage, appeared remote indeed.

Two factors are changing this attitude. In the first place, many 6-j symbols have now been tabulated; calculations in which tensor operators are used are no longer irksome, and the advantage of this method over the conventional approach becomes obvious. Second, the increasing availability of elements in the rare-earth and actinide series has stimulated research into the microwave and optical spectra of not only the free atoms but also the ions situated at lattice sites in crystals. A rare-earth or actinide atom, whether neutral or ionized, very often possesses several electrons outside closed shells, a characteristic feature being a configuration including a number of f electrons. For electrons with such high angular momenta, the methods of Condon and Shortley are often extremely involved and compare very unfavorably with the powerful and elegant techniques of Racah.

Several elementary texts on the theory of tensor operators are now available. However, the theory of continuous groups is not readily accessible in a convenient form. In attempting to remedy this deficiency, I have taken the opportunity to review the theory of tensor operators with particular emphasis on those aspects of greatest interest to atomic spectroscopists. Throughout, I have had in mind the experimentalist who wishes to grasp the principles of the theory and to learn how to apply them to physical problems, rather than the mathematician, whose interests seem to be well catered for by such writers as Lie, Cartan, and Weyl. I have included a large number of problems both in the text and at the end of the chapters to emphasize the physical interpretation of the mathematics. Most problems are drawn from configurations of the type f^n, where Racah's techniques show to their best advantage; needless to say, I have not attempted anything approaching a review of the optical and microwave spectra of the rare-earth or actinide atoms.

This book has evolved from a course of lectures given at Berkeley in 1960. The interest of the audience in the applications of the theory and the encouragement of Professor W. A. Nierenberg were greatly appreciated. My obligation to the pioneering work of Professor Racah will be apparent in the pages that follow; in particular, I am indebted to him and to the editors of the *Physical Review* for permission to include a number of tables that originally appeared in that journal. I wish to thank Professor G. F. Koster, Dr. F. R. Innes, and Dr. B. G. Wybourne for reading the manuscript and suggesting a number of improvements. My own interest in applying the theory of continuous groups to problems in atomic spectroscopy was aroused by lectures given by Dr. J. P. Elliott at Oxford in 1956, and to him a special word of thanks is due.

Brian R. Judd

1

CLASSICAL METHODS

1-1 INTRODUCTION

All but the lightest atoms are dynamic systems of great complexity. To analyze the properties of such systems, certain assumptions and approximations must first be made. The immediate aim is to simplify the mathematics, but it is important to be guided by physical considerations; for the purpose of the analysis is not just to account for the properties of a particular atom as closely as possible but also to gain insight into its structure and to discern features that are shared by other atomic systems. After the assumptions and approximations have been made, their implications must be rigorously worked out. The conventional methods for carrying out these two steps are described by Condon and Shortley[1]‡ in their classic work *The Theory of Atomic Spectra.* Although the original assumptions remain virtually unchanged today, the second step in the analysis, often the longer and more difficult of the two, has been changed almost beyond recognition by the pioneering efforts of Racah. An understanding of the more direct approach of Condon and Shortley to problems in atomic spectroscopy is nevertheless essential for an appreciation of the later developments of the theory. We shall therefore begin by outlining the fundamental assumptions, and in the working out of these we shall use those techniques of Condon and Shortley that are of the greatest value at the present time, when alternative and more powerful methods are often open to us.

‡ References to the literature are at the end of the book.

1-2 THE CENTRAL-FIELD APPROXIMATION

At the outset an atom is visualized as a point nucleus of infinite mass and charge Ze surrounded by N electrons, each of mass m and charge $-e$. The nonrelativistic Hamiltonian of such a system is

$$H = \sum_{i=1}^{N} \left(\frac{p_i^2}{2m} - \frac{Ze^2}{r_i} \right) + \sum_{i>j=1}^{N} \frac{e^2}{r_{ij}}$$

In this expression, r_i is the distance of electron i from the nucleus, r_{ij} its distance from electron j. The second summation in H prevents a separation of the variables in the Schrödinger equation; yet it is too large to be dropped and treated later by perturbation theory. The approximation that is made is to suppose that the potential in which an electron moves can be reproduced by a function $-U(r_i)/e$. The approximate Hamiltonian is therefore

$$E = \sum_i \left(\frac{p_i^2}{2m} + U(r_i) \right)$$

The difference $H - E$ is the perturbation potential. The Schrödinger equation

$$\sum_{i=1}^{N} (-\hbar^2\nabla_i^2/2m + U(r_i))\Psi = E'\Psi$$

can be separated by choosing a set of functions $\psi(k_i)$ and writing

$$\Psi = \prod_{i=1}^{N} \psi_i(k_i) \qquad E' = \sum_{i=1}^{N} E'(k_i)$$

On deleting the superfluous references to the specific electrons, we find that the resultant equations are all of the type

$$(-\hbar^2\nabla^2/2m + U(r))\psi(k) = E'(k)\psi(k) \tag{1-1}$$

The only difference between Eq. (1-1) and the Schrödinger equation for the hydrogen atom lies in the presence of $U(r)$ in place of $-e^2/r$. The normalized solutions for bound states can thus be written in the form

$$\psi(k) = r^{-1}R_{nl}(r)Y_{lm_l}(\theta,\phi) \tag{1-2}$$

where the spherical harmonics Y_{lm} are defined by

$$Y_{lm}(\theta,\phi) = (-1)^m \left\{\frac{(2l+1)(l-m)!}{4\pi(l+m)!}\right\}^{\frac{1}{2}} P_l^m(\cos\theta)e^{im\phi} \tag{1-3}$$

with
$$P_l^m(x) = \frac{(1-x^2)^{\frac{1}{2}m}}{2^l l!} \frac{d^{l+m}}{dx^{l+m}} (x^2-1)^l \tag{1-4}$$

The radial function $R_{nl}(r)$ depends on the central potential $U(r)$; the angular part of $\psi(k)$ is identical to that for a hydrogenic eigenfunction. The symbol k can be regarded as standing for the set of quantum numbers (nlm_l). As in the case of the hydrogen atom, the number of nodes (not counting the origin) in $R_{nl}(r)$ is $n - l - 1$; however, the permitted energies $E'(k)$ are degenerate with respect to m_l only, and not to l as well, as they are in the hydrogen atom.

The spin of the electron is taken into account by multiplying $\psi(k)$ by either α or β, corresponding to the two possible orientations $m_s = \pm\frac{1}{2}$ along the z axis. Although the substitution for Ψ implies that $\psi(k_1)$, $\psi(k_2)$, $\psi(k_3)$, . . . are functions of the coordinates of electrons 1, 2, 3, . . . , respectively, it is clear that any permutation of the electrons with respect to the functions also leads to a solution of the Schrödinger equation. Because of the existence of spin, each of the $N!$ products of the type Ψ must be multiplied by N spin functions, one for each electron. Of all the possible linear combinations of these grand products that can be constructed, only those which are totally antisymmetric with respect to the simultaneous interchange of the spatial and spin coordinates of any pair of electrons are acceptable as eigenfunctions that can occur in nature. This is the mathematical expression of the Pauli exclusion principle; it limits the basic states to determinantal product states. An example of such a state for an atom comprising three electrons is

$$\Psi' = (3!)^{-\frac{1}{2}} \begin{vmatrix} \psi_1(k_1)\alpha_1 & \psi_1(k_1)\beta_1 & \psi_1(k_2)\alpha_1 \\ \psi_2(k_1)\alpha_2 & \psi_2(k_1)\beta_2 & \psi_2(k_2)\alpha_2 \\ \psi_3(k_1)\alpha_3 & \psi_3(k_1)\beta_3 & \psi_3(k_2)\alpha_3 \end{vmatrix} \tag{1-5}$$

Fortunately it is not necessary to write out determinants of this kind in full, since all the relevant information is provided by the diagonal terms. Using plus and minus signs for the functions α and β, respectively, corresponding to $m_s = \pm\frac{1}{2}$, we can simplify Ψ' to

$$\Psi' = \{\overset{+}{k_1}\overset{-}{k_1}\overset{+}{k_2}\} \tag{1-6}$$

The brackets denote a determinantal product state. The properties of such states follow from Eq. (1-5); for example,

$$\{\overset{+}{k}_1\overset{-}{k}_1\overset{+}{k}_2\} = -\{\overset{-}{k}_1\overset{+}{k}_1\overset{+}{k}_2\} = \{\overset{-}{k}_1\overset{+}{k}_2\overset{+}{k}_1\} \tag{1-7}$$

When the general theory of determinantal product states is being studied, it is often convenient to incorporate the spin quantum number m_s into the set $k = (nlm_l)$, and we shall use the symbol K to represent the complete set of four quantum numbers. Furthermore, if $K_j = \overset{+}{k}_j$, for example, then $\psi_i(K_j)$ will be written for $\psi_i(k_j)\alpha_i$. Expanding the general determinantal product state, we obtain

$$\{K_1K_2 \cdots K_N\} = (N!)^{-\frac{1}{2}} \sum_P (-1)^p \psi_1(K_1)\psi_2(K_2) \cdots \psi_N(K_N) \tag{1-8}$$

where the sum extends over all permutations P of the electrons' coordinates among the states defined by the quantum numbers $K_1, K_2, \ldots, K_N$, and where p is the parity of the permutation; that is, p is $+1$ if the permutation is even and -1 if it is odd.

Since the eigenvalues $E'(k)$ are degenerate with respect to m_l and m_s, the eigenvalues E' are distinguished by the list of symbols

$$(n_1l_1)(n_2l_2) \cdots (n_Nl_N)$$

This sequence is said to specify the electronic *configuration*. Traditionally, the letters $s, p, d, f, g, \ldots$ are used for the values 0, 1, 2, 3, 4, . . . of l; the number of times a particular symbol (nl) occurs is indicated by a raised suffix. A closed shell results when the suffix is equal to $2(2l + 1)$. This corresponds to the occurrence of all the symbols $K = (nlm_sm_l)$ for a given n and l. The occupation of all the quantum numbers of spatial and spin orientation gives a closed shell an inert character, and it is usually unnecessary to write out the quantum numbers explicitly in a determinantal product state. This is a useful simplification, since closed shells are a common feature of configurations that are energetically low-lying. If the electrons outside closed shells are *equivalent*, that is, if they possess the same values of n and l, these quantum numbers may also be dropped from the determinantal product state. For example, the electron configuration of Nd^{3+} is

$$1s^22s^22p^63s^23p^63d^{10}4s^24p^64d^{10}5s^25p^64f^3$$

Only three $4f$ electrons are not in closed shells. The basic eigenfunctions can therefore be written as

$$\{\overset{+}{3}\overset{+}{2}\overset{-}{1}\},\ \{\overset{+}{3}\overset{+}{1}-\overset{+}{2}\},\ \{\overset{-}{0}-\overset{-}{1}-\overset{+}{3}\},\ \text{etc.} \tag{1-9}$$

The numbers specify values of m_l, it being understood that $(nl) \equiv (4f)$ for all three electrons.

1-3 THE PERTURBATION POTENTIAL

The Hamiltonian E gives rise to a series of energy levels called configurations. Usually they are highly degenerate; there are, for example, 364 determinantal product states of the type (1-9), and this is for a configuration in which all but three electrons are in closed shells. The degeneracy is partially removed by the perturbation potential

$$H - E = \sum_{i=1}^{N} \left(\frac{-Ze^2}{r_i} - U(r_i) \right) + \sum_{i>j=1}^{N} \frac{e^2}{r_{ij}}$$

The effect of the first summation is to shift a configuration as a whole; since we shall be interested almost entirely in the structure of a single configuration, these terms can be dropped and there remains

$$H_1 = \sum_{i>j=1}^{N} \frac{e^2}{r_{ij}} \tag{1-10}$$

Corrections to the simplified Hamiltonian H are also represented by contributions to the perturbation potential. The nonrelativistic restriction can be removed for a single electron moving in a central potential represented by $U(r)$ by using Dirac's equation. If the average velocity of the electron is small compared with c, the velocity of light, the relativistic corrections can be represented by additional terms in the nonrelativistic Hamiltonian.[2] For our purposes the most important of these is

$$\hbar \mathbf{s} \cdot (\text{grad } U \times \mathbf{p})/2m^2c^2 \tag{1-11}$$

where $\mathbf{p}$ is the momentum of the electron. The vector $\mathbf{s}$ stands for the spin angular momentum of the electron; its z component s_z possesses the eigenvalues $+\frac{1}{2}$ and $-\frac{1}{2}$ for states defined by the symbols $+$ and $-$ introduced in Sec. 1-2. Since U is a function solely of r,

$$\text{grad } U = \frac{\mathbf{r}}{r}\frac{dU}{dr}$$

and (1-11) becomes

$$\frac{\hbar^2}{2m^2c^2r}\frac{dU}{dr}\mathbf{s}\cdot\mathbf{l} \tag{1-12}$$

where $\mathbf{l}$, the orbital angular momentum of an electron in units of $\hbar$, is defined by $\mathbf{l} = (\mathbf{r} \times \mathbf{p})/\hbar$. For a many-electron atom, (1-12) is replaced by

$$H_2 = \sum_i \xi(r_i)\mathbf{s}_i \cdot \mathbf{l}_i \tag{1-13}$$

where

$$\xi(r) = \frac{\hbar^2}{2m^2c^2r}\frac{dU}{dr}$$

This expression for H_2 is equivalent to that given by Condon and Shortley for the spin-orbit interaction.

For free atoms and ions, the combination $H_1 + H_2$ is usually by far the largest contribution to the perturbation potential. Under its influence each configuration is broken up into *levels*. As a first approach to calculating the effect of $H_1 + H_2$ in detail, it is assumed that the eigenvalues E' are sufficiently well dispersed to permit each configuration to be examined separately. The problem is to apply the perturbation $H_1 + H_2$ to a degenerate energy level, taking as basic states those determinantal product states which, for a single sequence $(n_1l_1)(n_2l_2) \cdots (n_Nl_N)$, run over all possible values of m_s and m_l. The direct method is to label columns and rows by the determinantal product states, evaluate the matrix elements of $H_1 + H_2$ and solve the secular determinant. This procedure is feasible for only the most elementary configurations; in the case of $Nd^{3+}4f^3$, for example, $364 \times 364 = 132496$ matrix elements have to be evaluated. Apart from being tedious, this method gives no insight into the structure of the configuration, gives no labels for distinguishing the eigenvalues, and takes no advantage of the frequent recurrence of identical roots which arise when solving the secular equation.

1-4 ANGULAR-MOMENTUM OPERATORS

Before we can be in a position to remedy the deficiencies in the direct approach to the problem, we must have some knowledge of the properties of angular momentum operators. The fundamental commutation relations for the components J_x, J_y, and J_z of an angular-momentum vector $\mathbf{J}$ are

$$[J_x,J_y] = iJ_z \qquad [J_y,J_z] = iJ_x \qquad [J_z,J_x] = iJ_y \tag{1-14}$$

where

$$[A,B] = AB - BA$$

The orthonormal states, in the notation of Dirac,[3] are distinguished by the two quantum numbers J and M. As is well known,[1] the commutation relations (1-14) lead to the equations

$$\begin{aligned} J_z|J,M) &= M|J,M) \\ \mathbf{J}^2|J,M) &= J(J+1)|J,M) \\ J_+|J,M) &= e^{i\phi}[J(J+1) - M(M+1)]^{\frac{1}{2}}|J, M+1) \\ J_-|J,M) &= e^{-i\phi}[J(J+1) - M(M-1)]^{\frac{1}{2}}|J, M-1) \end{aligned} \tag{1-15}$$

The symbols J_+ and J_- stand for $J_x + iJ_y$ and $J_x - iJ_y$, respectively. Because they raise and lower M, they are sometimes called *shift* operators. The quantum number J is a multiple of $\frac{1}{2}$. For a given J, the quantum number M can take on the $2J + 1$ values $J, J - 1, \ldots, -J$. The arbitrary phase angle ϕ can have no effect on any results of physical significance, and for convenience we follow the convention of Condon and Shortley and set $\phi = 0$. Questions of phase arise continually in the theory of angular momentum; we shall follow Edmonds[4] in always making the same choice as Condon and Shortley. It is to be observed that the quantities raised to the power $\frac{1}{2}$ in the last two equations of the set (1-15) can be factorized; however, they are much more memorable in their present form, since in both cases the part which is subtracted from $J(J + 1)$ is a product of the values of M linked by the shift operator.

For the purpose of finding the eigenvalues of the Coulomb interaction H_1, it is convenient to introduce the total orbital and total spin angular momenta of the electrons, namely,

$$\mathbf{S} = \sum_i \mathbf{s}_i \qquad \text{and} \qquad \mathbf{L} = \sum_i \mathbf{l}_i$$

The states, which must be certain linear combinations of the determinantal product states, are labeled by the quantum numbers γSM_SLM_L. The symbol γ stands for any other quantum numbers that are needed when the set SM_SLM_L fails to define the states uniquely. The importance of $\mathbf{S}$ and $\mathbf{L}$ lies in the fact that they commute with H_1; this is obviously true of $\mathbf{S}$, and it is a simple matter to demonstrate that any component of **L**, for example,

$$L_z = \frac{1}{i} \sum_j \left(x_j \frac{\partial}{\partial y_j} - y_j \frac{\partial}{\partial x_j} \right)$$

commutes with

$$H_1 = e^2 \sum_{i>j} [(x_i - x_j)^2 + (y_i - y_j)^2 + (z_i - z_j)^2]^{-\frac{1}{2}}$$

Hence

$$\begin{aligned}(\gamma SM_SLM_L|H_1|\gamma' S'M'_SL'M'_L)M'_L &= (\gamma SM_SLM_L|H_1L_z|\gamma' S'M'_SL'M'_L) \\ &= (\gamma SM_SLM_L|L_zH_1|\gamma' S'M'_SL'M'_L) \\ &= M_L(\gamma SM_SLM_L|H_1|\gamma' S'M'_SL'M'_L)\end{aligned}$$

If $M_L \neq M'_L$, the matrix element vanishes. Using $\mathbf{L}^2$ in place of L_z, we deduce that all matrix elements for which $L \neq L'$ vanish. By analogous arguments it can be seen that all matrix elements that are nondiagonal with respect to both S and M_S are also zero. Furthermore,

$$\begin{aligned}&(\gamma SM_SLM_L|H_1|\gamma' SM_SLM_L) \\ &\quad = (\gamma SM_SLM_L|H_1L_+|\gamma' SM_SLM_L - 1)[L(L+1) - M_L(M_L - 1)]^{-\frac{1}{2}} \\ &\quad = (\gamma SM_SLM_L - 1|H_1|\gamma' SM_SLM_L - 1),\end{aligned}$$

showing that the matrix elements are independent of M_L and, by a similar line of reasoning, of M_S too. The rows and columns of the matrix of H_1 can therefore be labeled simply by γ, S, and L; the secular determinant breaks up into a series of smaller determinants, each labeled by the pair of quantum numbers SL. The eigenvalues are independent of M_S and M_L and are thus $(2L + 1)(2S + 1)$fold degenerate. The set of states belonging to a single eigenvalue is called a *term*. The value of L is denoted by a capital letter according to the following scheme:

$L =$ 0	1	2	3	4	5	6	7	8	9	10	11	12 $\cdots$
S	P	D	F	G	H	I	K	L	M	N	O	$Q \cdots$

A raised prefix gives the *multiplicity* $2S + 1$. Thus ^4I, which occurs in the configuration $4f^3$, stands for the 52 degenerate states for which $S = \frac{3}{2}$, $L = 6$.

The high degeneracy of the terms is partially removed by the spin-orbit interaction H_2, which commutes, not with **S** and **L** separately, but with their resultant $\mathbf{J} = \mathbf{S} + \mathbf{L}$. This can be proved for the component J_z of **J** by first writing

$$J_z \sum_i \mathbf{s}_i \cdot \mathbf{l}_i = \sum_{i \neq k} j_{kz} \mathbf{s}_i \cdot \mathbf{l}_i + \sum_i j_{iz} \mathbf{s}_i \cdot \mathbf{l}_i$$

The function $\xi(r_i)$, being independent of the angular variables, commutes with $\mathbf{J}$, and for this reason has been dropped. It is obvious that j_{kz} commutes with $\mathbf{s}_i \cdot \mathbf{l}_i$ when $i \neq k$. For a term in the second summation,

$$\begin{aligned} j_z(\mathbf{s} \cdot \mathbf{l}) &= (s_z + l_z)(s_x l_x + s_y l_y + s_z l_z) \\ &= (s_x s_z + i s_y) l_x + (s_y s_z - i s_x) l_y + s_z l_z s_z \\ &\qquad + s_x(l_x l_z + i l_y) + s_y(l_y l_z - i l_x) + s_z l_z l_z \\ &= (\mathbf{s} \cdot \mathbf{l}) j_z \end{aligned}$$

If $H_2 \ll H_1$, it is possible to give an adequate account of the energy-level scheme by treating each term as an isolated entity. The various values of J are indicated by subscripts to the term symbol; the term ^4I of f^3, for example, gives rise to the *multiplet* $^4I_{9/2}$, $^4I_{11/2}$, $^4I_{13/2}$, and $^4I_{15/2}$. However, if the matrix elements of H_2 between the different terms are comparable with the energy separations of the terms, it is no longer a good approximation to consider each term separately and the complete matrix of $H_1 + H_2$ must be constructed. Although it would be possible to use the states $|\gamma S M_S L M_L)$ as a basis, it is much more convenient to use the set $|\gamma SLJM_J)$. The reasons for this preference are similar to those which prompted us to replace the determinantal product states by the set $|\gamma S M_S L M_L)$; the secular determinant breaks up into smaller determinants, and the recalculation of many identical roots is avoided.

Since both H_1 and H_2 are functions of the coordinates of the individual electrons and cannot be expressed as functions of S, L, and J, it would appear that the actual calculation of the matrix elements of H_1 and H_2 can be performed only when the basic states $|\gamma SLJM_J)$ are expressed in terms of determinantal product states. Transformations of this kind are of frequent occurrence in the study of many-particle systems and deserve to be treated in detail.

1-5 THE 3-j SYMBOLS

The expansion of states of the type $|\gamma SLJM_J)$ in terms of determinantal product states is best accomplished by first transforming to SLM_SM_L quantization. As an example of the technique, we shall consider the lowest level of f^2, namely, $^3H_4$. Suppose that M_J has its maximum value. Then the three solutions of

$$M_S + M_L = M_J = 4$$

lead to the equation

$$|^3H_4, M_J = 4) = a|^3H, M_S = 1, M_L = 3) + b|^3H,0,4) + c|^3H,-1,5) \tag{1-16}$$

The coefficients a, b, and c are special cases of the vector-coupling (VC) coefficients $(SM_SLM_L|SLJM_J)$. They are also called the Clebsch-Gordan coefficients or Wigner coefficients. When M_J possesses its maximum value, the VC coefficients may be rapidly calculated to within a phase factor by operating with $J_+ = S_+ + L_+$ and insisting that the result be identically zero. For example,

$$J_+|^3H_4,4) = 0$$

leads to

$$a(5 \cdot 6 - 3 \cdot 4)^{\frac{1}{2}}|1,4) + b(1 \cdot 2 - 0 \cdot 1)^{\frac{1}{2}}|1,4) + b(5 \cdot 6 - 4 \cdot 5)^{\frac{1}{2}}|0,5) + c(1 \cdot 2 - (-1) \cdot 0)^{\frac{1}{2}}|0,5) = 0$$

Because of the orthonormality of the states $|\gamma SM_SLM_L)$,

$$a(18)^{\frac{1}{2}} + b(2)^{\frac{1}{2}} = 0 \qquad b(10)^{\frac{1}{2}} + c(2)^{\frac{1}{2}} = 0$$

The equation

$$aa^* + bb^* + cc^* = 1$$

where the asterisk denotes the complex conjugate, follows from

$$(^3H_4,4|^3H_4,4) = 1$$

and

$$(^3H,M_SM_L|^3H,M'_SM'_L) = \delta(M_S,M'_S)\delta(M_L,M'_L)$$

Thus

$$aa^*(1 + 9 + 45) = 1 = 55aa^*$$

and so $a = e^{i\omega}(55)^{-\frac{1}{2}}$. If we set $\omega = 0$, Eq. (1-16) becomes

$$|^3H_4,4) = (\tfrac{1}{55})^{\frac{1}{2}}|1,3) - (\tfrac{9}{55})^{\frac{1}{2}}|0,4) + (\tfrac{9}{11})^{\frac{1}{2}}|-1,5) \qquad (1\text{-}17)$$

Other states $|^3H_4,M_J)$ can be found by using the shift operator $J_- = S_- + L_-$. For example,

$$|^3H_4,3) = (\tfrac{3}{55})^{\frac{1}{2}}|1,2) - (\tfrac{16}{55})^{\frac{1}{2}}|0,3) + (\tfrac{36}{55})^{\frac{1}{2}}|-1,4)$$

This procedure can be expressed algebraically. Let j_1 and j_2 be two angular momenta coupled to a resultant j. The generalization of Eq. (1-17) is

$$|j_1j_2jj) = C \sum_{\mu_1,\mu_2} (-1)^{j_1-\mu_1}\delta(\mu_1 + \mu_2, j) \left[\frac{(j_1 + \mu_1)!(j_2 + \mu_2)!}{(j_1 - \mu_1)!(j_2 - \mu_2)!}\right]^{\frac{1}{2}} \times |j_1\mu_1j_2\mu_2) \qquad (1\text{-}18)$$

for it is easy to show that this expansion satisfies the equation

$$j_+|j_1j_2jj) = 0$$

It is a matter of elementary algebra to prove

$$\sum_{\mu_1,\mu_2} \frac{(j_1+\mu_1)!(j_2+\mu_2)!}{(j_1-\mu_1)!(j_2-\mu_2)!}\,\delta(\mu_1+\mu_2, j) = \frac{(j+j_1-j_2)!(j+j_2-j_1)!(j+j_1+j_2+1)!}{(2j+1)!(j_1+j_2-j)!} \tag{1-19}$$

We therefore choose

$$C = \left[\frac{(2j+1)!(j_1+j_2-j)!}{(j+j_1-j_2)!(j+j_2-j_1)!(j+j_1+j_2+1)!}\right]^{1/2}$$

to ensure that the sum of the squares of the coefficients in Eq. (1-18) is unity. The conventional choice of phase is made.

To find the general VC coefficient, we operate $j - m$ times on Eq. (1-18) with $j_- = j_{1-} + j_{2-}$. From Eq. (1-15) it is straightforward to show that

$$(J_-)^N|J,M) = \left[\frac{(J+M)!(J-M+N)!}{(J-M)!(J+M-N)!}\right]^{1/2} |J, M-N)$$

Hence

$$\begin{aligned}
(j_-)^{j-m}|j_1j_2jj) &= \left[\frac{(2j)!(j-m)!}{(j+m)!}\right]^{1/2} |j_1j_2jm) \\
&= C \sum_{\mu_1,\mu_2} (-1)^{j_1-\mu_1}\delta(\mu_1+\mu_2, j) \left[\frac{(j_1+\mu_1)!(j_2+\mu_2)!}{(j_1-\mu_1)!(j_2-\mu_2)!}\right]^{1/2} \\
&\qquad \times \sum_x j_{1-}{}^{x} j_{2-}{}^{j-m-x} \frac{(j-m)!}{x!(j-m-x)!} |j_1\mu_1j_2\mu_2) \\
&= C \sum_{\mu_1,\mu_2} (-1)^{j_1-\mu_1}\delta(\mu_1+\mu_2, j) \sum_x \frac{(j-m)!}{x!(j-m-x)!} \\
&\quad \times \left[\frac{(j_1+\mu_1)!(j_2+\mu_2)!(j_1+\mu_1)!(j_1-\mu_1+x)!(j_2+\mu_2)!(j_2-\mu_2+j-m-x)!}{(j_1-\mu_1)!(j_2-\mu_2)!(j_1-\mu_1)!(j_1+\mu_1-x)!(j_2-\mu_2)!(j_2+\mu_2-j+m+x)!}\right]^{1/2} \\
&\qquad \times |j_1 \quad \mu_1 - x \quad j_2 \quad \mu_2 - j + m + x)
\end{aligned}$$

The substitutions $\mu_1 = m_1 + x$, $\mu_2 = j - m - x + m_2$ are now made. We find that

$$|j_1j_2jm) = \sum_{m_1,m_2} (j_1m_1j_2m_2|j_1j_2jm)|j_1m_1j_2m_2)$$

where $(j_1m_1j_2m_2|j_1j_2jm)$

$$= \delta(m_1 + m_2, m) \left[\frac{(2j+1)(j_1+j_2-j)!(j_1-m_1)!(j_2-m_2)!(j+m)!(j-m)!}{(j_1+j_2+j+1)!(j+j_1-j_2)!(j+j_2-j_1)!(j_1+m_1)!(j_2+m_2)!} \right]^{\frac{1}{2}}$$

$$\times \sum_x (-1)^{j_1-m_1-x} \frac{(j_1+m_1+x)!(j_2+j-m_1-x)!}{x!(j-m-x)!(j_1-m_1-x)!(j_2-j+m_1+x)!} \qquad (1\text{-}20)$$

This expression exhibits a high degree of symmetry. On writing $x = j - m - y$, for example, we obtain the result

$$(j_1m_1j_2m_2|j_1j_2jm) = (-1)^{j_1+j_2-j}(j_2m_2j_1m_1|j_2j_1jm) \qquad (1\text{-}21)$$

The 3-j symbol, defined by

$$\begin{pmatrix} j_1 & j_2 & j_3 \\ m_1 & m_2 & m_3 \end{pmatrix} = (-1)^{j_1-j_2-m_3}(2j_3+1)^{-\frac{1}{2}}(j_1m_1j_2m_2|j_1j_2j_3 - m_3) \qquad (1\text{-}22)$$

is designed to display symmetry relations of this kind in a systematic and uniform way. Even permutations of the columns leave the numerical value unaltered, i.e.,

$$\begin{pmatrix} j_1 & j_2 & j_3 \\ m_1 & m_2 & m_3 \end{pmatrix} = \begin{pmatrix} j_2 & j_3 & j_1 \\ m_2 & m_3 & m_1 \end{pmatrix} = \begin{pmatrix} j_3 & j_1 & j_2 \\ m_3 & m_1 & m_2 \end{pmatrix}$$

whereas odd permutations of the columns introduce the phase factor $(-1)^{j_1+j_2+j_3}$,

$$\begin{pmatrix} j_1 & j_2 & j_3 \\ m_1 & m_2 & m_3 \end{pmatrix} = (-1)^{j_1+j_2+j_3} \begin{pmatrix} j_2 & j_1 & j_3 \\ m_2 & m_1 & m_3 \end{pmatrix}$$

It follows that all 3-j symbols possessing two identical columns are zero if $j_1 + j_2 + j_3$ is odd. It can also be shown that

$$\begin{pmatrix} j_1 & j_2 & j_3 \\ m_1 & m_2 & m_3 \end{pmatrix} = (-1)^{j_1+j_2+j_3} \begin{pmatrix} j_1 & j_2 & j_3 \\ -m_1 & -m_2 & -m_3 \end{pmatrix}$$

Further symmetry relations have been given by Regge[5] (see Prob. 1-6). The orthonormality conditions

$$(j_1m_1j_2m_2|j_1m_1'j_2m_2') = \delta(m_1,m_1')\delta(m_2,m_2')$$

and

$$(j_1j_2j_3m_3|j_1j_2j_3'm_3') = \delta(j_3,j_3')\delta(m_3,m_3')$$

impose on the 3-j symbols the restrictions

$$\sum_{j_3,m_3} (2j_3+1)\begin{pmatrix} j_1 & j_2 & j_3 \\ m_1 & m_2 & m_3 \end{pmatrix}\begin{pmatrix} j_1 & j_2 & j_3 \\ m_1' & m_2' & m_3 \end{pmatrix} = \delta(m_1,m_1')\delta(m_2,m_2') \tag{1-23}$$

and

$$\sum_{m_1,m_2} \begin{pmatrix} j_1 & j_2 & j_3 \\ m_1 & m_2 & m_3 \end{pmatrix}\begin{pmatrix} j_1 & j_2 & j_3' \\ m_1 & m_2 & m_3' \end{pmatrix} = (2j_3+1)^{-1}\delta(j_3,j_3')\delta(m_3,m_3') \tag{1-24}$$

Of the available tables for VC coefficients or 3-j symbols, that of Rotenberg, Bivins, Metropolis, and Wooten[6] possesses the advantage that the entries, although machine-computed, are expressed as the square roots of the ratios of integers.

1-6 DETERMINANTAL PRODUCT STATES

The transformation from SM_SLM_L quantization to sm_slm_l quantization can also be performed with the aid of shift operators. The general determinantal product state is denoted by $\{K_1K_2 \cdots K_N\}$, where $K_i = (n_il_im_{si}m_{li})$. From Eq. (1-8) it is clear that, for $N = 2$,

$$\begin{aligned} L_+\{K_1K_2\} &= (2)^{-\frac{1}{2}}(l_{1+} + l_{2+})[\psi_1(K_1)\psi_2(K_2) - \psi_1(K_2)\psi_2(K_1)] \\ &= (2)^{-\frac{1}{2}}[a_1\psi_1(K_1')\psi_2(K_2) - a_2\psi_1(K_2')\psi_2(K_1) \\ &\qquad + a_2\psi_1(K_1)\psi_2(K_2') - a_1\psi_1(K_2)\psi_2(K_1')] \\ &= a_1\{K_1'K_2\} + a_2\{K_1K_2'\} \end{aligned}$$

where

$$\begin{aligned} K_i' &= (n_il_im_{si}\ m_{li}+1) \\ a_i &= [l_i(l_i+1) - m_{li}(m_{li}+1)]^{\frac{1}{2}} \end{aligned}$$

The generalization to N electrons is as follows:

$$\begin{aligned} L_+\{K_1K_2 \cdots K_N\} &= a_1\{K_1'K_2 \cdots K_N\} \\ &\quad + a_2\{K_1K_2' \cdots K_N\} + \cdots + a_N\{K_1K_2 \cdots K_N'\} \end{aligned} \tag{1-25}$$

By a similar method it can be shown that

$$\begin{aligned} S_+\{K_1K_2 \cdots K_N\} &= b_1\{K_1''K_2 \cdots K_N\} \\ &\quad + b_2\{K_1K_2'' \cdots K_N\} + \cdots + b_N\{K_1K_2 \cdots K_N''\} \end{aligned} \tag{1-26}$$

where

$$\begin{aligned} K_i'' &= (n_il_i\ m_{si}+1\ m_{li}) \\ b_i &= [\tfrac{3}{4} - m_{si}(m_{si}+1)]^{\frac{1}{2}} = \delta(m_{si},-\tfrac{1}{2}) \end{aligned}$$

As an example, we shall calculate the coefficients a, b, c, and d in the expansion

$$|f^3\,{}^2I,\ M_S = \tfrac{1}{2},\ M_L = 6) = a\{\overset{+-+}{330}\} + b\{\overset{+-+}{321}\} + c\{\overset{++-}{321}\} + d\{\overset{-++}{321}\} \qquad (1\text{-}27)$$

Since M_L possesses its maximum value,

$$L_+|f^3\,{}^2I,\tfrac{1}{2},6) = 0$$

On making the substitution (1-27) and using (1-25), we obtain

$$[(12)^{\frac{1}{2}}a + (6)^{\frac{1}{2}}b - (6)^{\frac{1}{2}}d]\{\overset{+-+}{331}\} + [-(10)^{\frac{1}{2}}b + (10)^{\frac{1}{2}}c]\{\overset{++-}{322}\} = 0$$

Similarly, the equation

$$S_+|f^3\,{}^2I,\tfrac{1}{2},6) = 0$$

leads to

$$(b + c + d)\{\overset{+++}{321}\} = 0$$

The three equations

$$b - d + a(2)^{\frac{1}{2}} = 0 \qquad b = c \qquad b + c + d = 0$$

are now combined with

$$aa^* + bb^* + cc^* + dd^* = 1$$

and the coefficients are determined to within a phase factor. Making an arbitrary choice of phase, we find that Eq. (1-27) can be written as

$$|f^3\,{}^2I,\tfrac{1}{2},6) = (\tfrac{3}{7})^{\frac{1}{2}}\{\overset{+-+}{330}\} - (\tfrac{2}{21})^{\frac{1}{2}}\{\overset{+-+}{321}\} - (\tfrac{2}{21})^{\frac{1}{2}}\{\overset{++-}{321}\} + (\tfrac{8}{21})^{\frac{1}{2}}\{\overset{-++}{321}\} \qquad (1\text{-}28)$$

Sometimes this method fails to give sufficient equations to determine the coefficients. For example, operation with S_+ and L_+ on the normalized state

$$|f^3\,{}^2H,\tfrac{1}{2},5) = [\{\overset{++-}{320}\} - (1 + x)\{\overset{+-+}{320}\} + x\{\overset{-++}{320}\} + (2)^{-\frac{1}{2}}(2x + 1)\{\overset{+-}{33}\overset{+}{-1}\} + x(2)^{\frac{1}{2}}\{\overset{+-+}{221}\} - (\tfrac{6}{5})^{\frac{1}{2}}\{\overset{++-}{311}\}](6x^2 + 4x + \tfrac{37}{10})^{-\frac{1}{2}}$$

gives identically zero for all values of x. Now the occurrence of six determinantal product states for which $M_S = \frac{1}{2}$, $M_L = 5$ indicates that there are six terms in the configuration f^3 for which $S \geq \frac{1}{2}$, $L \geq 5$;

moreover Eq. (1-27) shows that there are four terms for which $S \geq \frac{1}{2}$, $L \geq 6$. We deduce that there are two H terms in f^3. However, since

$$|f^3\ {}^4I,\tfrac{3}{2},6) = \{\overset{+}{3}\overset{+}{2}\overset{+}{1}\}$$

and

$$|f^3\ {}^4I,\tfrac{3}{2},5) = \{\overset{+}{3}\overset{+}{2}\overset{+}{0}\}$$

it is impossible to construct a state $|f^3\ {}^4H,\frac{3}{2},5)$ that is orthogonal to $|f^3\ {}^4I,\frac{3}{2},5)$. Therefore both H terms correspond to $S = \frac{1}{2}$. The reason for our inability to define unambiguously the state $|f^3\ {}^2H,\frac{1}{2},5)$ is now clear: the specification $f^3SM_SLM_L$ is incomplete. In fact, we could pick at random a value of x, define one state, and then calculate its orthogonal companion. Such a pair of states could be used as a basis for calculating the matrix elements of $H_1 + H_2$. The states

$$|f^3\gamma_1\ {}^2H,\tfrac{1}{2},5) = -(\tfrac{1}{6})^{\frac{1}{2}}\{\overset{+}{3}\overset{-}{2}\overset{+}{0}\} + (\tfrac{1}{6})^{\frac{1}{2}}\{\overset{-}{3}\overset{+}{2}\overset{+}{0}\} + (\tfrac{1}{3})^{\frac{1}{2}}\{\overset{+}{3}\overset{-}{3}-\overset{+}{1}\} + (\tfrac{1}{3})^{\frac{1}{2}}\{\overset{+}{2}\overset{-}{2}\overset{+}{1}\} \tag{1-29}$$

and

$$\begin{aligned}|f^3\gamma_2\ {}^2H,\tfrac{1}{2},5) = {} & (\tfrac{30}{91})^{\frac{1}{2}}\{\overset{+}{3}\overset{+}{2}\overset{-}{0}\} - (\tfrac{40}{273})^{\frac{1}{2}}\{\overset{+}{3}\overset{-}{2}\overset{+}{0}\} \\ & - (\tfrac{10}{273})^{\frac{1}{2}}\{\overset{-}{3}\overset{+}{2}\overset{+}{0}\} + (\tfrac{5}{273})^{\frac{1}{2}}\{\overset{+}{3}\overset{-}{3}-\overset{+}{1}\} \\ & - (\tfrac{20}{273})^{\frac{1}{2}}\{\overset{+}{2}\overset{-}{2}\overset{+}{1}\} - (\tfrac{36}{91})^{\frac{1}{2}}\{\overset{+}{3}\overset{+}{1}\overset{-}{1}\}\end{aligned} \tag{1-30}$$

which correspond to $x = \infty$ and $x = -\frac{1}{3}$, are a suitable orthonormal pair.

1-7 MATRIX ELEMENTS

By carrying out the transformations symbolized by the sequence of quantizations

$$SLJM_J \to SM_SLM_L \to sm_slm_l$$

we can expand every state $|\gamma SLJM_J)$ as a linear combination of determinantal product states. The matrix elements

$$(\gamma SLJM_J|H_1 + H_2|\gamma' S'L'JM_J)$$

can therefore be expressed as a sum of integrals of the type

$$\int \{K_1K_2 \cdot \cdot \cdot K_N\}^*(H_1 + H_2)\{K_1'K_2' \cdot \cdot \cdot K_N'\}$$

where the integration is carried out over all spatial and spin variables. The method of evaluating such integrals is described in detail by

Condon and Shortley,[1] both for single-particle operators (like H_2) of the form $F = \sum_i f_i$ and for two-particle operators (like H_1) of the form $G = \sum_{i>j} g_{ij}$. The results, which may be proved by using the expansion (1-8), are summarized below:

1. When the two sets $K_1K_2 \cdots K_N$ and $K_1'K_2' \cdots K_N'$ can be ordered, by means of operations of the type (1-9), to be identical, so that $K_i = K_i'$ for all i, then

$$\int \{K_1K_2 \cdots K_N\}^* F \{K_1K_2 \cdots K_N\} = \sum_i \int \{K_i\}^* f_1 \{K_i\} = \sum_i \int \psi_1^*(K_i) f_1 \psi_1(K_i) \quad (1\text{-}31)$$

and
$$\int \{K_1K_2 \cdots K_N\}^* G \{K_1K_2 \cdots K_N\} = \sum_{i>j} \int \{K_iK_j\}^* g_{12} \{K_iK_j\}$$
$$= \sum_{i>j} \left[\int \psi_1^*(K_i)\psi_2^*(K_j) g_{12} \psi_1(K_i)\psi_2(K_j) - \int \psi_1^*(K_i)\psi_2^*(K_j) g_{12} \psi_2(K_i)\psi_1(K_j) \right] \quad (1\text{-}32)$$

Integrals in this summation preceded by a positive sign are called *direct integrals;* those preceded by a negative sign, *exchange integrals.*

2. If all the components K_i of the set $K_1K_2 \cdots K_N$ reappear in the set $K_1'K_2' \cdots K_N'$ with the exception of K_r, then, subject to a possible reordering of the type (1-7) to match the $N - 1$ identical components, the matrix element of a single-particle operator reduces to

$$\int \{K_1K_2 \cdots K_r \cdots K_N\}^* F \{K_1K_2 \cdots K_r' \cdots K_N\} = \int \{K_r\}^* f_1 \{K_r'\} = \int \psi_1^*(K_r) f_1 \psi_1(K_r')$$

and that of a two-particle operator to

$$\int \{K_1K_2 \cdots K_r \cdots K_N\}^* G \{K_1K_2 \cdots K_r' \cdots K_N\} = \sum_i \int \{K_iK_r\}^* g_{12} \{K_iK_r'\}$$

The 2×2 determinants can be expanded as in (1), but no special names attach to the various integrals in the sum.

3. If the components K_i of the set $K_1K_2 \cdots K_N$ reappear in the set $K_1'K_2' \cdots K_N'$ with the exception of K_r and K_t, then all matrix elements of F vanish; and under a possible reordering to match the

$N - 2$ identical components, the matrix element of G becomes

$$\int \{K_1K_2 \cdots K_r \cdots K_t \cdots K_N\}^* G \{K_1K_2 \cdots K_r' \cdots K_t' \cdots K_N\} = \int \{K_rK_t\}^* g_{12} \{K_r'K_t'\}$$

All matrix elements of F and G that do not fall into one of the three classes above are zero.

1-8 THE CONFIGURATION *sd*

Given the basic single-electron eigenfunctions, we now have at our disposal the necessary apparatus to solve any problem in atomic spectroscopy. The method will probably be clarified by working through a specific example and making a direct comparison with experiment. The evaluation of the matrix elements of H_1 requires the construction of subsidiary tables (Tables 1[6] and 2[6] of Condon and Shortley, or Tables A20-1 and A20-2 of Slater[1]), and since the developments in the theory of tensor operators render much of this machinery superfluous, we choose a configuration where the number of terms does not exceed the number of radial integrals connected with the evaluation of H_1 by more than one. In cases such as these no great advantage attaches to calculating the matrix elements of H_1 unless we are prepared to estimate the radial integrals. Since these integrals are often extracted from experiment or used as adjustable parameters, very little is lost by our present reluctance to evaluate the matrix elements of H_1.

The configuration *sd* is a suitable one for our purpose; it has been observed in many spectra and is a reasonably simple one to analyze. The 20 basic determinantal product states are

$$\begin{aligned}
&\{\overset{++}{20}\},\ \{\overset{++}{10}\},\ \{\overset{++}{00}\},\ \{-\overset{++}{10}\},\ \{-\overset{++}{20}\},\ \{\overset{+-}{20}\}\\
&\{\overset{+-}{10}\},\ \{\overset{+-}{00}\},\ \{-\overset{+-}{10}\},\ \{-\overset{+-}{20}\},\ \{\overset{-+}{20}\},\ \{\overset{-+}{10}\},\\
&\{\overset{-+}{00}\},\ \{-\overset{-+}{10}\},\ \{-\overset{-+}{20}\},\ \{\overset{--}{20}\},\ \{\overset{--}{10}\},\ \{\overset{--}{00}\},\\
&\{-\overset{--}{10}\},\ \{-\overset{--}{20}\}
\end{aligned} \tag{1-33}$$

in which the quantum numbers m_s and m_l of the d electron are placed before those of the s electron. The state with the maximum values of M_S and M_L is $\{\overset{++}{20}\}$, for which $M_S = 1$ and $M_L = 2$. This proves the existence of the term ^3D. Operating on

$$|^3D, M_S = 1, M_L = 2) = \{\overset{++}{20}\}$$

with S_-, we obtain

$$|^3D,0,2) = (\tfrac{1}{2})^{\frac{1}{2}}\{\overset{+-}{20}\} + (\tfrac{1}{2})^{\frac{1}{2}}\{\overset{-+}{20}\}$$

The orthogonal state cannot belong to ^3D and can only derive from ^1D,

$$|^1D,0,2) = (\tfrac{1}{2})^{\frac{1}{2}}\{\overset{+-}{20}\} - (\tfrac{1}{2})^{\frac{1}{2}}\{\overset{-+}{20}\}$$

Since $3 \times 5 + 1 \times 5 = 20$, the terms ^3D and ^1D are the only ones in the configuration. We write

$$(^1D|H_1|\,^1D) - (^3D|H_1|^3D) = G$$

In order to calculate the matrix elements of H_2, we use the techniques of Secs. (1-5) and (1-6) to prove

$$\begin{aligned}
|^3D_3,3) &= |^3D,1,2) \\
|^3D_2,2) &= (\tfrac{1}{3})^{\frac{1}{2}}|^3D,1,1) - (\tfrac{2}{3})^{\frac{1}{2}}|^3D,0,2) \\
|^3D_1,1) &= (\tfrac{1}{10})^{\frac{1}{2}}|^3D,1,0) - (\tfrac{3}{10})^{\frac{1}{2}}|^3D,0,1) + (\tfrac{3}{5})^{\frac{1}{2}}|^3D,-1,2)
\end{aligned}$$

$$\begin{aligned}
|^3D,1,1) &= \{\overset{++}{10}\} \\
|^3D,0,1) &= (\tfrac{1}{2})^{\frac{1}{2}}\{\overset{+-}{10}\} + (\tfrac{1}{2})^{\frac{1}{2}}\{\overset{-+}{10}\} \\
|^3D,-1,2) &= \{\overset{--}{20}\}
\end{aligned}$$

Thus
$$(^3D_3|H_2|\,^3D_3) = \int\{\overset{++}{20}\}^*H_2\{\overset{++}{20}\} = \zeta_d$$

where, in general,
$$\zeta_{nl} = \int R_{nl}(r)\xi(r)R_{nl}(r)\,dr \tag{1-34}$$

In practice, either or both of the quantum numbers n and l are often dropped from ζ_{nl} if it is clear from the context what n and l are or if the formulas in which ζ_{nl} appears are valid for all n or l. For the level $^3D_2$ we find

$$\begin{aligned}
(^3D_2|H_2|\,^3D_2) &= (\tfrac{1}{3})(^3D,1,1|H_2|\,^3D,1,1) - 2(\tfrac{2}{9})^{\frac{1}{2}}(^3D,1,1|H_2|\,^3D,0,2) \\
&\qquad + (\tfrac{2}{3})(^3D,0,2|H_2|\,^3D,0,2) \\
&= -\tfrac{1}{2}\zeta_d
\end{aligned}$$

and, continuing in this manner, we get

$$\begin{aligned}
(^3D_1|H_2|\,^3D_1) &= \frac{-3\zeta_d}{2} \\
(^3D_2|H_2|\,^1D_2) &= (\tfrac{3}{2})^{\frac{1}{2}}\zeta_d
\end{aligned}$$

If the ^3D term is taken as the zero on the energy scale, the perturbation matrix takes the form

	$^3D_3$	$^3D_2$	$^1D_2$	$^3D_1$
$^3D_3$	ζ	0	0	0
$^3D_2$	0	$-\zeta/2$	$(\frac{3}{2})^{\frac{1}{2}}\zeta$	0
$^1D_2$	0	$(\frac{3}{2})^{\frac{1}{2}}\zeta$	G	0
$^3D_1$	0	0	0	$-3\zeta/2$

The subscript on ζ has been dropped. If we were to include all M_J components of the levels in the labeling of the rows and columns, the first entry, ζ, would occur in all seven times on the diagonal, the 2×2 central matrix would occur five times, and $-3\zeta/2$ three times. The total number of roots is 20, which agrees with the number of determinantal product states. Denoting the energy of a level with angular momentum J by $\epsilon(J)$, we find

$$\epsilon(3) = \zeta \qquad \epsilon(1) = \frac{-3\zeta}{2}$$

$$\epsilon(2) = \frac{G}{2} - \frac{\zeta}{4} \pm \left[\left(\frac{G}{2} + \frac{\zeta}{4}\right)^2 + \frac{3\zeta^2}{2}\right]^{\frac{1}{2}}$$

As a convenient way of displaying the dependence of $\epsilon(J)$ on G and ζ, we plot

$$\eta = \left(\epsilon - \frac{G}{4}\right)\left[G^2 + \left(\frac{5\zeta}{2}\right)^2\right]^{-\frac{1}{2}}$$

against $$\xi = \frac{\chi}{1 + \chi}$$

where $$\chi = \frac{5\zeta}{2G}$$

This is done in Fig. 1-1. The use of the functions η and ξ gives the figure the following properties:

1. The energy levels in both the limits $G = 0$ and $\zeta = 0$ can be accommodated on the same diagram.
2. The difference between the values of η corresponding to the highest and lowest levels is unity in both limits.
3. The center of gravity of the levels, calculated by giving each level a weight equal to its degeneracy $2J + 1$, lies on the line $\eta = 0$ for all ξ in the range $0 \leq \xi \leq 1$.
4. For any ξ, the difference between the values of η are proportional to the energy separations of the levels.

From the many examples of the configuration sd in the literature,[7] five have been selected and compared directly with the theory. It can be seen from Fig. 1-1 that the agreement is extremely good. The coincidence of pairs of levels in the limit $G = 0$ is somewhat unexpected. This is because we chose to couple the angular momenta $\mathbf{s}$ and $\mathbf{l}$ of the

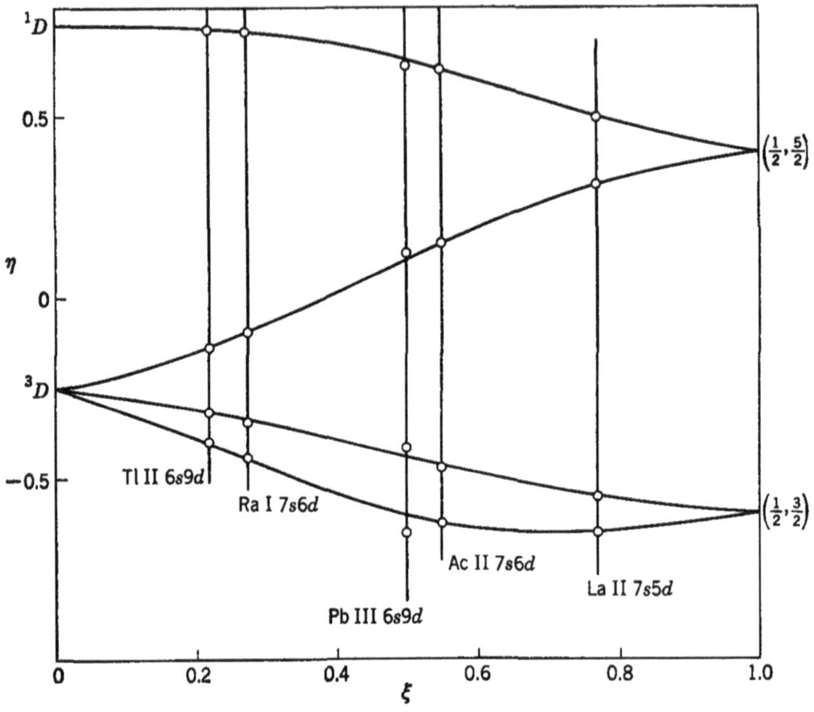

FIG. 1-1 The four levels of the configuration sd are plotted in intermediate coupling and compared with experiment. The energy levels in LS coupling (corresponding to $H_1 \gg H_2$) are shown on the left; the situation for jj coupling (for which $H_2 \gg H_1$) is shown on the right. Deviations between experiment and theory are due to configuration interaction.

individual electrons separately to $\mathbf{S}$ and $\mathbf{L}$ and then form the resultant $\mathbf{J}$,

$$\sum_i \mathbf{s}_i = \mathbf{S} \qquad \sum_i \mathbf{l}_i = \mathbf{L} \qquad \mathbf{S} + \mathbf{L} = \mathbf{J} \tag{1-35}$$

However, if $H_2 \gg H_1$, it is more appropriate to couple up each $\mathbf{s}_i$ and $\mathbf{l}_i$ and then combine the separate resultants,

$$\mathbf{s}_i + \mathbf{l}_i = \mathbf{j}_i \qquad \sum_i \mathbf{j}_i = \mathbf{J} \tag{1-36}$$

The coupling schemes represented by (1-35) and (1-36) are called *LS* coupling [or Russell-Saunders (RS) coupling] and *jj* coupling, respectively. If we adopt the second scheme, the 20 determinantal product states (1-33) are replaced by the set of 12 defining the z components m_j of the pair of j values $(\frac{5}{2},\frac{1}{2})$, namely,

$$\{\tfrac{5}{2}\,\tfrac{1}{2}\},\{\tfrac{3}{2}\,\tfrac{1}{2}\},\{\tfrac{1}{2}\,\tfrac{1}{2}\},\{-\tfrac{1}{2}\,\tfrac{1}{2}\},\{-\tfrac{3}{2}\,\tfrac{1}{2}\},\{-\tfrac{5}{2}\,\tfrac{1}{2}\},\{\tfrac{5}{2}\,-\tfrac{1}{2}\},\{\tfrac{3}{2}\,-\tfrac{1}{2}\},\{\tfrac{1}{2}\,-\tfrac{1}{2}\},$$
$$\{-\tfrac{1}{2}\,-\tfrac{1}{2}\},\{-\tfrac{3}{2}\,-\tfrac{1}{2}\},\{-\tfrac{5}{2}\,-\tfrac{1}{2}\} \qquad (1\text{-}37)$$

together with the following set of 8, which define the z components of $(\frac{3}{2},\frac{1}{2})$,

$$\{\tfrac{3}{2}\,\tfrac{1}{2}\},\{\tfrac{1}{2}\,\tfrac{1}{2}\},\{-\tfrac{1}{2}\,\tfrac{1}{2}\},\{-\tfrac{3}{2}\,\tfrac{1}{2}\},\{\tfrac{3}{2}\,-\tfrac{1}{2}\},\{\tfrac{1}{2}\,-\tfrac{1}{2}\},\{-\tfrac{1}{2}\,-\tfrac{1}{2}\},\{-\tfrac{3}{2}\,-\tfrac{1}{2}\} \qquad (1\text{-}38)$$

Since the eigenvalues of

$$\sum_i \mathbf{s}_i \cdot \mathbf{l}_i = \frac{1}{2}\sum_i (\mathbf{j}_i^2 - \mathbf{s}_i^2 - \mathbf{l}_i^2)$$

are

$$\frac{1}{2}\sum_i [j_i(j_i+1) - s_i(s_i+1) - l_i(l_i+1)]$$

the matrix of H_2 is now diagonal. The states (1-37) and (1-38), corresponding to $j_1 = \frac{5}{2}, j_2 = \frac{1}{2}$ and $j_1 = \frac{3}{2}, j_2 = \frac{1}{2}$, respectively, belong to the eigenvalues ζ and $-3\zeta/2$.

1-9 DEFICIENCIES IN THE METHOD

The approach of the previous section can in principle be extended to deal with any configuration; however, a blind application of the simple techniques that have so far been discussed overlooks the possibility of important simplifications in the mathematics. For example, when analyzing the configuration *sd*, we calculated the matrix elements

$$({}^{2S+1}D_J|H_2|\,{}^{2S+1}D_J)$$

for each J separately; however, the results satisfy the equation

$$({}^3D_J|H_2|\,{}^3D_J) - ({}^3D_{J-1}|H_2|\,{}^3D_{J-1}) = \kappa J \qquad (1\text{-}39)$$

where κ is independent of J. This equation can be generalized to any multiplet in any configuration; however, we shall not use the approach of Condon and Shortley to prove this but instead shall wait until we have tensor operator techniques at our disposal. Even when equations such as Eq. (1-39) are available, there are other reasons for supposing that the theory is not in so simple or so appropriate a form as it might be. For example, the determinantal product states introduce the eigenvalues m_s and m_l of the z components of **s** and **l**, and yet many perturbations do not specify an axis in any way. Again, in studying the properties of terms with small values of S and L in complex configurations (such as f^3), the transition from SM_SLM_L to sm_slm_l quantization often involves a large number of determinantal product states, prefaced by coefficients in which quite high prime numbers make their appearance; yet on calculating the matrix elements of various perturbations, the coefficients often combine in a striking and unexpected way to give a simple result. We shall see later how the theory of tensor operators, used in conjunction with group theory and the concept of fractional parentage, removes much of the superfluous mathematics and gives a deeper insight into the arrangement of the energy levels of a complex atom.

PROBLEMS

1-1. Prove that, in a configuration of n equivalent electrons, no term exists possessing maximum multiplicity and a value of L that is one unit less than the maximum for that multiplicity.

1-2. Show that, for all configurations of two equivalent electrons, the sum of the quantum numbers S and L characterizing any term is an even number.

1-3. Show that in f^3 the terms ^2D, ^2F, ^2G, and ^2H occur twice.

1-4. Obtain the following formulas for certain 3-j symbols:

$$\begin{pmatrix} j & j & 0 \\ m & -m & 0 \end{pmatrix} = (-1)^{j-m}(2j+1)^{-\frac{1}{2}}$$

$$\begin{pmatrix} j & k & j \\ -j & 0 & j \end{pmatrix} = \frac{(2j)!}{[(2j-k)!(2j+k+1)!]^{\frac{1}{2}}}$$

1-5. Prove that Eqs. (1-14) and (1-15) remain valid if the following substitutions are made:

$$J_+ \to -\eta\frac{\partial}{\partial\xi} \qquad J_- \to -\xi\frac{\partial}{\partial\eta}$$

$$J_z \to -\frac{1}{2}\left(\xi\frac{\partial}{\partial\xi} - \eta\frac{\partial}{\partial\eta}\right)$$

$$|J,M) \to f_M{}^J(\xi,\eta) = (-1)^{J-M}\xi^{J-M}\eta^{J+M}[(J-M)!(J+M)!]^{-\frac{1}{2}}$$

Find the linear combinations of $f_{m_1}{}^{j_1}(\xi_1,\eta_1)f_{m_2}{}^{j_2}(\xi_2,\eta_2)$ with fixed j_1 and j_2 that are eigenstates of

$$j_{1z} + j_{2z} = -\frac{1}{2}\left(\xi_1\frac{\partial}{\partial\xi_1} + \xi_2\frac{\partial}{\partial\xi_2} - \eta_1\frac{\partial}{\partial\eta_1} - \eta_2\frac{\partial}{\partial\eta_2}\right)$$

with eigenvalues $m = m_1 + m_2$ and of $(\mathbf{j}_1 + \mathbf{j}_2)^2$ with eigenvalues $j(j+1)$. Hence obtain the equation

$$
\begin{aligned}
(j_1m_1j_2m_2|j_1j_2jm) &= \delta(m_1+m_2, m)\left[\frac{(2j+1)(j_1+j_2-j)!(j_1-j_2+j)!(j_2-j_1+j)!}{(j_1+j_2+j+1)!}\right]^{\frac{1}{2}} \\
&\quad \times [(j_1+m_1)!(j_1-m_1)!(j_2+m_2)!(j_2-m_2)!(j+m)!(j-m)!]^{\frac{1}{2}} \\
\times \sum_z (-1)^z &\frac{1}{[z!(j_1+j_2-j-z)!(j_1-m_1-z)!(j_2+m_2-z)!\,(j-j_2+m_1+z)!(j-j_1-m_2+z)!]}
\end{aligned}
$$

and show that it is equivalent to Eq. (1-20). (This method is due to Sharp.[8])

1-6. Use the formula of Prob. 1-5 to prove that the coefficient of

$$u_1^{j_1-m_1}u_2^{j_1+m_1}u_3^{j_2+j_3-j_1}v_1^{j_2-m_2}v_2^{j_2+m_2}v_3^{j_3+j_1-j_2}w_1^{j_3-m_3}w_2^{j_3+m_3}w_3^{j_1+j_2-j_3}$$

in the expansion of

$$\begin{vmatrix} u_1 & v_1 & w_1 \\ u_2 & v_2 & w_2 \\ u_3 & v_3 & w_3 \end{vmatrix}^J$$

is

$$
\begin{aligned}
(-1)^J &\begin{pmatrix} j_1 & j_2 & j_3 \\ m_1 & m_2 & m_3 \end{pmatrix} (J+1)^{\frac{1}{2}}(J!)^{\frac{1}{2}} \\
&\times [(j_1-m_1)!(j_1+m_1)!(j_2+j_3-j_1)!(j_2-m_2)!(j_2+m_2)!(j_1+j_3-j_2)! \\
&\qquad (j_3-m_3)!(j_3+m_3)!(j_1+j_2-j_3)!]^{-\frac{1}{2}}
\end{aligned}
$$

Deduce that the symbol defined by

$$\begin{bmatrix} j_2+j_3-j_1 & j_3+j_1-j_2 & j_1+j_2-j_3 \\ j_1-m_1 & j_2-m_2 & j_3-m_3 \\ j_1+m_1 & j_2+m_2 & j_3+m_3 \end{bmatrix} = \begin{pmatrix} j_1 & j_2 & j_3 \\ m_1 & m_2 & m_3 \end{pmatrix}$$

possesses the following properties:

1. Even permutations of columns or rows leave the symbol unchanged.
2. Odd permutations of columns or rows introduce the phase factor $(-1)^{j_1+j_2+j_3}$.
3. Transposition leaves the symbol unchanged (see Regge[5]).

1-7. Prove that

$$|f^3\,{}^2K, M_S = \tfrac{1}{2}, M_L = 7) = (\tfrac{3}{8})^{\frac{1}{2}}\{\overset{+}{3}\overset{-}{3}\overset{+}{1}\} + (\tfrac{5}{8})^{\frac{1}{2}}\{\overset{+}{3}\overset{+}{2}\overset{-}{2}\}$$

Use this expansion to show that

$$({}^2K_{15/2}|H_2|\,{}^2K_{15/2}) = \frac{9\zeta_f}{8}$$

1-8. Make use of Eqs. (1-29) and (1-30) to prove that

$$(f^3\gamma_1\,{}^2H_{11/2}|H_2|f^3\gamma_1\,{}^2H_{11/2}) = 0$$

and

$$(f^3\gamma_2\,{}^2H_{11/2}|H_2|f^3\gamma_1\,{}^2H_{11/2}) = 0$$

1-9. The ground configuration of PuI comprises six $5f$ electrons. Prove that three ^5D terms occur, and complete the expansion

$$|\gamma(x,y)\ {}^5D, M_S = M_L = 2)$$
$$= n\{\overset{+}{3}\overset{+}{2}\overset{+}{1}\overset{+}{0}\,\overset{+}{-1}\,\overset{-}{-3}\} + nx\{\overset{+}{3}\overset{+}{2}\overset{+}{1}\overset{+}{0}\,\overset{-}{-1}\,\overset{+}{-3}\} + ny\{\overset{+}{3}\overset{+}{2}\overset{-}{1}\overset{+}{0}\,\overset{+}{-1}\,\overset{+}{-3}\} + \cdots$$

Show that $\gamma_1 = \gamma(0,0)$, $\gamma_2 = \gamma(-\frac{3}{2},0)$, and $\gamma_3 = \gamma(-\frac{1}{8},-\frac{7}{18})$ define three orthogonal states. Suppose that the actual states existing in the plutonium atom are $\gamma_i = \gamma(x_i,y_i)$, where $i = 1, 2, 3$. Show that the six equations

$$(\gamma_i|\gamma_j) = 0 \qquad (\gamma_i|H_1|\gamma_j) = 0$$

for $i \neq j$ can be regarded as defining the vertices (x_i,y_i) of the common self-conjugate triangle with respect to two conics $S = 0$ and $S' = 0$. Prove that the eigenvalues of H_1 are given by the values of λ that ensure that the conic $S' - \lambda S = 0$ is a pair of straight lines. (The expansion in determinantal product states has been used in the analysis of the ^5D multiplet of EuIV.[9])

1-10. Show that the transformation from $SLJM_J$ to SM_SLM_L quantization may be performed when $J \neq M_J$ by means of the equivalence

$$(2S_zL_z + S_+L_- + S_-L_+)|SLJM_J) = [J(J+1) - S(S+1) - L(L+1)]|SLJM_J)$$

Use this method to check Eq. (1-17). Show that the analogous operators to use for the transformation to sm_slm_l quantization are

$$\sum_{i\neq j} \mathbf{l}_i \cdot \mathbf{l}_j \qquad \text{and} \qquad \sum_{i\neq j} \mathbf{s}_i \cdot \mathbf{s}_j$$

and check the expansion of Prob. 1-7. (These techniques are due to Stevens.[10])

1-11. Prove that the only terms of f^7 with $S \geq \frac{5}{2}$ are ^8S, ^6P, ^6D, ^6F, ^6G, ^6H, and ^6I. Prove that

$$(f^7\ {}^8S_{\frac{7}{2}}|H_2|f^7\ {}^6P_{\frac{7}{2}}) = (14)^{\frac{1}{2}}\zeta$$

2

CRYSTAL FIELDS

2-1 THE CONTRIBUTION TO THE HAMILTONIAN

It often happens that the electrons of an ion that is situated in a crystal lattice are sufficiently localized to allow the effect of the environment of the ion to be treated as a perturbation on the system of configurations of the free ion. The absorption and fluorescence spectra of salts for which this situation obtains vary considerably from salt to salt: sometimes, as for crystals containing ions of the rare-earth series, the lines are quite sharp, and the energy level structures of the free ions can be deduced from the spectra. It is fairly simple to arrange matters so that all the constituent components of a crystal except the ion under investigation are spectroscopically inert within a certain range of wavelengths. Ions can often be substituted into inert lattices; in this way interactions between spectroscopically active ions in neighboring cells of the lattice can be virtually eliminated, and the analyses of the spectra give directly the energy-level system of a single ion subject to the influence of the crystal lattice. In what follows we shall fix our attention on one such ion, taking its nucleus as the origin of the coordinate system. The theory of crystal fields and in particular the interpretation of their effects in terms of the theory of groups are due to Bethe.[11]

As a first approximation, the presence of the surrounding atoms and ions can be allowed for by supposing their effect to be purely electrostatic. If $V(r,\theta,\phi)$ is the electric potential produced by this complex system of atoms and ions, the contribution to the Hamiltonian is

$$H_3 = -e \sum_i V(r_i,\theta_i,\phi_i)$$

where the sum extends over all the electrons of the central ion. It is usual to expand this in a series of spherical harmonics,[12]

$$H_3 = \sum_{i,k,q} B_k{}^q r_i{}^k Y_{kq}(\theta_i,\phi_i) \tag{2-1}$$

As will be seen, the number of terms in this series that need to be considered is often quite small and in many cases the constants $B_k{}^q$ can be treated as adjustable parameters. When the structure of the crystal lattice is known, they can in principle be calculated. The potential at (r,θ,ϕ) produced by a charge $-ge$ at (ρ,α,β) is

$$V = -ge/|\mathbf{r} - \boldsymbol{\rho}| = -ge \sum_k r^k P_k(\cos \omega)/\rho^{k+1}$$

where ω is the angle between the vectors $\mathbf{r}$ and $\boldsymbol{\rho}$ and P_k is a Legendre polynomial. The expansion is valid in the region $r < \rho$. With the aid of the addition theorem for spherical harmonics,[1] namely

$$P_k(\cos \omega) = \frac{4\pi}{2k+1} \sum_q Y^*_{kq}(\alpha,\beta) Y_{kq}(\theta,\phi) \tag{2-2}$$

it is easy to show that, for a number of charges $-g_je$,

$$\begin{aligned} B_k{}^q &= \frac{4\pi}{2k+1} \sum_j \frac{g_je^2}{\rho_j{}^{k+1}} Y^*_{kq}(\alpha_j,\beta_j) \\ &= (-1)^q \left\{\frac{4\pi(k-q)!}{(2k+1)(k+q)!}\right\}^{\frac{1}{2}} \sum_j \frac{g_je^2}{\rho_j{}^{k+1}} P_k{}^q(\cos \alpha_j)e^{-iq\beta_j} \end{aligned} \tag{2-3}$$

As an example, we consider the highly symmetrical configuration of charges $-ge$ at the points $(\pm\rho,0,0)$, $(0,\pm\rho,0)$, and $(0,0,\pm\rho)$. (If the charges are negative, $g > 0$.) With the aid of the equation

$$Y_{k,-q}(\theta,\phi) = (-1)^q Y^*_{kq}(\theta,\phi) \tag{2-4}$$

we get

$$\begin{aligned} B_k{}^q &= (-1)^q B_k{}^{-q} \\ &= (-1)^q \frac{ge^2}{\rho^{k+1}} \left\{\frac{4\pi(k-q)!}{(2k+1)(k+q)!}\right\}^{\frac{1}{2}} [P_k{}^q(1) + 4P_k{}^q(0) + P_k{}^q(-1)] \end{aligned}$$

provided $\frac{1}{4}q$ is an integer or zero. Coefficients $B_k{}^q$, for which this is not the case, all vanish. Substituting back into Eq. (2-1), and limiting the expansion to terms for which $k \leq 6$, we obtain

$$\begin{aligned} H_3' = A_4{}^0\Sigma(35z^4 - 30z^2r^2 + 3r^4) + A_4{}^4\Sigma(x^4 - 6x^2y^2 + y^4) \\ + A_6{}^0\Sigma(231z^6 - 315z^4r^2 + 105z^2r^4 - 5r^6) \\ + A_6{}^4\Sigma(11z^2 - r^2)(x^4 - 6x^2y^2 + y^4) \end{aligned} \quad (2\text{-}5)$$

where

$$A_4{}^0 = \frac{7ge^2}{16\rho^5} \qquad A_4{}^4 = \frac{35ge^2}{16\rho^5}$$
$$A_6{}^0 = \frac{3ge^2}{64\rho^7} \qquad A_6{}^4 = \frac{-63ge^2}{64\rho^7} \quad (2\text{-}6)$$

For simplicity, the running suffix i has not been attached to the coordinates of the electrons in Eq. (2-5).

2-2 FINITE GROUPS

The effect of the crystalline electric field on the energy-level structure of an ion embedded in the crystal lattice can be allowed for by including H_3 in the Hamiltonian of the free ion. For ions of the rare-earth series, $H_1 \gg H_2 \gg H_3$, and as a first approximation each level, characterized by the quantum number J and possessing a $(2J + 1)$fold degeneracy, can be treated as an isolated entity. The rows and columns of the matrix of H_3 are labeled by M_J; the entries are found by expanding the states in determinantal product states and using the fact that H_3 is an operator of the type F. (It will be seen later that this procedure can be greatly simplified.) The eigenvalues are obtained by solving the secular equation. For ions of the iron group, $H_3 \gg H_2$: the states $|\gamma SM_SLM_L)$ form a more convenient basis, and the spin-orbit coupling H_2 is added after the Coulomb interaction and crystal field effects have been allowed for.

There is no suggestion in this procedure that the theory is incomplete. In practice, however, it is found that many calculations give unexpected results; for example, the eigenvalues often exhibit a surprising degeneracy, especially in cases where the symmetry at an ion is high. Matrix elements of simple operators $\left(\text{such as } L_z, \sum_i z_i, \text{etc.}\right)$ between eigenstates of $H_1 + H_2 + H_3$ sometimes vanish for no apparent reason. The appropriate branch of mathematics for giving an understanding of these properties is the theory of groups. It would be too great a digression to derive here the well-known results of elementary group theory; however, an appreciation of the principal concepts and results of the theory is very helpful for an understanding of the properties of semi-simple continuous groups, which will be discussed in detail later. We shall therefore limit ourselves to a review of the elementary theory; the reader is referred to the works of Lomont[13] and Wigner[14] for extensive accounts of the subject.

A finite group $\mathcal{G}$ is a set of g distinct elements $\mathcal{R}_1, \mathcal{R}_2, \ldots, \mathcal{R}_g$ that satisfy the following conditions:

1. Every pair of elements $\mathcal{R}_\sigma$, $\mathcal{R}_\rho$ can be combined as a product $\mathcal{R}_\sigma\mathcal{R}_\rho$ that again belongs to the set.

2. The associative law is valid, i.e.,

$$\mathcal{R}_\tau(\mathcal{R}_\sigma\mathcal{R}_\rho) = (\mathcal{R}_\tau\mathcal{R}_\sigma)\mathcal{R}_\rho$$

3. The set contains an identity element $\mathcal{I}$ such that

$$\mathcal{R}_\sigma\mathcal{I} = \mathcal{I}\mathcal{R}_\sigma = \mathcal{R}_\sigma$$

for all σ.

4. For each element $\mathcal{R}_\sigma$ there exists an inverse $\mathcal{R}_\sigma^{-1}$ belonging to the set such that

$$\mathcal{R}_\sigma\mathcal{R}_\sigma^{-1} = \mathcal{R}_\sigma^{-1}\mathcal{R}_\sigma = \mathcal{I}$$

Two groups $\mathcal{G}$ and $\mathcal{H}$ are said to be *isomorphic* if a unique correspondence can be drawn between the elements $\mathcal{R}_1, \mathcal{R}_2, \ldots$ of $\mathcal{G}$ and $\mathcal{P}_1, \mathcal{P}_2, \ldots$ of $\mathcal{H}$ such that the products $\mathcal{R}_\sigma\mathcal{R}_\rho$ correspond to $\mathcal{P}_\sigma\mathcal{P}_\rho$ for all σ and ρ. If all the elements of $\mathcal{G}$ commute, then $\mathcal{G}$ is called an *Abelian* group. The element $\mathcal{R}_\rho\mathcal{R}_\sigma\mathcal{R}_\rho^{-1} = \mathcal{R}_\tau$ is said to be *conjugate* to $\mathcal{R}_\sigma$. A set of elements that are conjugate to one another is called a *class;* two classes cannot have an element in common, and hence the elements of $\mathcal{G}$ can be completely subdivided into distinct classes. The identity element always forms a class by itself.

If the only interpretation of the elements $\mathcal{G}$ is that they are the elements of a group, then the group is said to be *abstract*: if they are quantities of a special kind, for example, if they are matrices or permutations, then they are said to form a *representation* of the corresponding abstract group. For the matrices $R_1, R_2, \ldots, R_\sigma, \ldots, R_g$ to form a representation of the group $\mathcal{G}$, every element $\mathcal{R}_\sigma$ must correspond to a nonsingular matrix R_σ such that $\mathcal{R}_\sigma\mathcal{R}_\rho$ corresponds to $R_\sigma R_\rho$. The correspondence is not necessarily unique; for example, every $\mathcal{R}_\sigma$ could correspond to a unit matrix. Unlike the abstract group, the representations need not be composed of distinct entities. If the elements of the group can be interpreted as operators that carry equivalent points of a certain space into one another, it is usually a simple matter to construct matrix representations. Suppose that under the operations of the group the n linearly independent functions $\phi_1, \phi_2, \ldots, \phi_n$

transform among themselves according to the equation

$$\mathcal{R}_\sigma \phi_i = \sum_j r_{ji}(\sigma)\phi_j$$

Under the action of a second operator $\mathcal{R}_\rho$, we get

$$\begin{aligned}\mathcal{R}_\rho \mathcal{R}_\sigma \phi_i &= \sum_j r_{ji}(\sigma)\mathcal{R}_\rho \phi_j \\ &= \sum_{j,k} r_{ji}(\sigma) r_{kj}(\rho)\phi_k = \sum_k (R_\rho R_\sigma)_{ki}\phi_k\end{aligned}$$

provided we interpret the quantity $r_{ji}(\sigma)$ as the element in row j and column i of the matrix R_σ; that is, as the quantity $(R_\sigma)_{ji}$. This demonstrates that $\mathcal{R}_\rho \mathcal{R}_\sigma$ corresponds to $R_\rho R_\sigma$, and so the matrices R_σ form an n-dimensional representation of the group. The functions ϕ_j are said to form a *basis* for the representation; it is usually advantageous to select an orthonormal set, in which case the matrices are unitary.

If the matrices R_σ form a representation of $\mathcal{G}$, then so do the matrices $P_\sigma = SR_\sigma S^{-1}$, where S is a nonsingular matrix. The representations Γ_R and Γ_P, comprising the matrices R_1, R_2, . . . and P_1, P_2, . . . , respectively, are called *equivalent* representations. A representation Γ_R is said to be *reducible* if a similarity transformation can be found that throws all the matrices R_σ (assumed unitary) into the form

$$\begin{pmatrix} R'_\sigma & 0 \\ 0 & R''_\sigma \end{pmatrix}$$

where R'_σ and R''_σ are square unitary matrices whose dimensions are independent of σ. If it is impossible to find such a transformation, the representation is *irreducible*. The decomposition of the reducible representation into its constituent representations $\Gamma_{R'}$ and $\Gamma_{R''}$ is symbolized by

$$\Gamma_R = \Gamma_{R'} + \Gamma_{R''}$$

The development of the theory of representations rests upon an important theorem known as Schur's lemma. Weyl[15] has expressed the content of this theorem in the following form:

1. If R_σ and Q_σ form two inequivalent irreducible matrix representations of a group, and if a matrix A (possibly rectangular) exists such that

$$R_\sigma A = AQ_\sigma$$

for all σ, then $A = 0$.

2. If the matrices R_σ form an irreducible representation of a group, and if

$$R_\sigma A = AR_\sigma$$

for all σ, then A is a constant times the unit matrix.

From (1) it can be shown that, if the matrices R_σ and Q_σ of two inequivalent irreducible representations are unitary, then their elements satisfy the equation

$$\sum_\sigma r_{ij}(\sigma) q_{lk}^*(\sigma) = 0 \tag{2-7}$$

for all i, j, k, and l. Part 2 leads to the equation

$$\sum_\sigma r_{ij}(\sigma) r_{lk}^*(\sigma) = g\delta(i,l)\delta(j,k)/n_R \tag{2-8}$$

provided that the matrices R_σ, of dimension n_R, are unitary and form an irreducible representation of the group.

The character $\chi^R(\sigma)$ of the matrix R_σ is defined as the sum of its diagonal elements,

$$\chi^R(\sigma) = \sum_i r_{ii}(\sigma)$$

Since the trace of a matrix is invariant under a similarity transformation, the characters of all the matrices R_σ belonging to a given class are the same. For the same reason, the decomposition

$$\Gamma_W = \sum_R c_R \Gamma_R$$

which simply states that the irreducible representation Γ_R occurs c_R times in the decomposition of Γ_W, can be expressed algebraically by the equation

$$\chi^W(\sigma) = \sum_R c_R \chi^R(\sigma)$$

for all σ.

Suppose that the n functions $\phi_1, \phi_2, \cdots, \phi_n$ form a basis for an n-dimensional representation Γ_X of $\mathcal{G}$ and that the m functions $\psi_1, \psi_2, \ldots, \psi_m$ form a basis for an m-dimensional representation Γ_Y of $\mathcal{G}$. Under the operations of the group, the mn functions $\phi_i\psi_k$ transform into linear combinations of themselves and therefore form a basis for a representation Γ_W of $\mathcal{G}$. This representation is called the *direct product*, or *Kronecker product*, of the representations Γ_X and Γ_Y

and is denoted by $\Gamma_X \times \Gamma_Y$. If

$$\mathcal{R}_\sigma \phi_i = \sum_j r_{ji}(\sigma)\phi_j$$

and

$$\mathcal{R}_\sigma \psi_k = \sum_l q_{lk}(\sigma)\psi_l$$

then

$$\mathcal{R}_\sigma(\phi_i\psi_k) = \sum_{j,l} r_{ji}(\sigma)q_{lk}(\sigma)(\phi_j\psi_l)$$

Since

$$\sum_{i,k} r_{ii}(\sigma)q_{kk}(\sigma) = \left(\sum_i r_{ii}(\sigma)\right)\left(\sum_k q_{kk}(\sigma)\right)$$

the characters of the representations obey the equation

$$\chi^W(\sigma) = \chi^X(\sigma)\chi^Y(\sigma) \tag{2-9}$$

In most cases Γ_W is reducible.

2-3 THE OCTAHEDRAL GROUP

As an illustration of the preceding summary and to prepare the ground for subsequent applications, we examine the group O of rotations that send an octahedron into itself. Every such rotation carries the cube whose vertices are the centers of the faces into itself, and it is clear that the octahedral and cubic groups are isomorphic. The four fixed axes passing through opposite vertices of the cube are labeled 1, 2, 3, and 4, as shown in Fig. 2-1. If a rotation of the group carries the

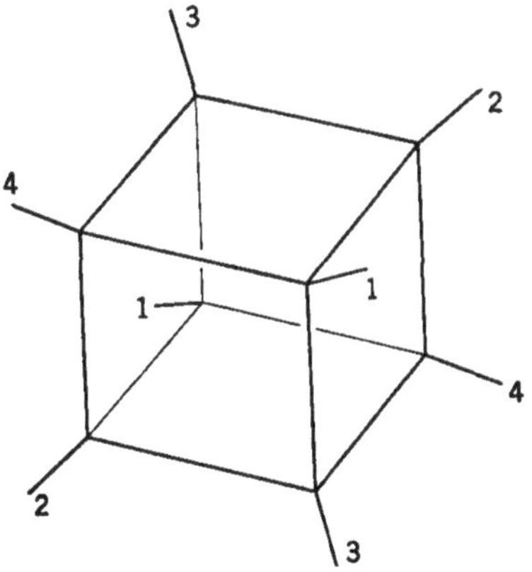

FIG. 2-1 The axes 1, 2, 3, and 4 pass through opposite vertices of the cube and are to be regarded as fixed in space. A rotation which sends the cube into itself can be defined by stating that the vertices which formerly lay on axis a now lie on axis 1, those which formerly lay on axis b now lie on axis 2, etc. The rotation is represented by the permutation

$$\begin{pmatrix} 1234 \\ abcd \end{pmatrix}$$

vertices lying on axis a into positions on axis 1, vertices on axis b into positions on axis 2, etc., the operation is conveniently denoted by the symbol

$$\begin{pmatrix}1234\\abcd\end{pmatrix}$$

The upper row is useful as a guide to the eye in combining more than one rotation,

$$\begin{pmatrix}1&\cdot&\cdot&\cdot\\a&\cdot&\cdot&\cdot\end{pmatrix}\begin{pmatrix}\cdot&a&\cdot&\cdot\\\cdot&e&\cdot&\cdot\end{pmatrix}\begin{pmatrix}\cdot&e&\cdot&\cdot\\\cdot&f&\cdot&\cdot\end{pmatrix}=\begin{pmatrix}1&\cdot&\cdot&\cdot\\f&\cdot&\cdot&\cdot\end{pmatrix}$$

Although operations on the right are performed first (i.e., the vertices on axis f are moved to axis e, then to axis a, then to axis 1), it is convenient to construct the final permutation by beginning at the left and working to the right (i.e., observe that a lies beneath 1 in the first permutation, e lies beneath a in the second, and f lies beneath e in the third: thus f lies beneath 1 in the result). For example,

$$\begin{pmatrix}1234\\2341\end{pmatrix}\begin{pmatrix}1234\\1243\end{pmatrix}=\begin{pmatrix}1234\\2431\end{pmatrix}$$

The operations that send the cube into itself are set out below.

1. The identity operation

$$\begin{pmatrix}1234\\1234\end{pmatrix}$$

2. Rotations by π about axes joining the midpoints of opposite edges of the cube,

$$\begin{pmatrix}1234\\1243\end{pmatrix},\begin{pmatrix}1234\\1432\end{pmatrix},\begin{pmatrix}1234\\1324\end{pmatrix},\begin{pmatrix}1234\\4231\end{pmatrix},\begin{pmatrix}1234\\3214\end{pmatrix},\begin{pmatrix}1234\\2134\end{pmatrix}$$

3. Rotations by π about axes passing through the centers of opposite faces of the cube,

$$\begin{pmatrix}1234\\3412\end{pmatrix},\begin{pmatrix}1234\\2143\end{pmatrix},\begin{pmatrix}1234\\4321\end{pmatrix}$$

4. Rotations by $2\pi/3$ about axes 1, 2, 3 and 4,

$$\begin{pmatrix}1234\\1342\end{pmatrix},\begin{pmatrix}1234\\1423\end{pmatrix},\begin{pmatrix}1234\\4213\end{pmatrix},\begin{pmatrix}1234\\3241\end{pmatrix},\begin{pmatrix}1234\\4132\end{pmatrix},\begin{pmatrix}1234\\2431\end{pmatrix},\begin{pmatrix}1234\\3124\end{pmatrix},\begin{pmatrix}1234\\2314\end{pmatrix}$$

5. Rotations by $\frac{1}{2}\pi$ about axes passing through the centers of opposite faces of the cube,

$$\begin{pmatrix}1234\\2341\end{pmatrix}, \begin{pmatrix}1234\\4123\end{pmatrix}, \begin{pmatrix}1234\\4312\end{pmatrix}, \begin{pmatrix}1234\\3421\end{pmatrix}, \begin{pmatrix}1234\\2413\end{pmatrix}, \begin{pmatrix}1234\\3142\end{pmatrix}$$

These 24 operations account for all the permutations on four objects: the octahedral group O is clearly isomorphic with the symmetric group of the fourth degree, S_4. By examining triple products of the type $\mathcal{R}_\rho \mathcal{R}_\sigma \mathcal{R}_\rho^{-1}$, for example,

$$\begin{pmatrix}1234\\2341\end{pmatrix}\begin{pmatrix}1234\\1243\end{pmatrix}\begin{pmatrix}1234\\4123\end{pmatrix} = \begin{pmatrix}1234\\1324\end{pmatrix}$$

it can be shown that the sets of elements 1, 2, . . . , 5 above form five classes, which we denote by C_1, C_2, . . . , C_5. The characters χ_j^R for the matrices R_σ belonging to the class C_j and the irreducible representation Γ_R are given in Table 2-1.

Table 2-1 The Characters of the Octahedral Group O

	C_1	C_2	C_3	C_4	C_5
Γ_1	1	1	1	1	1
Γ_2	1	−1	1	1	−1
Γ_3	2	0	2	−1	0
Γ_4	3	−1	−1	0	1
Γ_5	3	1	−1	0	−1

Bases for the irreducible representations may be easily found. Take x, y, and z axes parallel to the edges of the cube and passing through the centers of its faces. The points

$$(x,y,z),\ (y,x,-z),\ (x,-y,-z),\ (z,x,y),\ (x,-z,y)$$

can be obtained from the point (x,y,z) by means of a single operation from each of the five classes. The traces of the three-dimensional matrices that represent these operations are 3, −1, −1, 0, 1, respectively. From the character table we conclude that the functions x, y, and z form a basis for the irreducible representation Γ_4 of O. Under the same operations, the triple product xyz transforms into

$$xyz,\quad -yxz,\quad xyz,\quad zxy,\quad -xzy$$

Hence the function xyz forms by itself a basis for the irreducible representation Γ_2. Of course, the representations derived from using functions of x, y, and z as bases are not necessarily irreducible; for example, x^2, y^2, z^2 transform into

$$(x^2,y^2,z^2),\ (y^2,x^2,z^2),\ (x^2,y^2,z^2),\ (z^2,x^2,y^2),\ (x^2,z^2,y^2)$$

and the sequence of characters 3, 1, 3, 0, 1 does not appear in the table. It can, however, be reproduced by the sum $\Gamma_1 + \Gamma_3$, showing that the functions x^2, y^2, and z^2 can be combined into an invariant form (namely, $x^2 + y^2 + z^2 = r^2$) and into two linearly independent expressions ($x^2 - y^2$ and $3z^2 - r^2$, for example) that form a basis for the representation Γ_3 of O.

As an illustration of a Kronecker product, we take the functions x_1, y_1, z_1 and x_2, y_2, z_2, the coordinates of two electrons. Each triad separately forms a basis for the irreducible representation Γ_4 of O. From Table 2-1 and Eq. (2-9), we find that the nine functions x_1x_2, x_1y_2, . . . , z_1z_2 form a basis for a representation of O whose characters are 9, 1, 1, 0, 1. The decomposition of this representation is expressed by the equation

$$\Gamma_4 \times \Gamma_4 = \Gamma_1 + \Gamma_3 + \Gamma_4 + \Gamma_5$$

The bases ϕ_i for Γ_1, Γ_3, Γ_4, and Γ_5 may be easily constructed by inspection. A convenient orthogonal set, normalized in the same way as the original functions, is the following:

$$\begin{aligned}
\Gamma_1\colon \phi_1 &= (\tfrac{1}{3})^{\frac{1}{2}}(x_1x_2 + y_1y_2 + z_1z_2)\\
\Gamma_3\colon \phi_2 &= (\tfrac{1}{2})^{\frac{1}{2}}(x_1x_2 - y_1y_2)\\
\phi_3 &= (\tfrac{1}{6})^{\frac{1}{2}}(2z_1z_2 - x_1x_2 - y_1y_2)\\
\Gamma_5\colon \phi_4 &= (\tfrac{1}{2})^{\frac{1}{2}}(y_1z_2 + z_1y_2)\\
\phi_5 &= (\tfrac{1}{2})^{\frac{1}{2}}(z_1x_2 + x_1z_2)\\
\phi_6 &= (\tfrac{1}{2})^{\frac{1}{2}}(x_1y_2 + y_1x_2)\\
\Gamma_4\colon \phi_7 &= (\tfrac{1}{2})^{\frac{1}{2}}(y_1z_2 - z_1y_2)\\
\phi_8 &= (\tfrac{1}{2})^{\frac{1}{2}}(z_1x_2 - x_1z_2)\\
\phi_9 &= (\tfrac{1}{2})^{\frac{1}{2}}(x_1y_2 - y_1x_2)
\end{aligned} \tag{2-10}$$

2-4 THE ROTATION GROUP IN THREE DIMENSIONS

The treatment of the previous section can be repeated for any group that sends a geometrical figure into itself. A natural extension of the theory is to the rotation group in three dimensions, R_3, since it can be regarded as the group comprising the operations that send a sphere into itself. Apart from the importance it enjoys in its own right, it serves as a link between the finite groups on the one hand and, on the other, the continuous groups, of which it is a particularly simple example.

It is well known that, under a rotation of axes through the Euler angles α, β, γ, a spherical harmonic $Y_{lm}(\theta, \phi)$ transforms into a function that can be expressed as a linear combination of the spherical harmonics of rank l. Carrying over the language of the theory of finite groups, we say that the $2l + 1$ harmonics form a basis for a representation of R_3. It can be shown that this representation, which is denoted by $\mathfrak{D}_l$, is irreducible. For the octahedral group O, all the operations of a class correspond to rotations through the same angle. This is also true for R_3; furthermore, all rotations through the same angle belong to a single class, irrespective of the axis about which any of the rotations is made. A simple proof of this statement can be constructed by considering the successive positions of a point on the surface of a sphere. To find the character common to all elements of a class, we choose the rotation to be through an angle α about the z axis. Since

$$Y_{lm}(\theta, \phi + \alpha) = e^{im\alpha}Y_{lm}(\theta,\phi) \tag{2-11}$$

the representation matrix is diagonal, with a trace given by

$$\begin{aligned}\chi_\alpha{}^l &= e^{il\alpha} + e^{i(l-1)\alpha} + \cdots + e^{-il\alpha} \\ &= \frac{e^{i(l+\frac{1}{2})\alpha} - e^{-i(l+\frac{1}{2})\alpha}}{e^{i\frac{1}{2}\alpha} - e^{-i\frac{1}{2}\alpha}} = \frac{\sin \frac{1}{2}(2l+1)\alpha}{\sin \frac{1}{2}\alpha}\end{aligned} \tag{2-12}$$

The octahedral group is a *subgroup* of R_3: the elements of O are contained in R_3 and satisfy the group postulates among themselves. The functions $Y_{lm}(\theta,\phi)$ form a basis for the irreducible representation $\mathfrak{D}_l$ of R_3 and also for a representation of O, which, in general, is reducible. To discover how the representations $\mathfrak{D}_l$ decompose when considered as representations of O, the characters of $\mathfrak{D}_l$ are evaluated for the angles of rotation occurring in O, and the sequence of numbers is then expressed as a linear combination of the rows of the character table for O. The angles of rotation associated with the classes C_1, C_2, C_3, C_4, and C_5 of O are 0, π, π, $2\pi/3$, and $\pi/2$, respectively. It is easy to show from Eq. (2-12) that

$$\chi_0{}^3 = 7 \qquad \chi_\pi{}^3 = -1 \qquad \chi^3_{2\pi/3} = 1 \qquad \chi^3_{\pi/2} = -1$$

By inspection, the sequence of numbers 7, -1, -1, $1, -1$ can be reproduced by the sum of the characters of the irreducible representations Γ_2, Γ_4, and Γ_5 of O. We say that, under the reduction $R_3 \rightarrow O$, the irreducible representation $\mathfrak{D}_3$ decomposes according to

$$\mathfrak{D}_3 \rightarrow \Gamma_2 + \Gamma_4 + \Gamma_5$$

This is an example of a *branching rule.* A set of rules for $l \leq 6$ is given in Table 2-2.

TABLE 2-2 BRANCHING RULES FOR THE REDUCTION $R_3 \rightarrow O$

$$\begin{aligned}
&\mathfrak{D}_0 \rightarrow \Gamma_1\\
&\mathfrak{D}_1 \rightarrow \Gamma_4\\
&\mathfrak{D}_2 \rightarrow \Gamma_3 + \Gamma_5\\
&\mathfrak{D}_3 \rightarrow \Gamma_2 + \Gamma_4 + \Gamma_5\\
&\mathfrak{D}_4 \rightarrow \Gamma_1 + \Gamma_3 + \Gamma_4 + \Gamma_5\\
&\mathfrak{D}_5 \rightarrow \Gamma_3 + 2\Gamma_4 + \Gamma_5\\
&\mathfrak{D}_6 \rightarrow \Gamma_1 + \Gamma_2 + \Gamma_3 + \Gamma_4 + 2\Gamma_5
\end{aligned}$$

The decomposition of the Kronecker product $\mathfrak{D}_j \times \mathfrak{D}_k$ is not difficult to obtain. Substituting the characters $\chi_\alpha{}^j$ and $\chi_\alpha{}^k$ as given by Eq. (2-12), into Eq. (2-9), we find

$$\begin{aligned}
\chi_\alpha{}^j\chi_\alpha{}^k &= [e^{ij\alpha} + e^{i(j-1)\alpha} + \cdots + e^{-ij\alpha}]\\
&\qquad\qquad\qquad \times [e^{ik\alpha} + e^{i(k-1)\alpha} + \cdots + e^{-ik\alpha}]\\
&= [e^{i(j+k)\alpha} + e^{i(j+k-1)\alpha} + \cdots + e^{-i(j+k)\alpha}]\\
&\qquad + [e^{i(j+k-1)\alpha} + e^{i(j+k-2)\alpha} + \cdots + e^{-i(j+k-1)\alpha}]\\
&\qquad\qquad + \cdots + [e^{i|j-k|\alpha} + e^{i(|j-k|-1)\alpha} + \cdots + e^{-i|j-k|\alpha}]\\
&= \chi_\alpha^{j+k} + \chi_\alpha^{j+k-1} + \cdots + \chi_\alpha^{|j-k|}
\end{aligned}$$

Therefore $$\mathfrak{D}_j \times \mathfrak{D}_k = \mathfrak{D}_{j+k} + \mathfrak{D}_{j+k-1} + \cdots + \mathfrak{D}_{|j-k|} \qquad (2\text{-}13)$$

2-5 EIGENFUNCTIONS AND OPERATORS

We are now in a position to begin applying group theory to problems in atomic spectroscopy. Suppose that the linearly independent functions ϕ_i are the eigenfunctions of an operator H corresponding to the eigenvalue E,

$$H\phi_i = E\phi_i \qquad i = 1, 2, \ldots, n \qquad (2\text{-}14)$$

If H is invariant under the operations of a group $\mathcal{G}$, then the various functions $\mathcal{R}_\sigma\phi_i$ are eigenfunctions of H corresponding to the eigenvalue E. Accordingly, any one of them must be expressible as a linear combination of the original functions ϕ_i, and the eigenfunctions ϕ_i therefore form a basis for an n-dimensional representation Γ_R of $\mathcal{G}$. The symbol Γ_R may be used to label the n-fold degenerate level; if Γ_R is reducible, the degeneracy is said to be *accidental.* In this case, two or more sets of linear combinations of the eigenfunctions ϕ_i can be constructed, each set forming a basis for an irreducible representation of $\mathcal{G}$. Group theory, in determining what irreducible representations of $\mathcal{G}$ can arise from a given set of basis functions, fixes the minimum amount of degeneracy that can occur; at the same time, it makes no guarantee that this extreme case obtains, though instances of accidental degeneracy are extremely rare (see Problem 2-3).

For the octahedral group O, we are able to investigate the transformation properties of functions of x, y, and z by the simple expedient of making replacements of the type $x \to y$, $y \to x$, $z \to -z$. To study the transformation properties of operators and eigenfunctions under general spatial rotations, we introduce operators of the type $e^{i\alpha J_z}$. Let (x,y,z) be the coordinates of an electron in a complex atom whose total angular momentum is $\mathbf{J}$. By expanding the operator $e^{i\alpha J_z}$ in a power series, it can be shown that[16]

$$\begin{aligned} e^{-i\alpha J_z}(x + iy)e^{i\alpha J_z} &= (x + iy)e^{-i\alpha} \\ e^{-i\alpha J_z} z e^{i\alpha J_z} &= z \end{aligned} \tag{2-15}$$

The components of the linear momentum $\mathbf{p}$ of an electron and the components of all angular momenta are transformed in a similar way, e.g.,

$$\begin{aligned} e^{-i\alpha J_z} s_+ e^{i\alpha J_z} &= s_+ e^{-i\alpha} \\ e^{-i\alpha J_z} s_z e^{i\alpha J_z} &= s_z \end{aligned} \tag{2-16}$$

The effect of the operator can be regarded as referring the coordinates and momenta of the electrons to a new coordinate frame, whose x and y axes are set forward an angle α from their former directions; alternatively, we may interpret the transformations (2-15) and (2-16) as rotating the entire atom through an angle $-\alpha$ about the z axis.‡ Any operator H, constructed from the coordinates and momenta of the electrons, transforms into the operator H', where

$$H' = e^{-i\alpha J_z} H e^{i\alpha J_z} \tag{2-17}$$

The equation

$$e^{-i\alpha J_z}|J,M_J) = e^{-i\alpha M_J}|J,M_J) \tag{2-18}$$

indicates that the eigenstate $|J,M_J)$ becomes multiplied by the factor $e^{-i\alpha M_J}$.

‡ This argument has neglected the structure of the nucleus, but it can easily be allowed for by replacing $\mathbf{J}$ with $\mathbf{F} = \mathbf{J} + \mathbf{I}$, where $\mathbf{I}$ is the spin of the nucleus. By replacing $e^{i\alpha J_z}$ by $e^{i\alpha F_z}$, both electrons and nucleons are rotated through an angle $-\alpha$ about the z axis.

Since the operators J_x, J_y and J_z, acting on $|J,M_J)$, produce states with the same quantum number J, it follows that under any spatial rotation the state $|J,M_J)$ transforms into some linear combination of these $2J + 1$ states. This is to be expected, since $H_1 + H_2$ is invariant under the operations of R_3, and the eigenfunctions of every J level must form a basis for a representation of R_3. The character χ_α^J of the representation can be easily derived from Eq. (2-18); in fact, comparing this equation with Eq. (2-11), we see that we need only replace l with J in Eq. (2-12) to get χ_α^J. Thus the $2J + 1$ states $|J,M_J)$ form a basis for the irreducible representation $\mathfrak{D}_J$ of R_3.

Unlike l, the quantum number J can be half-integral. The first of the half-integral representations of R_3 is $\mathfrak{D}_{\frac{1}{2}}$, whose most elementary basis is provided by the spin functions α and β, corresponding to $m_s = \pm\frac{1}{2}$. When J is half-integral, the factor $e^{-2\pi i M_J}$ is -1, showing that, under a rotation by 2π about the z axis, $|J,M_J)$ becomes multiplied by -1. It is clearly impossible to decompose $\mathfrak{D}_J$ into representations of O, since the latter possess the property that rotations by 2π and 0 are equivalent. On the face of it, this implies that when J is half-integral we cannot construct linear combinations of the states $|J,M_J)$ that display octahedral symmetry. To overcome this difficulty, O is augmented with an element $\mathcal{E}$ that commutes with all elements of the group, obeys the equation $\mathcal{E}^2 = \mathcal{I}$, and possesses the property that, if the element $\mathcal{R}$ formerly satisfied $\mathcal{R}^n = \mathcal{I}$, it now satisfies $\mathcal{R}^n = \mathcal{E}$. Groups that have been extended in this way are called *double* groups. In general they do not have twice as many classes as the original group, since a class consisting of rotations through π is often not doubled. The characters of the double octahedral group, as found by Bethe[11], are given in Table 2-3. Some branching rules for $R_3 \to O$ when J is half-integral are set out in Table 2-4.

TABLE 2-3 THE CHARACTERS OF THE DOUBLE OCTAHEDRAL GROUP

	C_1	C_1'	C_2	C_3	C_4	C_4'	C_5	C_5'
Γ_1	1	1	1	1	1	1	1	1
Γ_2	1	1	−1	1	1	1	−1	−1
Γ_3	2	2	0	2	−1	−1	0	0
Γ_4	3	3	−1	−1	0	0	1	1
Γ_5	3	3	1	−1	0	0	−1	−1
Γ_6	2	−2	0	0	1	−1	$(2)^{\frac{1}{2}}$	$-(2)^{\frac{1}{2}}$
Γ_7	2	−2	0	0	1	−1	$-(2)^{\frac{1}{2}}$	$(2)^{\frac{1}{2}}$
Γ_8	4	−4	0	0	−1	1	0	0

TABLE 2-4 BRANCHING RULES FOR HALF-INTEGRAL J

$\mathfrak{D}_{\frac{1}{2}}$	$\rightarrow \Gamma_6$
$\mathfrak{D}_{\frac{3}{2}}$	$\rightarrow \Gamma_8$
$\mathfrak{D}_{\frac{5}{2}}$	$\rightarrow \Gamma_7 + \Gamma_8$
$\mathfrak{D}_{\frac{7}{2}}$	$\rightarrow \Gamma_6 + \Gamma_7 + \Gamma_8$
$\mathfrak{D}_{\frac{9}{2}}$	$\rightarrow \Gamma_6 + 2\Gamma_8$
$\mathfrak{D}_{\frac{11}{2}}$	$\rightarrow \Gamma_6 + \Gamma_7 + 2\Gamma_8$

2-6 THE WIGNER-ECKART THEOREM

A knowledge of the transformation properties of eigenfunctions and operators finds its most useful application in the evaluation of matrix elements such as

$$\int \phi_i^* h_j \theta_k \, d\tau$$

Suppose that the following conditions are satisfied:

1. The p functions ϕ_i $(i = 1, 2, \ldots, p)$ form an orthonormal basis for the irreducible representation Γ_P of a group $\mathcal{G}$.

2. The q functions ψ_j $(j = 1, 2, \ldots, q)$ form an orthonormal basis for the irreducible representation Γ_Q of $\mathcal{G}$.

3. The r functions θ_k $(k = 1, 2, \ldots, r)$ form an orthonormal basis for the irreducible representation Γ_R of $\mathcal{G}$.

4. Under the operations of $\mathcal{G}$, the q operators h_j $(j = 1, 2, \ldots, q)$ transform in an identical manner to the basis functions ψ_j. In detail, if $\psi_j \rightarrow \psi_j'$ and $h_j \rightarrow h_j'$ under an operation $\mathcal{R}_\sigma$ of the group, where

$$\psi_j' = \sum_l r_{lj}(\sigma)\psi_l$$

then

$$h_j' = \sum_l r_{lj}(\sigma)h_l$$

for any σ.

The qr functions $\psi_j\theta_k$ form a basis for the representation $\Gamma_Q \times \Gamma_R$ of $\mathcal{G}$. Corresponding to the decomposition

$$\Gamma_Q \times \Gamma_R = \sum_S c_S \Gamma_S$$

sets of linear combinations of the products $\psi_j\theta_k$ can be constructed, each set forming an orthonormal basis for an irreducible representation of $\mathcal{G}$. We write

$$\rho_l(\Gamma_S\beta) = \sum_{j,k} (\Gamma_Q j;\Gamma_R k|\Gamma_S \beta l)\psi_j\theta_k \tag{2-19}$$

where l runs from 1 to s, for the s basis functions of an irreducible representation Γ_S of $\mathcal{G}$. The symbol β serves to distinguish equivalent irreducible representations that arise when $c_S > 1$. Equations (2-10) are examples of Eq. (2-19). Considerable freedom exists in the actual choice of the orthonormal basis functions $\rho_l(\Gamma_S\beta)$ for a given Γ_S; once the choice has been made, the coefficients in Eq. (2-19) are fixed. Care is taken to ensure that the functions $\rho_i(\Gamma_P\beta)$ transform in an identical manner to the functions ϕ_i for all β.

Owing to the unitarity of the transformation (2-19),

$$\psi_j\theta_k = \sum_{l,S,\beta} (\Gamma_S\beta l|\Gamma_Q j;\Gamma_R k)\rho_l(\Gamma_S\beta) \tag{2-20}$$

where

$$(\Gamma_S\beta l|\Gamma_Q j;\Gamma_R k) = (\Gamma_Q j;\Gamma_R k|\Gamma_S\beta_l)^*$$

Taking precisely the same linear combinations of the quantities $h_j\theta_k$ as are taken of the functions $\psi_j\theta_k$, we may construct functions $\lambda_l(\Gamma_S\beta)$ that transform in an identical manner to the functions $\rho_l(\Gamma_S\beta)$. In analogy with Eq. (2-20),

$$h_j\theta_k = \sum_{\beta,S,l} (\Gamma_S\beta l|\Gamma_Q j;\Gamma_R k)\lambda_l(\Gamma_S\beta)$$

We multiply both sides of this equation by ϕ_i^* and integrate over the entire coordinate space. The only surviving terms in the sum are those for which $\Gamma_S = \Gamma_P$ and $l = i$; moreover it is also possible to show that the integrals

$$\int \phi_i^*\lambda_i(\Gamma_P\beta)\, d\tau \tag{2-21}$$

are independent of i and can therefore be written as A_β.‡ Hence

$$\int \phi_i^* h_j\theta_k\, d\tau = \sum_\beta A_\beta(\Gamma_P\beta i|\Gamma_Q j;\Gamma_R k) \tag{2-22}$$

This equation constitutes a general statement of the *Wigner-Eckart* theorem, so named after its original proponents.[14,17]

‡ A formal proof of the condition $\Gamma_S = \Gamma_P$ can be readily constructed for a finite group with g elements. The equations

$$\mathcal{R}_\sigma\phi_i^* = \sum_m p_{mi}^*(\sigma)\phi_m^* \qquad \mathcal{R}_\sigma\lambda_l = \sum_n s_{nl}(\sigma)\lambda_n$$

indicate that

$$\begin{aligned}\int \phi_i^*\lambda_l\, d\tau &= g^{-1}\sum_\sigma \int (\mathcal{R}_\sigma\phi_i)^*(\mathcal{R}_\sigma\lambda_l)\, d\tau \\ &= g^{-1}\sum_{m,n}\sum_\sigma p_{mi}^*(\sigma)s_{nl}(\sigma)\int \phi_m^*\lambda_n\, d\tau\end{aligned}$$

According to Eq. (2-7), the sum over σ vanishes for all i, l, m, and n if Γ_P and Γ_S are inequivalent irreducible representations; it follows that, for the integral to be nonzero, $\Gamma_S = \Gamma_P$. The condition $l = i$ and the invariance of the integrals (2-21) can be derived in a similar fashion with the aid of Eq. (2-8).

Sometimes other bases for Γ_P and Γ_R or other operators that transform according to Γ_Q can be found. We use a superscript to distinguish quantities that possess identical transformation properties: for example, $(\theta_1^1, \theta_2^1, \ldots, \theta_r^1)$ and $(\theta_1^2, \theta_2^2, \ldots, \theta_r^2)$ are two sets of functions that separately transform according to the equation

$$\mathcal{R}_\sigma \theta_k^c = \sum_l r_{lk}(\sigma)\theta_l^c$$

Equation (2-22) is now written

$$\int \phi_i^{a*} h_j^b \theta_k^c \, d\tau = \sum_\beta A_\beta(a,b,c)(\Gamma_P\beta i|\Gamma_Q j;\Gamma_R k) \qquad (2\text{-}23)$$

The sum over β comprises c_P terms, where c_P is the number of times Γ_P occurs in the decomposition of $\Gamma_Q \times \Gamma_R$. Hence every set of pqr integrals

$$\int \phi_i^{a*} h_j^b \theta_k^c \, d\tau$$

for a particular choice of a, b, and c can be expressed as a sum over at most c_P linearly independent sets of integrals. Put another way, coefficients f_{uvw}^{abc} can be found such that

$$\int \phi_i^{a*} h_j^b \theta_k^c \, d\tau = \Sigma f_{uvw}^{abc} \int \phi_i^{u*} h_j^v \theta_k^w \, d\tau \qquad (2\text{-}24)$$

for all i, j, and k, where the sum extends over at most c_P terms, provided that the sets of pqr integrals for different u, v, and w are linearly independent. Equation (2-24) is usually a more convenient form for the Wigner-Eckart theorem if the coefficients $(\Gamma_P\beta i|\Gamma_Q j;\Gamma_R k)$ have not been tabulated. Needless to say, if $c_P = 0$, all matrix elements vanish.

2-7 TENSOR OPERATORS

The most familiar application of the Wigner-Eckart theorem is to the group R_3. The functions ϕ_i^* and θ_k are replaced by $(\gamma K M_K|$ and $|\gamma' K' M_K')$, respectively, where $\mathbf{K}$ and $\mathbf{K}'$ are any pair of angular-momentum vectors of the same kind (for example, $\mathbf{J}$ and $\mathbf{J}'$). In order that the Wigner-Eckart theorem may be applied, the operators h_j must transform in an identical fashion to some set of states $|\gamma'' kq)$, where $q = k, k-1, \ldots, -k$. This is guaranteed if the operators, which we now write as $T_q^{(k)}$, transform like the spherical harmonics Y_{kq}, or, what is equivalent, like the quantities

$$C_q^{(k)} = \left(\frac{4\pi}{2k+1}\right)^{\frac{1}{2}} Y_{kq}(\theta,\phi) = (-1)^q \left[\frac{(k-q)!}{(k+q)!}\right]^{\frac{1}{2}} P_k^q(\cos\theta)e^{iq\phi} \qquad (2\text{-}25)$$

The $2k+1$ operators $T_q^{(k)}$, where $q = k, k-1, \ldots, -k$, are said to form the components of a *tensor operator* $\mathbf{T}^{(k)}$ of rank k.

According to Eq. (2-17), the transformation properties of an operator under rotations about the z axis are completely determined by the commutation relations of the operator with J_z. For the operators $T_q^{(k)}$ and $C_q^{(k)}$ to transform in an identical fashion under any rotation, it is necessary only that the commutation relations of $T_q^{(k)}$ with J_x, J_y, and J_z be the same as those of $C_q^{(k)}$. The only part of $\mathbf{J}$ that does not commute with $C_q^{(k)}$ is the orbital angular momentum $\mathbf{l}$; on making the substitutions

$$l_z = \frac{1}{i}\frac{\partial}{\partial\phi} \qquad l_\pm = e^{\pm i\phi}\left(\pm\frac{\partial}{\partial\theta} + i\cot\theta\,\frac{\partial}{\partial\phi}\right)$$

it is straightforward to prove

$$\begin{aligned}[J_z,C_q^{(k)}] &= qC_q^{(k)}\\ [J_\pm,C_q^{(k)}] &= [k(k+1) - q(q\pm1)]^{\frac{1}{2}}C_{q\pm1}^{(k)}\end{aligned}$$

Accordingly, the quantities $T_q^{(k)}$ possess the same transformation properties as the harmonics $C_q^{(k)}$ if they satisfy the commutation relations

$$\begin{aligned}[J_z,T_q^{(k)}] &= qT_q^{(k)}\\ [J_\pm,T_q^{(k)}] &= [k(k+1) - q(q\pm1)]^{\frac{1}{2}}T_{q\pm1}^{(k)}\end{aligned} \tag{2-26}$$

These equations were taken by Racah[18] as the starting point for the theory of tensor operators.

From Eq. (2-13) it is clear that $\mathfrak{D}_K$ does not occur more than once in the decomposition of the Kronecker product $\mathfrak{D}_k \times \mathfrak{D}_{K'}$. Equation (2-22) therefore runs

$$(\gamma KM_K|T_q^{(k)}|\gamma'K'M_K') = A(KM_K|kq;K'M_K')$$

where A is independent of M_K, q, and M_K'. The coefficient on the right is simply a VC coefficient; on converting it to a 3-j symbol and expanding the description of A, we get

$$(\gamma KM_K|T_q^{(k)}|\gamma'K'M_K') = (-1)^{K-M_K}\begin{pmatrix} K & k & K' \\ -M_K & q & M_K' \end{pmatrix}(\gamma K\|T^{(k)}\|\gamma'K') \tag{2-27}$$

The final factor on the right is called a *reduced* matrix element. The representation $\mathfrak{D}_K$ occurs in the decomposition of $\mathfrak{D}_k \times \mathfrak{D}_{K'}$ if the *triangular condition* is satisfied, that is, if a triangle can be formed having sides of lengths K, k, and K'. If the triangular condition is not satisfied, the 3-j symbol automatically vanishes.

Each term $\sum_i r_i^k Y_{kq}(\theta_i,\phi_i)$ in Eq. (2-1), being a sum of spherical harmonics, is a tensor operator. If H_3 is treated as a perturbation, the first-order approximation requires the evaluation of matrix elements for which $K = K'$. For this case, we can construct operators that involve K_x, K_y, and K_z and that transform like the spherical harmonics Y_{kq}; consequently the matrix elements of these operators are proportional to those of the tensor operators $\sum_i r_i^k Y_{kq}(\theta_i,\phi_i)$. The following are examples of such operator equivalences for a configuration of equivalent electrons:

$$\sum_i (3z_i^2 - r_i^2) \equiv \alpha\langle r^2\rangle[3K_z^2 - K(K+1)] \tag{2-28}$$

$$\begin{aligned}\sum_i (35z_i^4 - 30z_i^2r_i^2 + 3r_i^4)& \\ \equiv \beta\langle r^4\rangle[35K_z^4 &- 30K(K+1)K_z^2 + 25K_z^2 \\ &- 6K(K+1) + 3K^2(K+1)^2]\end{aligned} \tag{2-29}$$

$$\begin{aligned}\sum_i (231z_i^6 - 315z_i^4r_i^2 + 105z_i^2r_i^4 - 5r_i^6)& \\ \equiv \gamma\langle r^6\rangle[231K_z^6 &- 315K(K+1)K_z^4 + 735K_z^4 \\ + 105K^2(K+1)^2K_z^2 &- 525K(K+1)K_z^2 + 294K_z^2 \\ - 5K^3(K+1)^3 &+ 40K^2(K+1)^2 - 60K(K+1)]\end{aligned} \tag{2-30}$$

where
$$\langle r^k\rangle = \int_0^\infty R_{nl}(r)r^kR_{nl}(r)\,dr$$

By separating out $\langle r^k\rangle$, *the operator equivalent factors* α, β, and γ become pure numbers, independent of the radial eigenfunction. The expressions on the right of Eqs. (2-28) to (2-30) are most easily obtained by replacing M_K by K_z in the explicit expressions for the corresponding 3-j symbols; an alternative method is to take the appropriate spherical harmonics and substitute K_x, K_y, and K_z for x, y, and z, respectively. Due regard must be paid to the fact that K_x, K_y, and K_z, unlike x, y, and z, do not commute. Thus, in setting up an operator to reproduce the eigenvalues of $\sum_i (35z_i^4 - 30z_i^2r_i^2 + 3r_i^4)$, we must replace $\sum_i r_i^4$, not with $(\mathbf{K}^2)^2$, whose eigenvalues are $K^2(K+1)^2$, but with

$$\frac{1}{6}\sum_{i,j=x,y,z}(K_i^2K_j^2 + K_j^2K_i^2 + K_iK_jK_iK_j + K_jK_iK_jK_i + K_iK_j^2K_i + K_jK_i^2K_j)$$

Repeated use of the commutation relations (1-15) reduces this to an operator the eigenvalues of which are

$$K^2(K+1)^2 - K(K+1)/3$$

but the calculation is very tedious. The introduction of the operator equivalences (2-28) to (2-30) is due to Stevens.[10] Examples of their use in treating the hyperfine interactions have been given by Ramsey.[19]

Since the calculation for any tensor operator $T_q^{(k)}$ must reduce ultimately to an integral between single-particle states, the triangular condition on l, k, and l indicates that the number of terms that need be included in H_3 is limited. For f electrons, all tensor operators $T_q^{(k)}$ for which $k > 6$ have vanishing matrix elements, and the expansion for H_3' given in Eq. (2-5) is sufficient for all configurations of the type f^n. It is obvious that there is no need to include terms of odd parity in H_3 if all matrix elements are taken between states of a single configuration. In addition, all components $T_q^{(k)}$ of a tensor operator $\mathbf{T}^{(k)}$ share the same operator equivalent factor, since the latter, being a product of a reduced matrix element and some factors involving K, is independent of q. These considerations, taken together, show that the three operator equivalent factors α, β, and γ are the only ones that have to be evaluated when crystal field effects in configurations of the type f^n are being studied. Stevens[10] has tabulated α, β, and γ for the ground levels of all triply ionized rare-earth atoms.

2-8 AN EXTERNAL MAGNETIC FIELD

To illustrate the power of the Wigner-Eckart theorem even in the weaker form (2-24), we consider the effect of an external magnetic field $\mathbf{H}$ on the energy-level system of a rare-earth ion in a crystal lattice. Dirac's equation indicates that for a single electron the appropriate addition to the Hamiltonian is $\beta\mathbf{H}\cdot(\mathbf{l}+2\mathbf{s})$, where $\beta = e\hbar/2mc$ is the Bohr magneton. As a first (and extremely good) approximation for the many-electron case, we therefore include

$$H_4 = \sum_i \beta\mathbf{H}\cdot(\mathbf{l}_i + 2\mathbf{s}_i) = \beta\mathbf{H}\cdot(\mathbf{L}+2\mathbf{S}) \qquad (2\text{-}31)$$

in the Hamiltonian. The inequalities $H_1 \gg H_2 \gg H_3 \gg H_4$, which obtain for rare-earth ions, indicate that H_4 can be treated as a perturbation on the energy levels whose eigenfunctions are of the type $|f^n\gamma SLJ\Gamma_R t)$, where t distinguishes the various functions that together form a basis for Γ_R. If the symmetry at the nucleus of the rare-earth ion is octahedral, Γ_R denotes an irreducible representation of the group O. Suppose we want to find the effect of an external magnetic field on a doublet Γ_3. The quantity $\mathbf{L} + 2\mathbf{S}$ is a vector, and under simple rotations its three components transform among themselves like the coordinates (x,y,z) of a point. Thus, for the group O, the components of $\mathbf{L} + 2\mathbf{S}$ transform according to Γ_4. From Table 2-1 it is easy to prove $\Gamma_4 \times \Gamma_3 = \Gamma_4 + \Gamma_5$. The absence of Γ_3 in the decomposition immediately reveals that all matrix elements of $\mathbf{L} + 2\mathbf{S}$ between states transforming like Γ_3 vanish; hence the doublet remains degenerate (to first order) when an external magnetic field is applied.

Examples of Eq. (2-24) with $c_P = 1$ are easy to find. The states $|J,M_J)$ form a basis for the irreducible representation $\mathfrak{D}_J$ of R_3. A vector transforms like $\mathfrak{D}_1$, and $\mathfrak{D}_1 \times \mathfrak{D}_J$ contains in its decomposition the representation $\mathfrak{D}_J$ once. Thus the matrix elements of the components of $\mathbf{L} + 2\mathbf{S}$ must be proportional to those of the corresponding components of any other vector. Since the states are labeled by J, it is very convenient to choose $\mathbf{J}$ for this role. Equation (2-24) now runs

$$(JM_J|L_j + 2S_j|JM_J') = g(JM_J|J_j|JM_J') \tag{2-32}$$

for all M_J, j, and M_J'. The symbol g is called the Landé g factor and depends on the detailed nature of the states. Equation (2-32) corresponds to $a = u$, $b \neq v$, $c = w$ in Eq. (2-24); the subscript j on L, S, and J stands for x, y, or z and corresponds to the same symbol in Eq. (2-24).

Instances where $a \neq u$, $c \neq v$ are not difficult to find. For example, the two eigenstates $|S' = \frac{1}{2}, M_S' = \pm\frac{1}{2})$ form a basis for the irreducible representation Γ_6 of O. The two basis functions for any other doublet of the type Γ_6 can be chosen to be in a one-to-one correspondence with the components $M_S' = \pm\frac{1}{2}$; and since Γ_6 occurs once in the decomposition of $\Gamma_4 \times \Gamma_6$, the matrix elements of $\beta\mathbf{H} \cdot (\mathbf{L} + 2\mathbf{S})$ can be reproduced by $g'\beta\mathbf{H} \cdot \mathbf{S}'$ operating between the states for which $M_S' = \pm\frac{1}{2}$, where g' is a constant. This shows that the splitting produced by an external magnetic field in a doublet of the type Γ_6 is independent of the orientation of $\mathbf{H}$ relative to the cubic axes. Experimentalists working in the field of paramagnetic resonance find that they can almost invariably summarize their observations in a succinct manner by quoting numbers (e.g., g' above) which appear in the *spin-Hamiltonian*, that is, the Hamiltonian in which the effective spin $\mathbf{S}'$ enters. In fact, they can often use spin-Hamiltonians for many more cases than are allowed by arguments of a strictly group theoretical nature. This point has been discussed by Koster and Statz.[20]

As an example of $c_P = 2$, we cite the action of $\mathbf{H}$ on an energy level corresponding to the irreducible representation Γ_8 of the group O.

The states $|S' = \frac{3}{2}, M'_S)$ form a basis for Γ_8, and since Γ_8 occurs twice in the reduction of $\Gamma_4 \times \Gamma_8$ (as can be verified by using Table 2-3), we need two independent sets of functions involving $\mathbf{S}'$ that transform according to Γ_4 in order to reproduce the effect of H_4. It is easy to see that $(S'_x)^d$, $(S'_y)^d$, and $(S'_z)^d$, where d is an odd positive integer, possess the appropriate transformation properties, and hence a possible form for the spin-Hamiltonian is[21]

$$\beta H_x(gS'_x + fS'^3_x) + \beta H_y(gS'_y + fS'^3_y) + \beta H_z(gS'_z + fS'^3_z)$$

The constants g and f correspond to f^{abc}_{uvw} and $f^{abc}_{uv'w}$ in the notation of Eq. (2-24).

2-9 NEPTUNIUM HEXAFLUORIDE

To link together the various topics of this chapter, we consider the problem of a single f electron in an octahedral field. The Hamiltonian is simply $H_2 + H'_3$. The first part, H_2, represents the spin-orbit coupling and is given by Eq. (1-13). The second part, H'_3, is the interaction energy of the electron and the octahedral field, which, for f electrons, is adequately represented by Eq. (2-5). A direct approach is to take the 14 states $|sm_slm_l)$, for which $s = \frac{1}{2}$, $m_s = \pm\frac{1}{2}$, $l = 3$, $m_l = 3, 2, \ldots, -3$, and evaluate the complete matrix of $H_2 + H'_3$. It is much better, however, to take as basic states those linear combinations of $|sm_slm_l)$ that diagonalize either H_2 or H'_3, for then the problem of factorizing the 14×14 secular determinant is avoided. If we decide to diagonalize H'_3, we have to evaluate the matrix elements of tensor operators between the states $|sm_slm_l)$. We follow the current practice of using the operator-equivalent approach rather than applying the Wigner-Eckart theorem in the form of Eq. (2-27). From Eq. (2-29),

$$35z^4 - 30z^2r^2 + 3r^4 \equiv \beta\langle r^4\rangle[35l_z^4 - 30l(l+1)l_z^2 + 25l_z^2 - 6l(l+1) + 3l^2(l+1)^2]$$

The substitutional method for deriving operator equivalences gives at once

$$x^4 - 6x^2y^2 + y^4 = \tfrac{1}{2}[(x+iy)^4 + (x-iy)^4] \equiv \tfrac{1}{2}\beta\langle r^4\rangle(l_+^4 + l_-^4)$$

Similar equivalences hold for the spherical harmonics of rank 6. The matrix elements are easily written down from the tables of Stevens[10] and of Baker, Bleaney, and Hayes;[22] the two operator-equivalent factors β and γ can be found by evaluating a single matrix element by means of equations such as

$$\begin{aligned}\beta\langle r^4\rangle(f, m_l = 0|35l_z^4 - 30l(l+1)l_z^2 + \cdots |f, m_l = 0) \\ = 360\beta\langle r^4\rangle = (f, 0|(35z^4 - 30z^2r^2 + 3r^4)|f, 0) \\ = 28\langle r^4\rangle \int_{-1}^{1} P_3(\mu)P_4(\mu)P_3(\mu)\,d\mu = \tfrac{18}{11}\langle r^4\rangle\end{aligned}$$

The secular determinant for H_3' factorizes into a number of two-dimensional and one-dimensional determinants, each determinant being characterized by those values of m_l that label the rows and columns. Equating the determinants to zero, we obtain

$$\begin{vmatrix} 3a + b - \epsilon & (a - 7b)(15)^{\frac{1}{2}} \\ (a - 7b)(15)^{\frac{1}{2}} & a + 15b - \epsilon \end{vmatrix} = 0$$

for m_l admixtures of the type 3, -1 and also for admixtures of the type -3, 1;

$$\begin{vmatrix} -7a - 6b - \epsilon & 5a + 42b \\ 5a + 42b & -7a - 6b - \epsilon \end{vmatrix} = 0$$

for admixtures of the type 2, -2; and

$$6a - 20b - \epsilon = 0$$

for $m_l = 0$. The constants a and b are given by

$$\begin{aligned} a &= 60\beta A_4{}^0\langle r^4\rangle = 8A_4{}^0\langle r^4\rangle/33 \\ b &= 180\gamma A_6{}^0\langle r^6\rangle = -80A_6{}^0\langle r^6\rangle/429 \end{aligned} \qquad (2\text{-}33)$$

In view of the dissimilarity between the secular equations, we might anticipate five distinct roots for ϵ, arising from the two quadratics and the linear form. However, the branching rule $\mathfrak{D}_3 \to \Gamma_2 + \Gamma_4 + \Gamma_5$ indicates that there can be at most three distinct roots, and this is immediately verified on actually solving for ϵ. Energies and eigenstates (defined by m_l) are as follows:

$\epsilon_1 = 6a - 20b$: $|0)$
$(\tfrac{5}{8})^{\frac{1}{2}}|3) + (\tfrac{3}{8})^{\frac{1}{2}}|-1)$
$(\tfrac{5}{8})^{\frac{1}{2}}|-3) + (\tfrac{3}{8})^{\frac{1}{2}}|1)$

$\epsilon_2 = -2a + 36b$: $(\tfrac{1}{2})^{\frac{1}{2}}|2) + (\tfrac{1}{2})^{\frac{1}{2}}|-2)$
$(\tfrac{3}{8})^{\frac{1}{2}}|3) - (\tfrac{5}{8})^{\frac{1}{2}}|-1)$
$(\tfrac{3}{8})^{\frac{1}{2}}|-3) - (\tfrac{5}{8})^{\frac{1}{2}}|1)$

$\epsilon_3 = -12a - 48b$: $(\tfrac{1}{2})^{\frac{1}{2}}|2) - (\tfrac{1}{2})^{\frac{1}{2}}|-2)$

It is clear that the third energy level corresponds to Γ_2, since it is the only nondegenerate one. By considering rotations by $\frac{1}{2}\pi$ about the

z axis, it is easy to show that the first energy level, ϵ_1, corresponds to Γ_4, the second, ϵ_2, to Γ_5.

Spin may be included simply by putting either + or − above the symbols for m_l in the eigenstates above. However, since the states $|m_s = \pm\frac{1}{2})$ form a basis for the irreducible representation Γ_6 of the double octahedral group, and since

$$\Gamma_6 \times \Gamma_4 = \Gamma_6 + \Gamma_8$$
$$\Gamma_6 \times \Gamma_5 = \Gamma_7 + \Gamma_8$$
and
$$\Gamma_6 \times \Gamma_2 = \Gamma_7$$

it is more appropriate to construct those linear combinations of the states which transform like irreducible representations of the double group. Since the spin-orbit coupling is a scalar (i.e., transforms like Γ_1), these linear combinations may be found by considering each set of functions $\Gamma_6 \times \Gamma_i$ separately (where $i = 4, 5, 2$) and diagonalizing H_2 *within* one set. The eigenstates with spin are now as follows:

$$\epsilon_1 \left\{ \begin{array}{ll} \left.\begin{array}{l} (\frac{5}{8})^{\frac{1}{2}}|-\overset{+}{3}) + (\frac{3}{8})^{\frac{1}{2}}|\overset{+}{1}) \\ (\frac{5}{8})^{\frac{1}{2}}|\overset{-}{3}) + (\frac{3}{8})^{\frac{1}{2}}|-\overset{-}{1}) \\ (\frac{2}{3})^{\frac{1}{2}}|\overset{+}{0}) - (\frac{5}{24})^{\frac{1}{2}}|-\overset{-}{3}) - (\frac{1}{8})^{\frac{1}{2}}|\overset{-}{1}) \\ (\frac{2}{3})^{\frac{1}{2}}|\overset{-}{0}) - (\frac{5}{24})^{\frac{1}{2}}|\overset{+}{3}) - (\frac{1}{8})^{\frac{1}{2}}|-\overset{+}{1}) \end{array} \right\} & \Gamma_8 \\ \left.\begin{array}{l} (\frac{1}{3})^{\frac{1}{2}}|\overset{+}{0}) + (\frac{5}{12})^{\frac{1}{2}}|-\overset{-}{3}) + (\frac{1}{2})|\overset{-}{1}) \\ (\frac{1}{3})^{\frac{1}{2}}|\overset{-}{0}) + (\frac{5}{12})^{\frac{1}{2}}|\overset{+}{3}) + (\frac{1}{2})|-\overset{+}{1}) \end{array} \right\} & \Gamma_6 \end{array} \right.$$

$$\epsilon_2 \left\{ \begin{array}{ll} \left.\begin{array}{l} (\frac{3}{8})^{\frac{1}{2}}|\overset{+}{3}) - (\frac{5}{8})^{\frac{1}{2}}|-\overset{+}{1}) \\ (\frac{3}{8})^{\frac{1}{2}}|-\overset{-}{3}) - (\frac{5}{8})^{\frac{1}{2}}|\overset{-}{1}) \\ (\frac{1}{3})^{\frac{1}{2}}|\overset{+}{2}) + (\frac{1}{3})^{\frac{1}{2}}|-\overset{+}{2}) - (\frac{1}{8})^{\frac{1}{2}}|\overset{-}{3}) + (\frac{5}{24})^{\frac{1}{2}}|-\overset{-}{1}) \\ (\frac{1}{3})^{\frac{1}{2}}|\overset{-}{2}) + (\frac{1}{3})^{\frac{1}{2}}|-\overset{-}{2}) - (\frac{1}{8})^{\frac{1}{2}}|-\overset{+}{3}) + (\frac{5}{24})^{\frac{1}{2}}|\overset{+}{1}) \end{array} \right\} & \Gamma_8 \\ \left.\begin{array}{l} (\frac{1}{6})^{\frac{1}{2}}|\overset{+}{2}) + (\frac{1}{6})^{\frac{1}{2}}|-\overset{+}{2}) + (\frac{1}{2})|\overset{-}{3}) - (\frac{5}{12})^{\frac{1}{2}}|-\overset{-}{1}) \\ (\frac{1}{6})^{\frac{1}{2}}|\overset{-}{2}) + (\frac{1}{6})^{\frac{1}{2}}|-\overset{-}{2}) + (\frac{1}{2})|-\overset{+}{3}) - (\frac{5}{12})^{\frac{1}{2}}|\overset{+}{1}) \end{array} \right\} & \Gamma_7 \end{array} \right.$$

$$\epsilon_3 \left\{ \begin{array}{l} (\frac{1}{2})^{\frac{1}{2}}|\overset{+}{2}) - (\frac{1}{2})^{\frac{1}{2}}|-\overset{+}{2}) \\ (\frac{1}{2})^{\frac{1}{2}}|\overset{-}{2}) - (\frac{1}{2})^{\frac{1}{2}}|-\overset{-}{2}) \end{array} \right\} \Gamma_7$$

The final step consists in evaluating the matrix elements of H_2 between the states above. Since H_2 is invariant under the operations of the octahedral group and at the same time commutes with J_z, the secular equation factorizes into four identical quadratics (for the four components of the two Γ_8 levels), two identical quadratics (for the two components of the two Γ_7 levels), and two linear equations (for the two components of Γ_6). They are, in order,

$$\begin{vmatrix} \epsilon_2 + \dfrac{\zeta}{4} - \epsilon & \dfrac{3\zeta(5)^{\frac{1}{2}}}{4} \\ \dfrac{3\zeta(5)^{\frac{1}{2}}}{4} & \epsilon_1 - \dfrac{3\zeta}{4} - \epsilon \end{vmatrix} = 0$$

$$\begin{vmatrix} \epsilon_3 - \epsilon & \zeta(3)^{\frac{1}{2}} \\ \zeta(3)^{\frac{1}{2}} & \epsilon_2 - \dfrac{\zeta}{2} - \epsilon \end{vmatrix} = 0 \qquad (2\text{-}34)$$

and

$$\epsilon_1 + \frac{3\zeta}{2} - \epsilon = 0$$

These equations enable the energy levels and the eigenstates to be determined in terms of ζ, ϵ_1, ϵ_2, and ϵ_3.

We are now in a position to apply the theory to NpF_6. The neptunium ion Np^{6+} is surrounded by an octahedron of negatively charged fluorine ions. According to Eqs. (2-6), $A_4^0 > 0$, and $A_6^0 > 0$. From Eqs. (2-33), $a > 0$, and $b < 0$. The relative importance of sixth-rank and fourth-rank spherical harmonics in the Hamiltonian determines the ordering of the levels in the limit $H_2 = 0$. If $b = 0$, $\epsilon_1 > \epsilon_2 > \epsilon_3$, while if $a = 0$, $\epsilon_3 > \epsilon_1 > \epsilon_2$. The energies ϵ_1, ϵ_2, and ϵ_3 of the components Γ_4, Γ_5, and Γ_2 of the configuration f^1 are plotted out in Fig. 2-2 for all negative ratios of a/b. The figure is constructed to possess similar properties to Fig. 1-1; the abscissa is $\xi = \chi/(1 + \chi)$, where $\chi = -14b/3a$, and the ordinate η ensures that the center of gravity lies on the line $\eta = 0$ and that the extent of the energy levels is unity in both limits. If a particular value of ξ is selected, the ratio $(\epsilon_1 - \epsilon_2)/(\epsilon_2 - \epsilon_3)$ is fixed and Eqs. (2-34) can be used to plot out the energy levels for different values of the spin-orbit coupling constant ζ. This is done in Fig. 2-3 for $\xi = 0.253$. By choosing this value the absorption spectrum[23] of NpF_6 is fitted, and at the same time the magnetic properties[24] of the lowest doublet are accounted for. The value of ξ', the parameter defining the relative strengths of the crystal field and the spin-orbit interaction, is 0.725. The analysis presented above is essentially a simplification of that of Eisenstein and Pryce,[24] who include covalent effects in their treatment.

Axe[25] has studied the properties of Pa^{4+} substituted in small quantities for Zr^{4+} in Cs_2ZrCl_6; the protoactinium ion possesses a single $5f$ electron, and the foregoing analysis is applicable. The crystal field parameters differ considerably from those for NpF_6, as

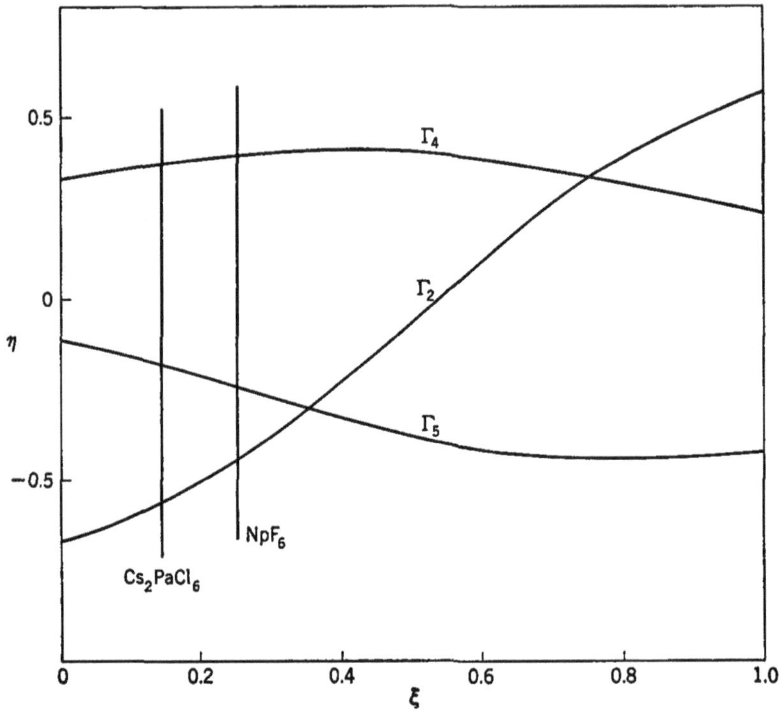

FIG. 2-2 The energy-level system of an $l = 3$ level in a crystal field of octahedral symmetry is shown for various relative strengths of the tensor operators of rank 4 and rank 6 in the Hamiltonian H_3. On the extreme left, tensor operators of rank 4 only contribute to the splitting; on the right, those of rank 6 only. The actual mixtures that are required to reproduce the observed energy-level schemes for Cs_2PaCl_6 and NpF_6 both lie to the left of the diagram, indicating that tensor operators of rank 4 are more important.

can be seen from Table 2-5. The predominance of $A_4^0\langle r^4\rangle$ over $A_6^0\langle r^6\rangle$ in both cases is to be expected when the inequality $\langle r^4\rangle/\rho^4 > \langle r^6\rangle/\rho^6$ is combined with Eqs. (2-6). It is clear from Fig. 2-2 that the spherical harmonics of the fourth rank are of greater relative importance for Cs_2PaCl_6 compared with NpF_6. However, the values of ξ are close enough for us to include Axe's data in Fig. 2-3 without grossly misrepresenting the situation. The agreement is improved by raising the Γ_5 level in the limit of $H_2 = 0$, as indicated in Fig. 2-2.

TABLE 2-5 PARAMETERS FOR TWO OCTAHEDRALLY COORDINATED IONS
(All values are in cm^{-1}.)

	Cs_2PaCl_6	NpF_6
$A_4^0\langle r^4\rangle$	888	5738
$A_6^0\langle r^6\rangle$	41.9	540.5
ζ	1490	2405

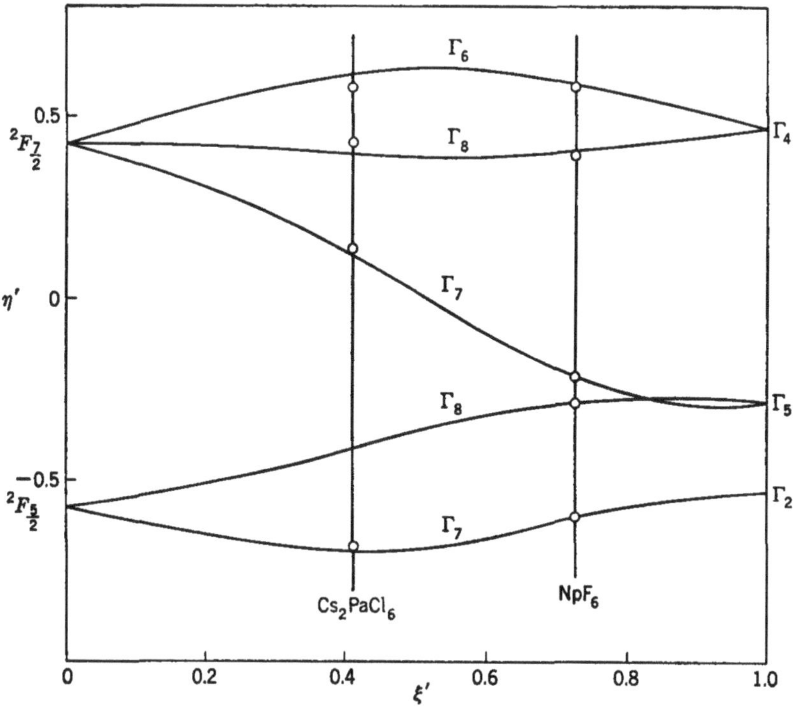

FIG. 2-3 The energy levels for a single f electron in an octahedral crystal field are plotted for various relative strengths of H_2 and H_3. On the left, $H_3 = 0$; the arrangement of energy levels in Fig. 2-2 for NpF_6 is taken as the right extreme. Strictly, the levels for Cs_2PaCl_6 should be based on a slightly different right extreme, corresponding to the arrangement in Fig. 2-2 for Cs_2PaCl_6; however, the situations are sufficiently similar to permit both sets of observations to be represented on the same diagram.

PROBLEMS

2-1. Prove that the subgroup of O comprising the permutations

$$\begin{pmatrix}1234\\1234\end{pmatrix}, \begin{pmatrix}1234\\2341\end{pmatrix}, \begin{pmatrix}1234\\3412\end{pmatrix}, \begin{pmatrix}1234\\4123\end{pmatrix}$$

is Abelian and that each element forms a class by itself. Show that the group is isomorphic with the group of planar rotations C_4 that send a square into itself and that the angles of rotation corresponding to the four permutations are 0, $\pi/2$, π, and $3\pi/2$. Show that the functions $x^2 + y^2$, xy, $x + iy$, $x - iy$ form four one-dimensional bases for irreducible representations of C_4, and construct the character table. Obtain the branching rules for the decomposition of irreducible representations Γ_R of O into irreducible representations γ_R of C_4 in the form

$$\begin{aligned}&\Gamma_1 \to \gamma_1\\&\Gamma_2 \to \gamma_2\\&\Gamma_3 \to \gamma_1 + \gamma_2\\&\Gamma_4 \to \gamma_1 + \gamma_3 + \gamma_4\\&\Gamma_5 \to \gamma_2 + \gamma_3 + \gamma_4\end{aligned}$$

2-2. Prove that

$$C_0^{(1)} = z/r \qquad C_{\pm 1}^{(1)} = \mp(\tfrac{1}{2})^{\frac{1}{2}}(x \pm iy)/r$$

and that

$$C_0^{(2)} = (3z^2 - r^2)/2r^2 \qquad C_{\pm 1}^{(2)} = \mp(\tfrac{3}{2})^{\frac{1}{2}}z(x \pm iy)/r^2$$
$$C_{\pm 2}^{(2)} = (\tfrac{3}{8})^{\frac{1}{2}}(x \pm iy)^2/r^2$$

2-3. Eight charges $-eg$ are each a distance ρ from the nucleus of an ion and lie at the vertices of a cube. Prove that the contribution to the Hamiltonian can be written in the form of Eq. (2-5), where

$$A_4^0 = -\frac{7ge^2}{18\rho^5} \qquad A_4^4 = -\frac{35ge^2}{18\rho^5}$$
$$A_6^0 = \frac{ge^2}{9\rho^7} \qquad A_6^4 = -\frac{7ge^2}{3\rho^7}$$

The cerium ion Ce^{3+}, which possesses a single f electron outside closed shells, can be inserted in the lattice of CaF_2 so that its immediate neighbors are eight fluorine ions lying at the vertices of a cube. According to the experiments of Dvir and Low,[26] the two lowest energy levels are of the types Γ_7 and Γ_8 and almost exactly coincide. Show that if the limiting cases (1) $\zeta = 0$, $-7b > -a$, and (2) $a = b = 0$ are excluded, where a and b are defined by Eqs. (2-33), ζ must satisfy

$$\zeta = (10a + 84b)(9a + 14b)(2a - 14b)/(165a^2 + 539ab)$$

2-4. The ion $(NpO_2)^{2+}$ possesses the linear structure O—Np—O. The free neptunium atom has seven electrons ($5f^46d7s^2$) outside a radon core, four of which are lost in bonds with the oxygen atoms, while two go in forming the ion. On the assumption that second-rank spherical harmonics are the most important in the expansion of the electric potential produced by the axial column of negative charge, show that the seven degenerate orbital states of the remaining f electron break up into four components, and find their separations. Show that the splittings produced by a small spin-orbit coupling are in the ratio 3:2:1:0. Prove that the effect of a magnetic field on the lowest doublet can be reproduced by the spin-Hamiltonian

$$g_{\parallel}\beta H_z S_z + g_{\perp}\beta(H_x S_x + H_y S_y)$$

with $S = \frac{1}{2}$, $g_{\parallel} = 4$, $g_{\perp} = 0$. The experimental g values[27] for crystals of $(UO_2)Rb(NO_3)_3$ containing small amounts of Np substituted for U are $g_{\parallel} = 3.40 \pm 0.05$, $g_{\perp} = 1.0 \pm 0.5$. Suggest reasons for the deviations between experiment and theory.[28]

3

THE n-j SYMBOLS

3-1 THREE ANGULAR MOMENTA

We have seen that the coefficient of $|j_1m_1j_2m_2)$ in the expansion of $|j_1j_2jm)$ can be expressed as an algebraic function of j_1, m_1, j_2, m_2, j, and m. This gives a general solution to the problem of passing from $SLJM_J$ to SM_SLM_L quantization. It is clear from Sec. 1-6, however, that in general the transition to sm_slm_l quantization involves the study of coupling schemes of several angular momenta. As a first step toward putting the many-electron problem on an algebraic footing we therefore investigate the coupling of three angular momenta j_1, j_2, and j_3 to give a resultant, j. It is not enough to write $|j_1j_2j_3jm)$ for a state of the system, since it is easy to see that such a state cannot be unambiguously expressed as a linear combination of the states $|j_1m_1,j_2m_2,j_3m_3)$. However, on specifying the resultant of j_1 and j_2, say, j_{12}, a unique reduction can be found:

$$\begin{aligned}|(j_1j_2)j_{12},j_3,jm)\\ &= \sum_{m_{12},m_3} |(j_1j_2)j_{12}m_{12},j_3m_3)(j_{12}m_{12}j_3m_3|j_{12}j_3jm)\\ &= \sum_{m_{12},m_3,m_1,m_2} |j_1m_1,j_2m_2,j_3m_3)(j_{12}m_{12}j_3m_3|j_{12}j_3jm)\\ &\qquad\times (j_1m_1j_2m_2|j_1j_2j_{12}m_{12}) \qquad (3\text{-}1)\end{aligned}$$

But we can equally couple j_2 and j_3 to form j_{23}, adding the resultant to j_1. In this case

$$
\begin{aligned}
|j_1,(j_2j_3)j_{23},jm) &= \sum_{m_1,m_{23}} |j_1m_1,(j_2j_3)j_{23}m_{23})(j_1m_1j_{23}m_{23}|j_1j_{23}jm) \\
&= \sum_{m_1,m_{23},m_2,m_3} |j_1m_1,j_2m_2,j_3m_3)(j_1m_1j_{23}m_{23}|j_1j_{23}jm) \\
&\qquad \times (j_2m_2j_3m_3|j_2j_3j_{23}m_{23}) \qquad (3\text{-}2)
\end{aligned}
$$

The transformation which connects the states in these two coupling schemes can be written as

$$
|j_1,(j_2j_3)j_{23},jm) = \sum_{j_{12}} ((j_1j_2)j_{12},j_3jm|j_1,(j_2j_3)j_{23},jm)|(j_1j_2)j_{12},j_3,jm) \qquad (3\text{-}3)
$$

It can be seen that the transformation coefficients are independent of m by operating on both sides of Eq. (3-3) with j_+. Substituting (3-1) and (3-2) into (3-3), and equating the coefficients of $|j_1m_1,j_2m_2,j_3m_3)$, we get

$$
\begin{aligned}
&\sum_{m_{23}} (-1)^{j_2-j_3+m_{23}+j_1-j_{23}+m}[(2j_{23}+1)(2j+1)]^{\frac{1}{2}} \\
&\qquad \times \begin{pmatrix} j_2 & j_3 & j_{23} \\ m_2 & m_3 & -m_{23} \end{pmatrix} \begin{pmatrix} j_1 & j_{23} & j \\ m_1 & m_{23} & -m \end{pmatrix} \\
&\qquad = \sum_{j_{12},m_{12}} ((j_1j_2)j_{12},j_3,j|j_1,(j_2j_3)j_{23},j) \\
&\qquad \times (-1)^{j_1-j_2+m_{12}+j_{12}-j_3+m}[(2j_{12}+1)(2j+1)]^{\frac{1}{2}} \\
&\qquad\qquad \times \begin{pmatrix} j_1 & j_2 & j_{12} \\ m_1 & m_2 & -m_{12} \end{pmatrix} \begin{pmatrix} j_{12} & j_3 & j \\ m_{12} & m_3 & -m \end{pmatrix}
\end{aligned}
$$

We multiply both sides of this equation by

$$
(-1)^{-m_{12}'} \begin{pmatrix} j_1 & j_2 & j_{12}' \\ m_1 & m_2 & -m_{12}' \end{pmatrix}
$$

and sum over m_1 and m_2. In view of Eq. (1-24), the right-hand side becomes

$$
\begin{aligned}
&\sum_{j_{12},m_{12}} ((j_1j_2)j_{12},j_3,j|j_1,(j_2j_3)j_{23},j)\delta(j_{12},j_{12}') \\
&\quad \times \delta(m_{12},m_{12}')(-1)^{j_1-j_2+j_{12}-j_3+m+m_{12}-m'_{12}} \left(\frac{2j+1}{2j_{12}+1}\right)^{\frac{1}{2}} \begin{pmatrix} j_{12} & j_3 & j \\ m_{12} & m_3 & -m \end{pmatrix} \\
&\qquad = ((j_1j_2)j_{12}',j_3,j|j_1,(j_2j_3)j_{23},j) \\
&\qquad \times (-1)^{j_1-j_2+j'_{12}-j_3+m} \left(\frac{2j+1}{2j_{12}'+1}\right)^{\frac{1}{2}} \begin{pmatrix} j_{12}' & j_3 & j \\ m_{12}' & m_3 & -m \end{pmatrix}
\end{aligned}
$$

We now define the 6-j symbol by the equation

$$((j_1j_2)j'_{12},j_3,j|j_1,(j_2j_3)j_{23},j)$$
$$= [(2j'_{12}+1)(2j_{23}+1)]^{\frac{1}{2}}(-1)^{j_1+j_2+j_3+j}\begin{Bmatrix} j_3 & j & j'_{12} \\ j_1 & j_2 & j_{23} \end{Bmatrix} \tag{3-4}$$

Thus

$$(-1)^{j_1-j_2+j'_{12}-j_3+m-j_1-j_2-j_3-j}[(2j+1)(2j_{23}+1)]^{\frac{1}{2}}$$
$$\times \begin{Bmatrix} j_3 & j & j'_{12} \\ j_1 & j_2 & j_{23} \end{Bmatrix}\begin{pmatrix} j'_{12} & j_3 & j \\ m'_{12} & m_3 & -m \end{pmatrix}$$
$$= \sum_{m_{23},m_1,m_2} (-1)^{j_2-j_3+m_{23}+j_1-j_{23}+m-m'_{12}}[(2j+1)(2j_{23}+1)]^{\frac{1}{2}}$$
$$\times \begin{pmatrix} j_2 & j_3 & j_{23} \\ m_2 & m_3 & -m_{23} \end{pmatrix}\begin{pmatrix} j_1 & j_{23} & j \\ m_1 & m_{23} & -m \end{pmatrix}\begin{pmatrix} j_1 & j_2 & j'_{12} \\ m_1 & m_2 & -m'_{12} \end{pmatrix} \tag{3-5}$$

The phase factors can be combined to give

$$(-1)^{j_1+3j_2+j_3-j'_{12}+j+m_{23}-m'_{12}-j_{23}}$$

Equation (3-5) can be thrown into a more symmetrical form by interchanging columns 1 and 2 of the first 3-j symbol on the right, columns 2 and 3 of the second, and reversing the signs in the lower row of the third. This introduces the phase factor

$$(-1)^{-(j_2+j_3+j_{23})-(j_1+j_{23}+j)+(j_1+j_2+j'_{12})}$$

Since $m'_{12} = m_1 + m_2$, the total phase factor is

$$(-1)^{j_1+3j_2+j_3-j_{23}-j'_{12}+j+m_{23}-m_1-m_2-j_2-j_3-j_{23}-j_1-j_{23}-j+j_1+j_2+j'_{12}}$$
$$= (-1)^{j_1+3j_2-3j_{23}+m_{23}-m_1-m_2} = (-1)^{j_1+j_2+j_{23}-m_1+(2j_2-m_2)+m_{23}}$$
$$= (-1)^{j_1+j_2+j_{23}-m_1+m_2+m_{23}}$$

Equation (3-5) now reads

$$\begin{Bmatrix} j_3 & j & j'_{12} \\ j_1 & j_2 & j_{23} \end{Bmatrix}\begin{pmatrix} j'_{12} & j_3 & j \\ m'_{12} & m_3 & -m \end{pmatrix} = \sum_{m_{23},m_1,m_2} (-1)^{j_1+j_2+j_{23}-m_1+m_2+m_{23}}$$
$$\times \begin{pmatrix} j_3 & j_2 & j_{23} \\ m_3 & m_2 & -m_{23} \end{pmatrix}\begin{pmatrix} j_1 & j & j_{23} \\ m_1 & -m & m_{23} \end{pmatrix}\begin{pmatrix} j_1 & j_2 & j'_{12} \\ -m_1 & -m_2 & m'_{12} \end{pmatrix}$$

The symmetry is displayed better if the following substitutions are made:

$$\begin{array}{lll} j_3 \to j_1 & j_2 \to l_2 & m'_{12} \to m_3 \\ j \to j_2 & j_{23} \to l_3 & m_1 \to -\mu_1 \\ j'_{12} \to j_3 & m_3 \to m_1 & m_2 \to \mu_2 \\ j_1 \to l_1 & m \to -m_2 & m_{23} \to \mu_3 \end{array}$$

We get, finally,

$$\begin{Bmatrix} j_1 & j_2 & j_3 \\ l_1 & l_2 & l_3 \end{Bmatrix} \begin{pmatrix} j_1 & j_2 & j_3 \\ m_1 & m_2 & m_3 \end{pmatrix} = \sum_{\mu_1,\mu_2,\mu_3} (-1)^{l_1+l_2+l_3+\mu_1+\mu_2+\mu_3}$$
$$\times \begin{pmatrix} j_1 & l_2 & l_3 \\ m_1 & \mu_2 & -\mu_3 \end{pmatrix} \begin{pmatrix} l_1 & j_2 & l_3 \\ -\mu_1 & m_2 & \mu_3 \end{pmatrix} \begin{pmatrix} l_1 & l_2 & j_3 \\ \mu_1 & -\mu_2 & m_3 \end{pmatrix} \tag{3-6}$$

in agreement with Eq. (2.20) of Rotenberg et al.[6]

3-2 THE 6-j SYMBOL

Equation (3-6) expresses a 6-j symbol as a sum over 3-j symbols. Since the latter themselves involve a summation, direct substitution through Eqs. (1-20) and (1-22) gives a very complicated expression for the 6-j symbol. However, since the 6-j symbol is independent of m_1, m_2, and m_3, we can set $m_1 = j_1$, $m_2 = -j_2$, thereby reducing the summations in three of the four 3-j symbols in Eq. (3-6) to a single term. After an appreciable amount of algebraic manipulation, the 6-j symbol reduces to a summation over a single variable,

$$\begin{Bmatrix} j_1 & j_2 & j_3 \\ l_1 & l_2 & l_3 \end{Bmatrix} = \Delta(j_1j_2j_3)\Delta(j_1l_2l_3)\Delta(l_1j_2l_3)\Delta(l_1l_2j_3)$$
$$\times \sum_z \frac{(-1)^z(z+1)!}{\begin{matrix}[(z-j_1-j_2-j_3)!(z-j_1-l_2-l_3)!(z-l_1-j_2-l_3)! \\ (z-l_1-l_2-j_3)!(j_1+j_2+l_1+l_2-z)! \\ (j_2+j_3+l_2+l_3-z)!(j_3+j_1+l_3+l_1-z)!]\end{matrix}} \tag{3-7}$$

where

$$\Delta(abc) = [(a+b-c)!(a-b+c)!(b+c-a)!/(a+b+c+1)!]^{\frac{1}{2}}$$

The four triangular conditions for the nonvanishing of the 3-j symbols are contained in the 6-j symbol in the following way:

The 6-j symbol is related to the W function of Racah[18] by the equation

$$\begin{Bmatrix} j_1 & j_2 & j_3 \\ l_1 & l_2 & l_3 \end{Bmatrix} = (-1)^{j_1+j_2+l_1+l_2} W(j_1j_2l_2l_1;j_3l_3)$$

From Eq. (3-7) it is clear that the 6-j symbol is invariant under any permutation of the columns, e.g.,

$$\begin{Bmatrix} j_1 & j_2 & j_3 \\ l_1 & l_2 & l_3 \end{Bmatrix} = \begin{Bmatrix} j_2 & j_1 & j_3 \\ l_2 & l_1 & l_3 \end{Bmatrix} \tag{3-8}$$

It is also invariant under an interchange of the upper and lower arguments in each of any two of its columns, e.g.,

$$\begin{Bmatrix} j_1 & j_2 & j_3 \\ l_1 & l_2 & l_3 \end{Bmatrix} = \begin{Bmatrix} l_1 & l_2 & j_3 \\ j_1 & j_2 & l_3 \end{Bmatrix} \tag{3-9}$$

The absence of phase factors in equations like (3-8) and (3-9) makes the 6-j symbol a more convenient quantity to handle in actual calculations than the W function. Regge[29] has noted that

$$\begin{Bmatrix} a & b & e \\ d & c & f \end{Bmatrix} = \begin{Bmatrix} \frac{1}{2}(a+b+d-c) & \frac{1}{2}(a+b+c-d) & e \\ \frac{1}{2}(a+d+c-b) & \frac{1}{2}(d+b+c-a) & f \end{Bmatrix}$$

and this property has been examined in detail by Jahn and Howell.[30] Strikingly different 6-j symbols are thereby related, e.g.,

$$\begin{Bmatrix} 6 & 3 & 4 \\ 2 & 2 & 5 \end{Bmatrix} = \begin{Bmatrix} \frac{9}{2} & \frac{9}{2} & 4 \\ \frac{7}{2} & \frac{1}{2} & 5 \end{Bmatrix}$$

Any 6-j symbol whose arguments do not exceed 8 can be found from the tables of Rotenberg et al.[6] Sometimes it is convenient to retain the detailed form of the 6-j symbol, particularly if one of the arguments is small. Cases where the smallest argument is 0, $\frac{1}{2}$, or 1 occur frequently, and their values are set out below:

$$\begin{Bmatrix} a & b & c \\ 0 & c & b \end{Bmatrix} = (-1)^s \left[\frac{1}{(2b+1)(2c+1)}\right]^{\frac{1}{2}}$$

$$\begin{Bmatrix} a & b & c \\ \frac{1}{2} & c-\frac{1}{2} & b+\frac{1}{2} \end{Bmatrix} = (-1)^s \left[\frac{(s-2b)(s-2c+1)}{(2b+1)(2b+2)2c(2c+1)}\right]^{\frac{1}{2}}$$

$$\begin{Bmatrix} a & b & c \\ \frac{1}{2} & c-\frac{1}{2} & b-\frac{1}{2} \end{Bmatrix} = (-1)^s \left[\frac{(s+1)(s-2a)}{2b(2b+1)2c(2c+1)}\right]^{\frac{1}{2}}$$

$$\begin{Bmatrix} a & b & c \\ 1 & c-1 & b-1 \end{Bmatrix} = (-1)^s \left[\frac{s(s+1)(s-2a-1)(s-2a)}{(2b-1)2b(2b+1)(2c-1)2c(2c+1)}\right]^{\frac{1}{2}} \tag{3-10}$$

$$\begin{Bmatrix} a & b & c \\ 1 & c-1 & b \end{Bmatrix} = (-1)^s \left[\frac{2(s+1)(s-2a)(s-2b)(s-2c+1)}{2b(2b+1)(2b+2)(2c-1)2c(2c+1)}\right]^{\frac{1}{2}}$$

$$\begin{Bmatrix} a & b & c \\ 1 & c-1 & b+1 \end{Bmatrix} = (-1)^s \left[\frac{(s-2b-1)(s-2b)(s-2c+1)(s-2c+2)}{(2b+1)(2b+2)(2b+3)(2c-1)2c(2c+1)}\right]^{\frac{1}{2}} \tag{3-10}$$

$$\begin{Bmatrix} a & b & c \\ 1 & c & b \end{Bmatrix} = (-1)^s \frac{2[a(a+1) - b(b+1) - c(c+1)]}{[2b(2b+1)(2b+2)2c(2c+1)(2c+2)]^{\frac{1}{2}}}$$

where, in each equation, $s = a + b + c$.

Sums over products of 6-j symbols can be obtained by considering various recoupling procedures and using Eq. (3-4).

From $$(j_1,(j_2j_3)j_{23},jm|j_1,(j_2j_3)j'_{23},jm) = \delta(j_{23},j'_{23})$$

we obtain the equation

$$\sum_{j_{12}} (j_1,(j_2j_3)j_{23},j|(j_1j_2)j_{12},j_3,j)((j_1j_2)j_{12},j_3,j|j_1,(j_2j_3)j'_{23},j) = \delta(j_{23},j'_{23})$$

which leads to

$$\sum_{j_{12}} (2j_{23}+1)(2j_{12}+1) \begin{Bmatrix} j_3 & j & j_{12} \\ j_1 & j_2 & j_{23} \end{Bmatrix} \begin{Bmatrix} j_3 & j & j_{12} \\ j_1 & j_2 & j'_{23} \end{Bmatrix} = \delta(j_{23},j'_{23}) \tag{3-11}$$

A second equation involving 6-j symbols can be obtained from the recoupling scheme

$$\sum_{j_{23}} ((j_1j_2)j_{12},j_3,j|j_1,(j_2j_3)j_{23},j)(j_1,(j_2j_3)j_{23},j|j_2,(j_3j_1)j_{31},j) = ((j_1j_2)j_{12},j_3,j|j_2,(j_3j_1)j_{31},j)$$

We write $((j_2j_3)j_{23},j_1,j|$ for $(j_1,(j_2j_3)j_{23},j|$ in the second transformation coefficient. This introduces the phase factor $(-1)^{j_1+j_{23}-j}$, owing to Eq. (1-21). We also write the right-hand side as

$$((j_2j_1)j_{12},j_3,j|j_2,(j_1j_3)j_{31},j)$$

the phase factor in this case being $(-1)^{j_1+j_2-j_{12}+j_1+j_3-j_{31}}$. Eq. (3-4) gives

$$\sum_{j_{23}} (2j_{23}+1)(-1)^{j_{23}+j_1-j} \begin{Bmatrix} j_3 & j & j_{12} \\ j_1 & j_2 & j_{23} \end{Bmatrix} \begin{Bmatrix} j_1 & j & j_{23} \\ j_2 & j_3 & j_{31} \end{Bmatrix}$$

$$= (-1)^{j_1+j_2-j_{12}+j_1+j_3-j_{31}-(j+j_1+j_2+j_3)} \begin{Bmatrix} j_3 & j & j_{12} \\ j_2 & j_1 & j_{31} \end{Bmatrix} \qquad (3\text{-}12)$$

3-3 THE 9-j SYMBOL

A transformation of great importance connects states in LS and jj coupling. The transformation coefficient is

$$((s_1s_2)S,(l_1l_2)L,J|(s_1l_1)j_1,(s_2l_2)j_2,J)$$

or, in more general terms,

$$((j_1j_2)j_{12},(j_3j_4)j_{34},j|(j_1j_3)j_{13},(j_2j_4)j_{24},j)$$

This can be written as

$$\sum_{j'} ((j_1j_2)j_{12},j_{34},j|j_1,(j_2j_{34})j',j)(j_2,(j_3j_4)j_{34},j'|j_3,(j_2j_4)j_{24},j')$$

$$\times\ (j_1,(j_3j_{24})j',j|(j_1j_3)j_{13},j_{24},j)$$

The adjacent parts of neighboring transformation coefficients in this sum are identical; we have merely dropped superfluous quantum numbers to make the connection with 6-j symbols more apparent. Using Eq. (3-4), we find the sum to be

$$[(2j_{12}+1)(2j_{34}+1)(2j_{13}+1)(2j_{24}+1)]^{\frac{1}{2}}$$

$$\times \sum_{j'} (-1)^{2j'}(2j'+1) \begin{Bmatrix} j_1 & j_2 & j_{12} \\ j_{34} & j & j' \end{Bmatrix} \begin{Bmatrix} j_3 & j_4 & j_{34} \\ j_2 & j' & j_{24} \end{Bmatrix} \begin{Bmatrix} j_{13} & j_{24} & j \\ j' & j_1 & j_3 \end{Bmatrix}$$

On defining the 9-j symbol through the equation

$$((j_1j_2)j_{12},(j_3j_4)j_{34},j|(j_1j_3)j_{13},(j_2j_4)j_{24},j)$$

$$= [(2j_{12}+1)(2j_{34}+1)(2j_{13}+1)(2j_{24}+1)]^{\frac{1}{2}} \begin{Bmatrix} j_1 & j_2 & j_{12} \\ j_3 & j_4 & j_{34} \\ j_{13} & j_{24} & j \end{Bmatrix} \qquad (3\text{-}13)$$

we obtain, with a rather more symmetrical set of arguments in the

9-j symbol, the equation

$$\begin{Bmatrix} j_{11} & j_{12} & j_{13} \\ j_{21} & j_{22} & j_{23} \\ j_{31} & j_{32} & j_{33} \end{Bmatrix} = \sum_x (-1)^{2x}(2x+1) \times \begin{Bmatrix} j_{11} & j_{21} & j_{31} \\ j_{32} & j_{33} & x \end{Bmatrix} \begin{Bmatrix} j_{12} & j_{22} & j_{32} \\ j_{21} & x & j_{23} \end{Bmatrix} \begin{Bmatrix} j_{13} & j_{23} & j_{33} \\ x & j_{11} & j_{12} \end{Bmatrix} \quad (3\text{-}14)$$

The 9-j symbol can be expressed in terms of 3-j symbols by writing

$$\begin{aligned} |(j_1j_2)j_{12},(j_3j_4)j_{34},jm) &= \sum_{m_{12},m_{34}} (j_{12}m_{12}j_{34}m_{34}|j_{12}j_{34}jm) \\ &\qquad \times |(j_1j_2)j_{12}m_{12},(j_3j_4)j_{34}m_{34}) \\ &= \sum_{m_{12},m_{34},m_1,m_2,m_3,m_4} (j_{12}m_{12}j_{34}m_{34}|j_{12}j_{34}jm) \\ &\qquad \times (j_1m_1j_2m_2|j_1j_2j_{12}m_{12})(j_3m_3j_4m_4|j_3j_4j_{34}m_{34}) \\ &\qquad \times |j_1m_1,j_2m_2,j_3m_3,j_4m_4) \end{aligned}$$

A similar expression can be found for $((j_1j_3)j_{13},(j_2j_4)j_{24},j_{24},jm|$. On putting the two parts together and using Eq. (1-22), the transformation coefficient is obtained as a sum over six 3-j symbols. Since the transformation coefficient is independent of m, the effect of summing over this variable is equivalent to a multiplication by $2j+1$. Using Eq. (3-13), and transforming to a new set of arguments, we obtain the remarkably symmetrical expression

$$\begin{aligned} &\begin{Bmatrix} j_{11} & j_{12} & j_{13} \\ j_{21} & j_{22} & j_{23} \\ j_{31} & j_{32} & j_{33} \end{Bmatrix} \\ &= \sum_{\text{All } m\text{'s}} \begin{pmatrix} j_{11} & j_{12} & j_{13} \\ m_{11} & m_{12} & m_{13} \end{pmatrix} \begin{pmatrix} j_{21} & j_{22} & j_{23} \\ m_{21} & m_{22} & m_{23} \end{pmatrix} \begin{pmatrix} j_{31} & j_{32} & j_{33} \\ m_{31} & m_{32} & m_{33} \end{pmatrix} \\ &\quad \times \begin{pmatrix} j_{11} & j_{21} & j_{31} \\ m_{11} & m_{21} & m_{31} \end{pmatrix} \begin{pmatrix} j_{12} & j_{22} & j_{32} \\ m_{12} & m_{22} & m_{32} \end{pmatrix} \begin{pmatrix} j_{13} & j_{23} & j_{33} \\ m_{13} & m_{23} & m_{33} \end{pmatrix} \quad (3\text{-}15) \end{aligned}$$

The symmetry properties of the 9-j symbol may be readily found from this formula. An odd permutation of the rows or columns multiplies the 9-j symbol by $(-1)^R$, where R is the sum of the nine arguments of the symbol. An even permutation produces no change of phase, and neither does a transposition. Curiously enough, Regge's representation of the 3-j symbol possesses the same symmetry properties (see Prob. 1-6).

3-4 RELATIONS BETWEEN THE n-j SYMBOLS FOR $n > 3$

By considering recoupling schemes, any number of equations involving various n-j symbols can be established. The underlying symmetry in these equations, which, as in Eqs. (3-11) and (3-12), is not immediately apparent, can be emphasized by a careful choice of notation. A deeper appreciation of the equations can be obtained by extracting from the n-j symbols within a sum all those triangular conditions that do not involve an index of summation. Every condition can be represented geometrically by the junction of three branches, each branch being labeled by one of the three angular-momentum quantum numbers involved in the triangular condition. The coupling diagrams characterize the sums and make the symmetry properties obvious. A great deal of attention has been paid to the topological properties of such diagrams, particularly in connection with the celebrated four-color problem; however, the theory[31] appears to have been insufficiently developed to assist us materially in our analysis of the properties of the n-j symbols.

An appreciable number of relations between various n-j symbols have appeared in the literature. In order to include most of them here, their derivations, which in virtually all cases are quite straightforward, will not be given in detail. Factors of the type $2x + 1$ occur frequently, and the abbreviation

$$[x] = 2x + 1 \tag{3-16}$$

is made. Since we still require brackets for other purposes, it is to be understood that Eq. (3-16) is valid only when there is a single (possibly subscripted) symbol enclosed within the brackets.

We begin by rewriting Eqs. (3-11) and (3-12),

$$\sum_x [x] \begin{Bmatrix} a & b & x \\ c & d & p \end{Bmatrix} \begin{Bmatrix} c & d & x \\ a & b & q \end{Bmatrix} = \frac{\delta(p,q)}{[p]} \tag{3-17}$$

$$\sum_x [x](-1)^{p+q+x} \begin{Bmatrix} a & b & x \\ c & d & p \end{Bmatrix} \begin{Bmatrix} c & d & x \\ b & a & q \end{Bmatrix} = \begin{Bmatrix} c & a & q \\ d & b & p \end{Bmatrix} \tag{3-18}$$

The coupling diagrams are given in Figs. 3-1 and 3-2. Special cases of Eqs. (3-17) and (3-18) have been given by Elliott and Lane.[32] On setting $q = 0$ and insisting in turn that the second 6-j symbol in both

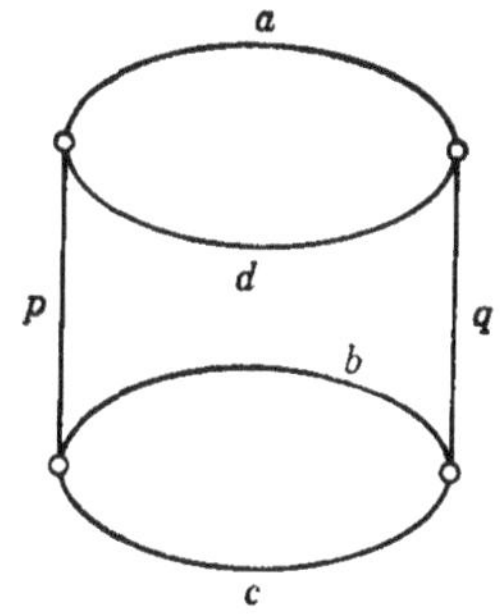

FIG. 3-1 The coupling diagram corresponding to Eq. (3-17).

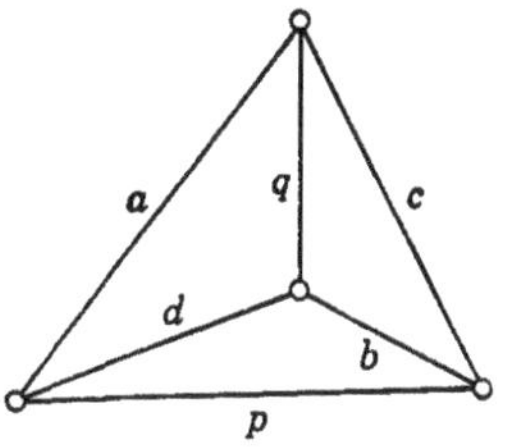

FIG. 3-2 The 6-j symbol.

sums does not vanish, we obtain, with the aid of Eq. (3-10),

$$\sum_x [x](-1)^{-x} \begin{Bmatrix} a & b & x \\ b & a & p \end{Bmatrix} = \delta(p,0)(-1)^{a+b}\{[a][b]\}^{\frac{1}{2}} \tag{3-19}$$

and

$$\sum_x [x] \begin{Bmatrix} a & b & x \\ a & b & p \end{Bmatrix} = (-1)^{2a+2b} \tag{3-20}$$

The Biedenharn-Elliott sum rule,[33,34] which can be derived by recoupling four angular momenta, may be regarded as an extension of Eq. (3-18):

$$\sum_x [x](-1)^{x+p+q+r+a+b+c+d+e+f} \begin{Bmatrix} a & b & x \\ c & d & p \end{Bmatrix} \begin{Bmatrix} c & d & x \\ e & f & q \end{Bmatrix} \begin{Bmatrix} e & f & x \\ b & a & r \end{Bmatrix}$$

$$= \begin{Bmatrix} p & q & r \\ e & a & d \end{Bmatrix} \begin{Bmatrix} p & q & r \\ f & b & c \end{Bmatrix} \tag{3-21}$$

The coupling diagram is given in Fig. 3-3. If the lower row in the third 6-j symbol in the sum is changed from bar to abr, and if the odd

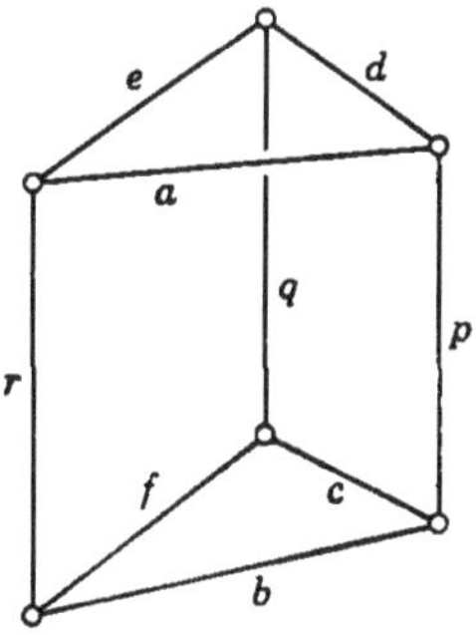

FIG. 3-3 The Biedenharn-Elliott sum rule.

multiple of x in the phase factor is eliminated, the equation becomes equivalent to the definition (3-14) of the 9-j symbol. It is convenient to write down at the same time a double sum over the products of two 9-j symbols; that it is equal to a 9-j symbol can be proved either by recoupling considerations or by algebraic manipulations with the 6-j symbols.

$$\sum_x [x](-1)^{2x} \begin{Bmatrix} a & b & x \\ c & d & p \end{Bmatrix} \begin{Bmatrix} c & d & x \\ e & f & q \end{Bmatrix} \begin{Bmatrix} e & f & x \\ a & b & r \end{Bmatrix}$$
$$= \sum_{x,y} [x][y](-1)^{2f+y+c-e} \begin{Bmatrix} a & f & r \\ q & d & e \\ x & y & b \end{Bmatrix} \begin{Bmatrix} a & q & x \\ d & f & y \\ p & c & b \end{Bmatrix}$$
$$= \begin{Bmatrix} a & f & r \\ d & q & e \\ p & c & b \end{Bmatrix} \tag{3-22}$$

The coupling diagram is given in Fig. 3-4 in two topologically equivalent forms. The first exhibits a threefold axis of symmetry; the second can be inscribed on a Möbius strip in such a way that no two

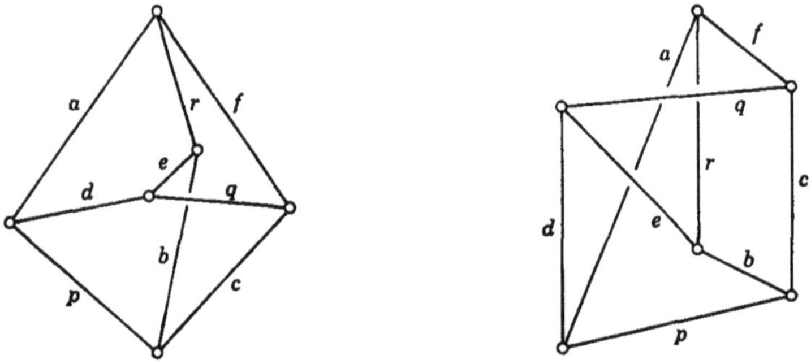

FIG. 3-4 Two topologically equivalent coupling diagrams for the 9-j symbol. That on the left exhibits a threefold axis of symmetry; that on the right can be inscribed on a Möbius strip.

branches cross each other. Higher n-j symbols can be easily generated by inserting extra segments in the strip; for example, the 12-j symbol[35,36] is defined by

$$\begin{Bmatrix} c & & b & & h & & e & \\ & p & & s & & r & & q \\ d & & a & & g & & f & \end{Bmatrix} = (-1)^{p+q+r+s+a+b+c+d+e+f+g+h}$$
$$\times \sum_x [x](-1)^{-x} \begin{Bmatrix} a & b & x \\ c & d & p \end{Bmatrix} \begin{Bmatrix} c & d & x \\ e & f & q \end{Bmatrix} \begin{Bmatrix} e & f & x \\ g & h & r \end{Bmatrix} \begin{Bmatrix} g & h & x \\ b & a & s \end{Bmatrix} \tag{3-23}$$

An equivalent definition is

$$\begin{Bmatrix} c & & b & & h & & e & \\ & p & & s & & r & & q \\ d & & a & & g & & f & \end{Bmatrix}$$

$$= \sum_x [x](-1)^{2x+b+f+a+e} \begin{Bmatrix} b & f & x \\ q & p & c \end{Bmatrix} \begin{Bmatrix} a & e & x \\ q & p & d \end{Bmatrix} \begin{Bmatrix} s & h & b \\ g & r & f \\ a & e & x \end{Bmatrix} \qquad (3\text{-}24)$$

Equations (3-23) and (3-24) possess the same coupling diagram, of course, and it is given in Fig. 3-5. A second 12-j symbol, for which

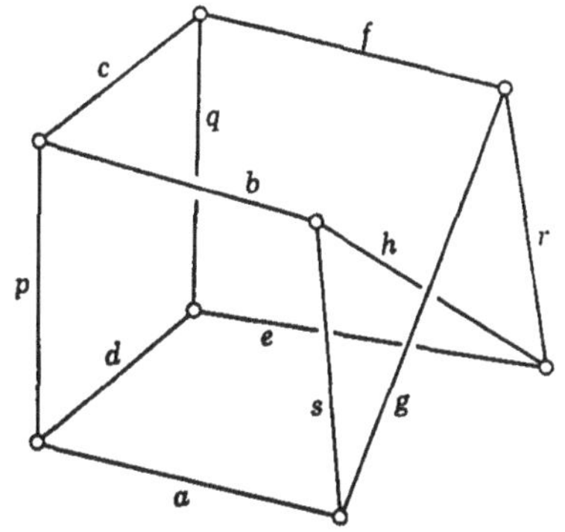

FIG. 3-5 The 12-j symbol of the first kind.

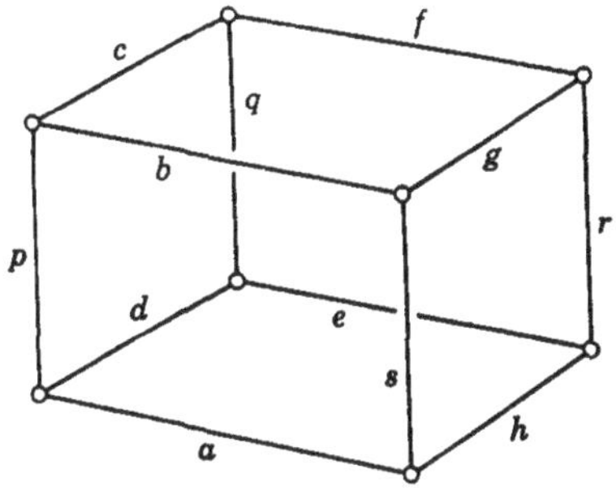

FIG. 3-6 The 12-j symbol of the second kind.

no special notation has yet come into general use, can be regarded as an extension of Eq. (3-21):

$$\sum_x [x] \begin{Bmatrix} a & b & x \\ c & d & p \end{Bmatrix} \begin{Bmatrix} c & d & x \\ e & f & q \end{Bmatrix} \begin{Bmatrix} e & f & x \\ g & h & r \end{Bmatrix} \begin{Bmatrix} g & h & x \\ a & b & s \end{Bmatrix}$$

$$= \sum_y [y] \begin{Bmatrix} d & q & e \\ p & c & b \\ a & f & y \end{Bmatrix} \begin{Bmatrix} h & r & e \\ s & g & b \\ a & f & y \end{Bmatrix} \qquad (3\text{-}25)$$

The coupling diagram, drawn in Fig. 3-6, can be inscribed on an untwisted strip.[37]

The orthonormality of the transformation coefficients of Sec. 3-3 leads to the equation

$$\sum_{x,y} [x][y] \begin{Bmatrix} a & f & x \\ d & q & y \\ p & c & b \end{Bmatrix} \begin{Bmatrix} a & f & x \\ d & q & y \\ p' & c' & b \end{Bmatrix} = \delta(p,p')\delta(c,c')/[p][c] \qquad (3\text{-}26)$$

and the coupling diagram is shown in Fig. 3-7. The following equation, which is often useful, is equivalent to the Biedenharn-Elliott sum rule:

$$\sum_x [x] \begin{Bmatrix} a & f & x \\ d & q & e \\ p & c & b \end{Bmatrix} \begin{Bmatrix} a & f & x \\ e & b & \lambda \end{Bmatrix} = (-1)^{2\lambda} \begin{Bmatrix} c & d & \lambda \\ e & f & q \end{Bmatrix} \begin{Bmatrix} a & b & \lambda \\ c & d & p \end{Bmatrix} \tag{3-27}$$

Innes and Ufford[38] have obtained the sum

$$\sum_{x,x',x''} [x][x'][x''] \begin{Bmatrix} e & e' & e'' \\ x & x' & x'' \end{Bmatrix} \begin{Bmatrix} b & a' & a'' \\ d & c' & c'' \\ e & x' & x'' \end{Bmatrix} \begin{Bmatrix} a & b' & a'' \\ c & d' & c'' \\ x & e' & x'' \end{Bmatrix} \begin{Bmatrix} a & a' & b'' \\ c & c' & d'' \\ x & x' & e'' \end{Bmatrix}$$
$$= \begin{Bmatrix} b & b' & b'' \\ a & a' & a'' \end{Bmatrix} \begin{Bmatrix} d & d' & d'' \\ c & c' & c'' \end{Bmatrix} \begin{Bmatrix} b & b' & b'' \\ d & d' & d'' \\ e & e' & e'' \end{Bmatrix} \tag{3-28}$$

corresponding to Fig. 3-8. Many relations can be derived as special cases of Eq. (3-28) by setting different angular-momentum quantum

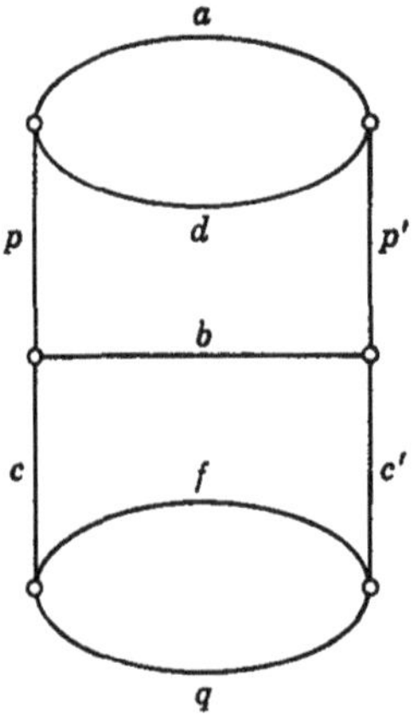

Fig. 3-7 The coupling diagram corresponding to Eq. (3-26).

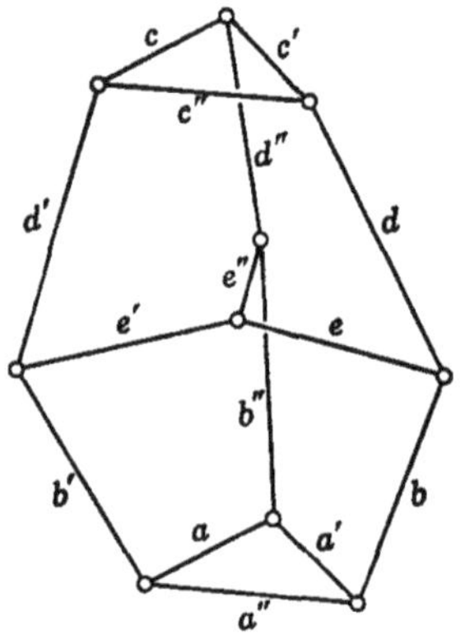

Fig. 3-8 The Innes-Ufford identity.

numbers to zero. It is clear that on putting $u = 0$ the coupling diagram in the region of the branch labeled by u simplifies as follows:

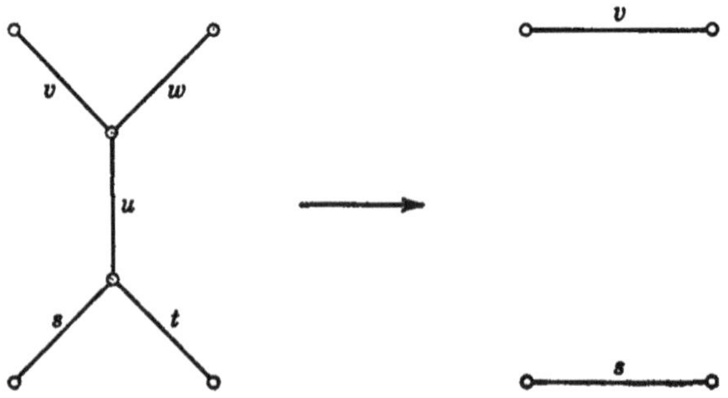

This reduction, when applied to Fig. 3-4, shows that a 9-j symbol with one of its arguments zero is equivalent to a 6-j symbol. In detail,

$$\begin{Bmatrix} a & b & e \\ c & d & e \\ f & f & 0 \end{Bmatrix} = \frac{(-1)^{b+c+e+f}}{\{[e][f]\}^{\frac{1}{2}}} \begin{Bmatrix} a & b & e \\ d & c & f \end{Bmatrix} \tag{3-29}$$

In Fig. 3-8, there are three essentially dissimilar branches. If any one of a, a', a'', c, c', or c'' is set equal to zero, one of the triangles collapses to a point and the identity of Arima, Horie, and Tanabe[39] is obtained. Thus, on putting $a'' = 0$, we get

$$\sum_{x,x'} [x][x'](-1)^{c+e+e''+d''-b-b'-d'-x'} \begin{Bmatrix} e'' & e' & e \\ c'' & x' & x \end{Bmatrix}$$

$$\times \begin{Bmatrix} b & d & e \\ c'' & x' & c' \end{Bmatrix} \begin{Bmatrix} b' & d' & e' \\ c'' & x & c \end{Bmatrix} \begin{Bmatrix} b & b' & b'' \\ c' & c & d'' \\ x' & x & e'' \end{Bmatrix}$$

$$= \begin{Bmatrix} d & d' & d'' \\ c & c' & c'' \end{Bmatrix} \begin{Bmatrix} b & b' & b'' \\ d & d' & d'' \\ e & e' & e'' \end{Bmatrix} \tag{3-30}$$

The coupling diagram is given in Fig. 3-9. It is easy to see that if c, c', or c'' is set equal to zero the figure for a 9-j symbol is produced, while if any one of b, b', b'', e, e', or e'' is taken to be zero the triangle in the diagram remains intact and we recover the Biedenharn-Elliott sum rule.

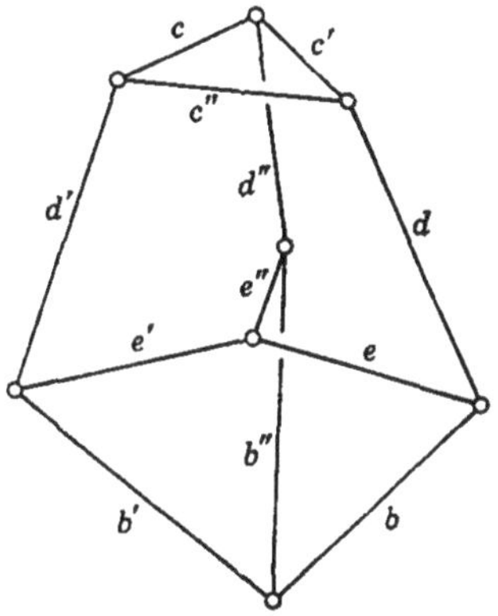

FIG. 3-9 The identity of Arima, Horie, and Tanabe.

Although the coupling diagrams, from the way they are constructed, exhibit the coupling schemes of the n-j symbols under the summation sign, a comparison between the diagrams and the formulas they represent makes it apparent that they determine, to within a phase factor, the result of actually carrying out the sum. All the results given above can be reproduced by taking the appropriate coupling diagram and decomposing it in steps, at each step including a factor according to the following rules:

1. Double links are removed by the substitution

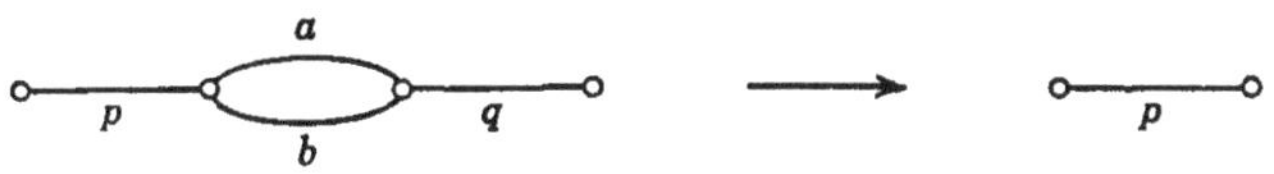

and the factor $\delta(p,q)/[p]$ is written down.

2. Triangles are eliminated according to

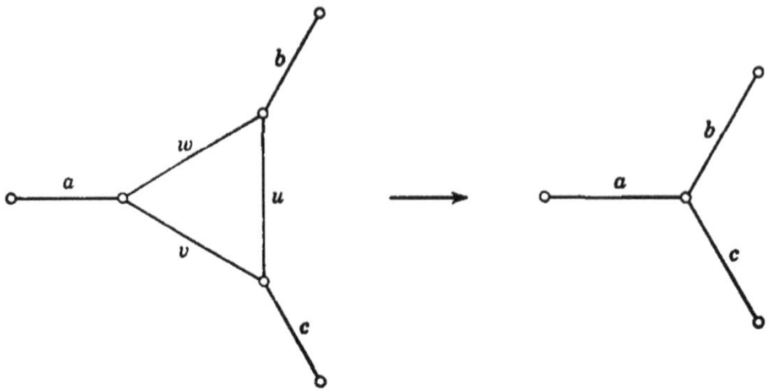

and the factor $\begin{Bmatrix} a & b & c \\ u & v & w \end{Bmatrix}$ is included.

3. Reductions of the type

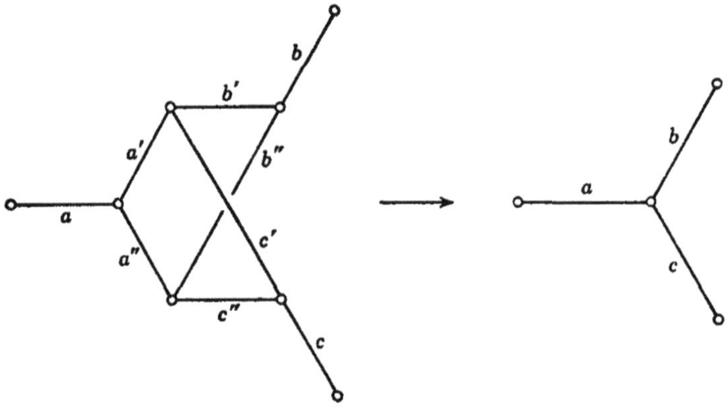

introduce as a factor the 9-j symbol

$$\begin{Bmatrix} a & a' & a'' \\ b & b' & b'' \\ c & c' & c'' \end{Bmatrix}$$

Eventually we arrive either at a coupling scheme involving an n-j symbol with $n > 9$ or else the triple linkage

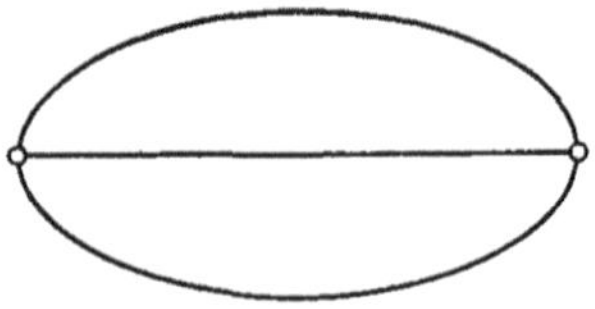

This indicates the termination of the reduction process. The rules given above are of use in anticipating the result of a complex sum over various n-j symbols, but the question of phase makes it necessary to study the sum in detail.

Suppose, for example, we are confronted with the triple sum

$$\sum_{x,y,z} [x][y][z] \begin{Bmatrix} q & e & d \\ c & g & x \\ f & y & z \end{Bmatrix} \begin{Bmatrix} x & z & d \\ c & b & p \\ g & g' & a \end{Bmatrix} \begin{Bmatrix} y & e & g \\ z & b & g' \\ f & r & a' \end{Bmatrix}$$

The coupling diagram, containing all the triangular conditions except those involving the running indices, is given in Fig. 3-10. The double link can be removed by (1) above and gives the factor $\delta(a,a')/[a]$. We may next apply the decomposition (3), though it is obvious that the coupling diagram corresponds to a 9-j symbol. We therefore expect the triple sum to evaluate to

$$\frac{\delta(a,a')}{[a]} \begin{Bmatrix} a & f & r \\ d & q & e \\ p & c & b \end{Bmatrix}$$

A detailed calculation shows this result to be correct with respect both to magnitude and, fortuitously, to phase.

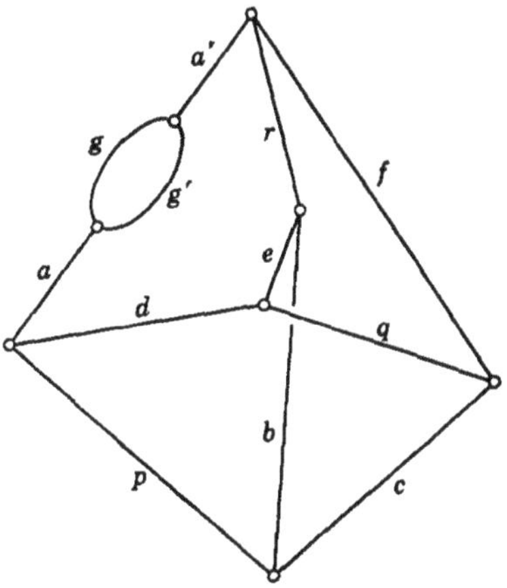

FIG. 3-10 A coupling diagram to exemplify the decomposition rules.

3-5 MIXED TENSOR OPERATORS

In the preceding chapter the tensor operator $\mathbf{T}^{(k)}$ is considered and the theory applied to the term H_3 in the Hamiltonian. It is evident, however, that other terms occur in the Hamiltonian, for instance, H_1 and H_2, which are not of this simple form. In order to extend the theory to deal with these contributions, we consider the quantities

$$X_Q^{(K)} = \sum_{q_1,q_2} T_{q_1}^{(k_1)} U_{q_2}^{(k_2)} (k_1 q_1 k_2 q_2 | k_1 k_2 K Q) \tag{3-31}$$

where $T_{q_1}^{(k_1)}$ and $U_{q_2}^{(k_2)}$ are the components of two tensor operators $\mathbf{T}^{(k_1)}$ and $\mathbf{U}^{(k_2)}$. Under the operations of R_3, the right-hand side of

Eq. (3-31) transforms in an identical manner to

$$\sum_{q_1,q_2} |k_1q_1k_2q_2)(k_1q_1k_2q_2|k_1k_2KQ)$$

and since this expression is equal to $|k_1k_2KQ)$, the quantities $X_Q{}^{(K)}$, where $Q = K, K - 1, \ldots, -K$, transform according to the irreducible representation $\mathfrak{D}_K$ of R_3. They must therefore be the components of a tensor operator $\mathbf{X}^{(K)}$ of rank K. This result may also be obtained by showing that

$$\begin{aligned}
[J_z, X_Q{}^{(K)}] &= QX_Q{}^{(K)} \\
[J_\pm, X_Q{}^{(K)}] &= [K(K+1) - Q(Q \pm 1)]^{\frac{1}{2}} X_{Q\pm 1}^{(K)}
\end{aligned}$$

Usually it is unnecessary to introduce the symbol X, and we write

$$X_Q{}^{(K)} \equiv \{\mathbf{T}^{(k_1)}\mathbf{U}^{(k_2)}\}_Q{}^{(K)} \tag{3-32}$$

It can be seen that

$$\begin{aligned}
\{\mathbf{T}^{(k)}\mathbf{U}^{(k)}\}_0{}^{(0)} &= \sum_q T_q{}^{(k)} U_{-q}{}^{(k)} (kqk - q|kk00) \\
&= \sum_q T_q{}^{(k)} U_{-q}{}^{(k)} \begin{pmatrix} k & k & 0 \\ q & -q & 0 \end{pmatrix} \\
&= (-1)^k (2k+1)^{-\frac{1}{2}} \sum_q (-1)^{-q} T_q{}^{(k)} U_{-q}{}^{(k)}
\end{aligned}$$

from Prob. 1-4. It is traditional to define a scalar product of two tensor operators by the equation

$$(\mathbf{T}^{(k)} \cdot \mathbf{U}^{(k)}) = \sum_q (-1)^q T_q{}^{(k)} U_{-q}{}^{(k)} \tag{3-33}$$

Thus $\qquad \{\mathbf{T}^{(k)}\mathbf{U}^{(k)}\}_0{}^{(0)} = (-1)^k (2k+1)^{-\frac{1}{2}} (\mathbf{T}^{(k)} \cdot \mathbf{U}^{(k)})$

The tensor $\{\mathbf{T}^{(k_1)}\mathbf{U}^{(k_2)}\}^{(K)}$ is conventionally written as $(\mathbf{T}^{(k_1)}\mathbf{U}^{(k_2)})^{(K)}$ when the tensors $\mathbf{T}^{(k_1)}$ and $\mathbf{U}^{(k_2)}$ are single-particle tensors of the same particle.[38] The absence of a dot and the presence of a superscript prevent confusion with the scalar product.

3-6 MATRIX ELEMENTS OF $X_Q{}^{(K)}$

Suppose that $\mathbf{T}^{(k_1)}$ and $\mathbf{U}^{(k_2)}$ operate on parts 1 and 2 of a system, respectively. Thus $\mathbf{T}^{(k_1)}$ might act on the orbit, $\mathbf{U}^{(k_2)}$ on the spin; or, in a two-particle system, $\mathbf{T}^{(k_1)}$ could be a function solely of the spin and positional coordinates of the first particle, while $\mathbf{U}^{(k_2)}$ held a similar position with respect to the second particle. The matrix element to be evaluated is

$$(\gamma j_1 j_2 J M_J | X_Q{}^{(K)} | \gamma' j_1' j_2' J' M_J')$$

where the subscripts 1 and 2 label the angular-momentum quantum numbers of the two parts of the system. Making use of the Wigner-Eckart theorem, we get

$$(\gamma j_1 j_2 J M_J | X_Q{}^{(K)} | \gamma' j_1' j_2' J' M_J') = (-1)^{J-M_J} \begin{pmatrix} J & K & J' \\ -M_J & Q & M_J' \end{pmatrix} (\gamma j_1 j_2 J \| X^{(K)} \| \gamma' j_1' j_2' J') \quad (3\text{-}34)$$

We want to allow $\mathbf{T}^{(k_1)}$ to act on part 1 and $\mathbf{U}^{(k_2)}$ on part 2; hence we uncouple the two parts and write $X_Q{}^{(K)}$ out in full:

$$\begin{aligned}
(\gamma j_1 j_2 J M_J | X_Q{}^{(K)} | \gamma' j_1' j_2' J' M_J') &= \sum_{m_1, m_2, m'_1, m'_2} \{[J][J']\}^{\frac{1}{2}} (-1)^{j_1 - j_2 + M_J + j'_1 - j'_2 + M'_J} \\
&\quad \times \begin{pmatrix} j_1 & j_2 & J \\ m_1 & m_2 & -M_J \end{pmatrix} \begin{pmatrix} j_1' & j_2' & J' \\ m_1' & m_2' & -M_J' \end{pmatrix} \sum_{q_1, q_2} \begin{pmatrix} k_1 & k_2 & K \\ q_1 & q_2 & -Q \end{pmatrix} \\
&\quad \times (-1)^{k_1 - k_2 + Q} [K]^{\frac{1}{2}} (\gamma j_1 m_1 j_2 m_2 | T_{q_1}{}^{(k_1)} U_{q_2}{}^{(k_2)} | \gamma' j_1' m_1' j_2' m_2') \\
&= \sum_{m_1, m_2, m'_1, m'_2, q_1, q_2, \gamma''} (-1)^{j_1 - j_2 + M_J + j'_1 - j'_2 + M'_J + k_1 - k_2 + Q + j_1 - m_1 + j_2 - m_2} \\
&\quad \times \begin{pmatrix} j_1 & j_2 & J \\ m_1 & m_2 & -M_J \end{pmatrix} \begin{pmatrix} j_1' & j_2' & J' \\ m_1' & m_2' & -M_J' \end{pmatrix} \begin{pmatrix} k_1 & k_2 & K \\ q_1 & q_2 & -Q \end{pmatrix} \\
&\quad \times \begin{pmatrix} j_1 & k_1 & j_1' \\ -m_1 & q_1 & m_1' \end{pmatrix} \begin{pmatrix} j_2 & k_2 & j_2' \\ -m_2 & q_2 & m_2' \end{pmatrix} \\
&\quad \times \{[J][J'][K]\}^{\frac{1}{2}} (\gamma j_1 \| T^{(k_1)} \| \gamma'' j_1)(\gamma'' j_2 \| U^{(k_2)} \| \gamma' j_2)
\end{aligned}$$

If Eq. (3-34) is multiplied by

$$(-1)^{J-M_J} \begin{pmatrix} J & K & J' \\ -M_J & Q & M_J' \end{pmatrix}$$

and both sides are summed over M_J, M_J', *and* Q, the right-hand side, from Eq. (1-24), becomes

$$(\gamma j_1 j_2 J \| X^{(K)} \| \gamma' j_1' j_2' J')$$

The left-hand side involves a sum over six 3-j symbols. On rearranging the rows and columns and changing the signs of their lower rows where necessary, the phase angle can be made to vanish. With the aid of Eq. (3-15) we immediately obtain

$$(\gamma j_1 j_2 J \| X^{(K)} \| \gamma' j_1' j_2' J') = \sum_{\gamma''} (\gamma j_1 \| T^{(k_1)} \| \gamma'' j_1')(\gamma'' j_2 \| U^{(k_2)} \| \gamma' j_2') \times \{[J][K][J']\}^{\frac{1}{2}} \begin{Bmatrix} j_1 & j_1' & k_1 \\ j_2 & j_2' & k_2 \\ J & J' & K \end{Bmatrix} \quad (3\text{-}35)$$

This equation is particularly valuable since three important special cases can be obtained from it. Setting $K = 0$, we find

$$\begin{aligned}(\gamma j_1 j_2 J M_J | (\mathbf{T}^{(k)} \cdot \mathbf{U}^{(k)}) | \gamma' j_1' j_2' J' M_J') &= (-1)^{k+J-M_J}(2k+1)^{\frac{1}{2}} \begin{pmatrix} J & 0 & J' \\ -M_J & 0 & M_J' \end{pmatrix} (\gamma j_1 j_2 J \| \{\mathbf{T}^{(k)}\mathbf{U}^{(k)}\}^{(0)} \| \gamma' j_1' j_2' J') \\ &= (-1)^{j_1'+j_2+J} \delta(J,J') \delta(M_J,M_J') \begin{Bmatrix} j_1' & j_2' & J \\ j_2 & j_1 & k \end{Bmatrix} \\ &\quad \times \sum_{\gamma''} (\gamma j_1 \| T^{(k)} \| \gamma'' j_1')(\gamma'' j_2 \| U^{(k)} \| \gamma' j_2') \end{aligned} \quad (3\text{-}36)$$

The two other equations are obtained by setting $k_2 = 0$ and then, for the final equation, $k_1 = 0$. We get

$$(\gamma j_1 j_2 J \| T^{(k)} \| \gamma' j_1' j_2' J') = \delta(j_2,j_2')(-1)^{j_1+j_2+J'+k}\{[J][J']\}^{\frac{1}{2}} \begin{Bmatrix} J & k & J' \\ j_1' & j_2 & j_1 \end{Bmatrix} (\gamma j_1 \| T^{(k)} \| \gamma' j_1') \quad (3\text{-}37)$$

for an operator $\mathbf{T}^{(k)}$ acting only on part 1, and

$$(\gamma j_1 j_2 J \| U^{(k')} \| \gamma' j_1' j_2' J') = \delta(j_1,j_1')(-1)^{j_1'+j_2'+J+k'}\{[J][J']\}^{\frac{1}{2}} \begin{Bmatrix} J & k' & J' \\ j_2' & j_1 & j_2 \end{Bmatrix} (\gamma j_2 \| U^{(k')} \| \gamma' j_2') \quad (3\text{-}38)$$

for an operator $\mathbf{U}^{(k')}$ acting only on part 2.

Sometimes it is not possible to regard $\mathbf{T}^{(k_1)}$ and $\mathbf{U}^{(k_2)}$ as acting on separate parts of a system. This is the case when the tensors are built from the same coordinates, for example. There is now no point in indicating that J is the resultant of j_1 and j_2 (or whatever the coupling scheme happens to be), and on dropping these quantum numbers the general matrix element simplifies as follows:

$$
\begin{aligned}
&(\gamma J M_J|X_Q^{(K)}|\gamma' J' M_J') \\
&= \sum_{q_1,q_2} (-1)^{k_1-k_2+Q}[K]^{\frac{1}{2}} \begin{pmatrix} k_1 & k_2 & K \\ q_1 & q_2 & -Q \end{pmatrix} (\gamma J M_J|T_{q_1}^{(k_1)} U_{q_2}^{(k_2)}|\gamma' J' M_J') \\
&= \sum_{q_1,q_2,J'',\gamma'',M_J''} (-1)^{k_1-k_2+Q}[K]^{\frac{1}{2}} \begin{pmatrix} k_1 & k_2 & K \\ q_1 & q_2 & -Q \end{pmatrix} \\
&\qquad \times (\gamma J M_J|T_{q_1}^{(k_1)}|\gamma'' J'' M_J'')(\gamma'' J'' M_J''|U_{q_2}^{(k_2)}|\gamma' J' M_J') \\
&= \sum_{q_1,q_2,J'',\gamma'',M_J''} [K]^{\frac{1}{2}}(-1)^{k_1-k_2+Q+J-M_J+J''-M_J''} \\
&\qquad \times \begin{pmatrix} k_1 & k_2 & K \\ q_1 & q_2 & -Q \end{pmatrix} \begin{pmatrix} J & k_1 & J'' \\ -M_J & q_1 & M_J'' \end{pmatrix} \begin{pmatrix} J'' & k_2 & J' \\ -M_J'' & q_2 & M_J' \end{pmatrix} \\
&\qquad \times (\gamma J\|T^{(k_1)}\|\gamma'' J'')(\gamma'' J''\|U^{(k_2)}\|\gamma' J') \\
&= \sum_{J'',\gamma''} [K]^{\frac{1}{2}}(-1)^{J+J'+K+J-M_J} \begin{pmatrix} J & K & J' \\ -M_J & Q & M_J' \end{pmatrix} \begin{Bmatrix} J & K & J' \\ k_2 & J'' & k_1 \end{Bmatrix} \\
&\qquad \times (\gamma J\|T^{(k_1)}\|\gamma'' J'')(\gamma'' J''\|U^{(k_2)}\|\gamma' J')
\end{aligned}
$$

The last step is made by rearranging the 3-j symbols and using Eq. (3-6) to evaluate the sum over q_1, q_2, and M_J''. From the Wigner-Eckart theorem,

$$
(\gamma J\|X^{(K)}\|\gamma' J') = [K]^{\frac{1}{2}}(-1)^{J+K+J'} \sum_{J'',\gamma''} \begin{Bmatrix} k_2 & K & k_1 \\ J & J'' & J' \end{Bmatrix} \times (\gamma J\|T^{(k_1)}\|\gamma'' J'')(\gamma'' J''\|U^{(k_2)}\|\gamma' J') \quad (3\text{-}39)
$$

Nothing in the derivation of this equation implies that $\mathbf{T}^{(k_1)}$ and $\mathbf{U}^{(k_2)}$ necessarily act on the same part of the system. Equation (3-39) is therefore equally valid for tensor operators that act on different parts of a system, but in this case it is a much weaker statement than Eq. (3-35). The latter can be derived from Eq. (3-39) by using Eqs. (3-37) and (3-38) to replace the reduced matrix elements and then summing over the three 6-j symbols by means of Eq. (3-14).

PROBLEMS

3-1. Derive the equation

$$
\begin{aligned}
&\sum_{m_3} (-1)^{l_1+l_2+\mu_1+\mu_2} \begin{pmatrix} j_1 & j_2 & j_3 \\ m_1 & m_2 & m_3 \end{pmatrix} \begin{pmatrix} l_1 & l_2 & j_3 \\ \mu_1 & -\mu_2 & m_3 \end{pmatrix} \\
&\quad = \sum_{l_3,\mu_3} (-1)^{l_3+\mu_3}(2l_3+1) \begin{Bmatrix} j_1 & j_2 & j_3 \\ l_1 & l_2 & l_3 \end{Bmatrix} \begin{pmatrix} l_1 & j_2 & l_3 \\ -\mu_1 & m_2 & \mu_3 \end{pmatrix} \begin{pmatrix} j_1 & l_2 & l_3 \\ m_1 & \mu_2 & -\mu_3 \end{pmatrix}
\end{aligned}
$$

(An equivalent equation has been given by Racah;[18] apart from a trivial rearrangement of the phase factors, the present form is due to Rotenberg et al.[6])

3-2. Verify the recursion relation

$$[(a+b+c+1)(b+c-a)(c+d+e+1)(c+d-e)]^{\frac{1}{2}}\begin{Bmatrix} a & b & c \\ d & e & f \end{Bmatrix}$$
$$= -2c[(b+d+f+1)(b+d-f)]^{\frac{1}{2}}\begin{Bmatrix} a & b-\frac{1}{2} & c-\frac{1}{2} \\ d-\frac{1}{2} & e & f \end{Bmatrix}$$
$$+ [(a+b-c+1)(a+c-b)(d+e-c+1)(c+e-d)]^{\frac{1}{2}}\begin{Bmatrix} a & b & c-1 \\ d & e & f \end{Bmatrix}$$

(See Edmonds[4]).

3-3. Show that the Möbius strip associated with the 12-j symbol

$$\begin{Bmatrix} p & e & b & h & \\ & r & a & d & s \\ q & g & c & f & \end{Bmatrix}$$

can be untwisted by multiplication by

$$[r]\begin{Bmatrix} e & p & r \\ g & q & \lambda \end{Bmatrix}$$

and summation over r but that multiplication by

$$(-1)^p[p]\begin{Bmatrix} \lambda & s & e \\ p & r & f \end{Bmatrix}$$

and summation over p gives a 12-j symbol that retains the twist.

3-4. Prove that

$$\sum_r [r]\begin{Bmatrix} p & e & b & h & \\ & r & a & d & s \\ q & g & c & f & \end{Bmatrix}\begin{Bmatrix} e & p & r \\ q & g & \lambda \end{Bmatrix}(-1)^{-r}$$
$$= (-1)^{p+q+e+g+a+b+c+d+h+f+s-\lambda}\begin{Bmatrix} b & c & \lambda \\ g & e & a \end{Bmatrix}\begin{Bmatrix} p & q & \lambda \\ h & f & s \end{Bmatrix}\begin{Bmatrix} h & f & \lambda \\ c & b & d \end{Bmatrix}$$

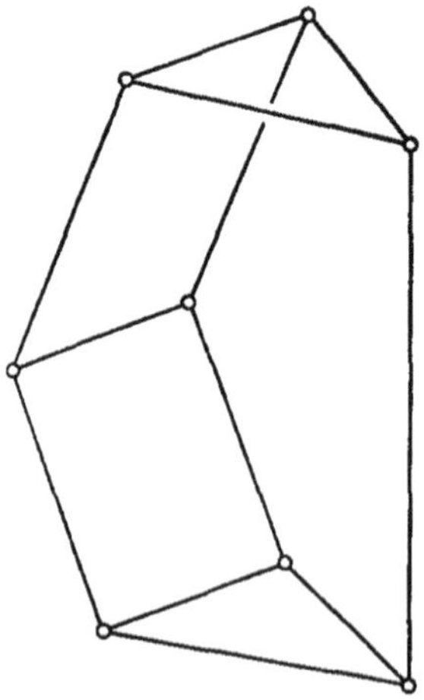

FIG. 3-11 The coupling diagram of Prob. 3-4.

Show that the coupling is represented by Fig. 3-11, and label the branches of the diagram.

3-5. Prove that

$$\sum_p [p] \left\{ \begin{array}{ccccccc} p & & e & & b & & h \\ & r & & a & & d & & s \\ q & & g & & c & & f \end{array} \right\} \begin{Bmatrix} \lambda & f & e \\ p & r & s \end{Bmatrix} = (-1)^{e+f-g-h} \begin{Bmatrix} r & g & q \\ h & s & \lambda \end{Bmatrix} \begin{Bmatrix} a & b & e \\ c & d & f \\ g & h & \lambda \end{Bmatrix}$$

and show that the coupling diagram is the same as Fig. 3-9. Deduce that the equation is equivalent to the identity of Arima et al. [Eq. (3-30)].

3-6. By considering the appropriate coupling diagrams, show that the 12-j symbol corresponding to a Möbius strip reduces to a 9-j symbol or to two 6-j symbols if one of its arguments is set equal to zero but that a 12-j symbol corresponding to an untwisted strip can reduce only to two 6-j symbols.

3-7. Examine the coupling diagrams that comprise 15 branches, and deduce that there are five species of 15-j symbols.

4

CONFIGURATIONS OF TWO ELECTRONS

4-1 INTRODUCTION

The formulas derived in Sec. 3-6, particularly Eq. (3-35) and its three special cases, can be applied to a large number of problems in atomic spectroscopy. To illustrate their use, the configuration f^2 will be considered in detail, though it will be clear that most of the methods could be illustrated equally well by other configurations comprising two electrons outside closed shells. For many years, $4f^2$ was known only as an excited configuration of LaII (see Condon and Shortley[1]), but later work[40] showed that it also occurs as the ground configuration of PrIV. The ion Pr^{3+} can be studied in compounds such as $PrCl_3$,[41] $Pr_2Mg_3(NO_3)_{12} \cdot 24H_2O$,[42] etc., where an examination of the term H_3 in the Hamiltonian is necessary for a complete elucidation of the spectra. Present indications are that more and more attention is being paid to the actinide ions, and the configuration $5f^2$ has been observed in ThIII,[43,44] UV,[45,46] and $(PuO_2)^{++}$.[47]

In this chapter, the following perturbations on the degenerate configuration f^2 are considered in detail:

1. H_1, the Coulomb interaction between the two electrons
2. H_2, the spin-orbit interaction
3. H_4, the term representing the effect of an external magnetic field
4. H_5, the magnetic hyperfine interaction
5. H_6, the magnetic interaction between the spins of the electrons

A general treatment of the hyperfine interaction H_5 by tensor-operator techniques has been given by Trees[48] and Schwartz[49]; most of the remaining interactions listed above were of sufficient interest to be analyzed before the advent of such methods. These interactions H_i do not form the entire Hamiltonian, but the general approach that is adopted should make it clear how to treat other contributions. The crystal field interaction H_3, the magnetic interaction between the spin of one electron and the orbit of the other, and the effect of the quadrupole moment of the nucleus are included in the problems at the end of the chapter.

All calculations proceed in three steps:

1. The contribution to the Hamiltonian is expressed in tensor-operator form.

2. The angular parts of the matrix elements are calculated by means of the formulas of Sec. 3-6.

3. The radial integrals are considered, and the theory is compared with experiment.

4-2 REDUCED MATRIX ELEMENTS

It is convenient to begin by evaluating the reduced matrix elements that are required in the calculations. The matrix element

$$(lm_l|l_z|lm_l) = m_l$$

can also be written as

$$\begin{aligned}(lm_l|l_z|lm_l) &= (-1)^{l-m_l}\begin{pmatrix} l & 1 & l \\ -m_l & 0 & m_l \end{pmatrix}(l\|l\|l) \\ &= m_l[l(l+1)(2l+1)]^{-\frac{1}{2}}(l\|l\|l)\end{aligned}$$

Thus

$$\begin{aligned}(l\|l\|l) &= [l(l+1)(2l+1)]^{\frac{1}{2}} \\ &= 2(21)^{\frac{1}{2}} \qquad \text{for } l = 3 \qquad (4\text{-}1)\end{aligned}$$

Similarly,

$$\begin{aligned}(s\|s\|s) &= [s(s+1)(2s+1)]^{\frac{1}{2}} \\ &= (\tfrac{3}{2})^{\frac{1}{2}} \qquad \text{for } s = \tfrac{1}{2} \qquad (4\text{-}2)\end{aligned}$$

We also need a reduced matrix element for a tensor involving the coordinates of the electrons. The tensors $\mathbf{C}^{(k)}$, introduced in Sec. 2-7, are widely used; for these, we have

$$\begin{aligned}(l, m_l = 0|C_0^{(k)}|l', m_l = 0) &= (-1)^l\begin{pmatrix} l & k & l' \\ 0 & 0 & 0 \end{pmatrix}(l\|C^{(k)}\|l') \\ &= \frac{1}{2}\int_{-1}^{1}[(2l+1)(2l'+1)]^{\frac{1}{2}}P_l(\mu)P_k(\mu)P_{l'}(\mu)\,d\mu\end{aligned}$$

It is advantageous to select $m_l = 0$ because closed expressions are available for 3-j symbols of the type

$$\begin{pmatrix} j_1 & j_2 & j_3 \\ 0 & 0 & 0 \end{pmatrix}$$

If the sum $j_1 + j_2 + j_3$, which is denoted here by J, is odd, the 3-j symbol vanishes; for a reversal of the signs of the lower row leaves the symbol unchanged and at the same time introduces the factor $(-1)^J$. If J is even, the 3-j symbol can be evaluated by an iterative procedure (see Edmonds[4]) and we obtain

$$\begin{pmatrix} j_1 & j_2 & j_3 \\ 0 & 0 & 0 \end{pmatrix} = (-1)^{\frac{1}{2}J} \left[\frac{(J-2j_1)!(J-2j_2)!(J-2j_3)!}{(J+1)!} \right]^{\frac{1}{2}} \times \frac{(\frac{1}{2}J)!}{(\frac{1}{2}J-j_1)!(\frac{1}{2}J-j_2)!(\frac{1}{2}J-j_3)!} \tag{4-3}$$

Gaunt[50] has given a general formula for the integral over the product of three Legendre polynomials. With the aid of Eq. (4-3) it can be expressed in the form

$$\frac{1}{2}\int_{-1}^{1} P_l(\mu)P_k(\mu)P_{l'}(\mu)\, d\mu = \begin{pmatrix} l & k & l' \\ 0 & 0 & 0 \end{pmatrix}^2$$

From these equations above we get

$$(l\|C^{(k)}\|l') = (-1)^l\{[l][l']\}^{\frac{1}{2}} \begin{pmatrix} l & k & l' \\ 0 & 0 & 0 \end{pmatrix} \tag{4-4}$$

Occasional use is made of the equation

$$(\mathbf{C}^{(k_1)}\mathbf{C}^{(k_2)})^{(K)} = (-1)^K[K]^{\frac{1}{2}} \begin{pmatrix} k_1 & K & k_2 \\ 0 & 0 & 0 \end{pmatrix} \mathbf{C}^{(K)} \tag{4-5}$$

where $\mathbf{C}^{(k_1)}$ and $\mathbf{C}^{(k_2)}$ are functions of the coordinates of the same electron. It can be proved by evaluating

$$(l\|(\mathbf{C}^{(k_1)}\mathbf{C}^{(k_2)})^{(K)}\|l')$$

by means of Eq. (3-39) and comparing the result with

$$(l\|C^{(K)}\|l')$$

When $k_1 = 1$, Eq. (4-5) reduces to

$$(\mathbf{C}^{(1)}\mathbf{C}^{(k)})^{(k+1)} = [(k+1)/(2k+1)]^{\frac{1}{2}}\mathbf{C}^{(k+1)} \tag{4-6}$$

and

$$(\mathbf{C}^{(1)}\mathbf{C}^{(k)})^{(k-1)} = -[k/(2k+1)]^{\frac{1}{2}}\mathbf{C}^{(k-1)} \tag{4-7}$$

The tensor $(\mathbf{C}^{(1)}\mathbf{C}^{(k)})^{(k)}$ is identically zero, since $k_1 + K + k_2$ is odd and the 3-j symbol in Eq. (4-5) vanishes.

4-3 THE COULOMB INTERACTION

To express H_1 in tensor-operator form we write

$$\frac{1}{r_{ij}} = \frac{1}{[r_i^2 + r_j^2 - 2r_ir_j \cos \omega]^{\frac{1}{2}}} = \sum_k \frac{r_<^k}{r_>^{k+1}} P_k(\cos \omega)$$

where $r_<$ is the lesser and $r_>$ the greater of r_i and r_j. Owing to the spherical-harmonic addition theorem [Eq. (2-2)],

$$\begin{aligned} P_k(\cos \omega) &= \frac{4\pi}{2k+1} \sum_q Y_{kq}^*(\theta_i,\phi_i) Y_{kq}(\theta_j,\phi_j) \\ &= \sum_q (-1)^q (C_{-q}^{(k)})_i (C_q^{(k)})_j = (\mathbf{C}_i^{(k)} \cdot \mathbf{C}_j^{(k)}) \end{aligned}$$

The subscript i to $\mathbf{C}^{(k)}$ indicates that it is a function of the coordinates of electron i. Thus

$$H_1 = e^2 \sum_k \frac{r_<^k}{r_>^{k+1}} (\mathbf{C}_i^{(k)} \cdot \mathbf{C}_j^{(k)}) \tag{4-8}$$

and the first step in the calculation is completed.

The general matrix element to work out is

$$(l^2SM_SLM_L|(\mathbf{C}_i^{(k)} \cdot \mathbf{C}_j^{(k)})|l^2S'M'_SL'M'_L)$$

The Pauli exclusion principle is automatically satisfied if $S + L$ is even (see Prob. 1-2). Since S can only be 0 or 1, it is obvious that a particular choice of L fixes S. Dropping superfluous quantum numbers, we get

$$\begin{aligned} &(l^2SM_SLM_L|(\mathbf{C}_i^{(k)} \cdot \mathbf{C}_j^{(k)})|l^2S'M'_SL'M'_L) \\ &\quad = \delta(S,S')\delta(M_S,M'_S)(llLM_L|(\mathbf{C}_i^{(k)} \cdot \mathbf{C}_j^{(k)})|llL'M'_L) \\ &\quad = \delta(S,S')\delta(M_S,M'_S)(-1)^{l+l+L}\delta(L,L')\delta(M_L,M'_L) \\ &\qquad\qquad \times \begin{Bmatrix} l & l & k \\ l & l & L \end{Bmatrix} (l\|C^{(k)}\|l)^2 \end{aligned}$$

from Eq. (3-36). For f electrons, this is

$$\delta(S,S')\delta(M_S,M'_S)\delta(L,L')\delta(M_L,M'_L)49(-1)^L \begin{Bmatrix} 3 & 3 & k \\ 3 & 3 & L \end{Bmatrix} \begin{pmatrix} 3 & k & 3 \\ 0 & 0 & 0 \end{pmatrix}^2$$

All nondiagonal matrix elements of H_1 therefore vanish. For the rest,

$$(f^2SM_SLM_L|H_1|f^2SM_SLM_L) = \sum_k 49F^{(k)}(-1)^L \begin{pmatrix} 3 & k & 3 \\ 0 & 0 & 0 \end{pmatrix}^2 \begin{Bmatrix} 3 & 3 & k \\ 3 & 3 & L \end{Bmatrix}$$

where the quantities $F^{(k)}$, the so-called *Slater integrals*, are given by

$$F^{(k)} = e^2 \int_0^\infty \int_0^\infty \frac{r_<^k}{r_>^{k+1}} [R_{nf}(r_i)R_{nf}(r_j)]^2 \, dr_i \, dr_j \qquad (4\text{-}9)$$

Since $3 + k + 3$ must be even for the 3-j symbols not to vanish, and since 3, k, and 3 must form a triangle, the sum extends over $k = 0, 2, 4,$ and 6 only. The values of the 3-j and 6-j symbols may be obtained from the tables of Rotenberg et al.,[6] and the following expressions are derived for the energies $E(^{2S+1}L)$ of the terms of f^2:

$$\begin{aligned}
E(^1S) &= F_0 + 60F_2 + 198F_4 + 1716F_6 \\
E(^3P) &= F_0 + 45F_2 + 33F_4 - 1287F_6 \\
E(^1D) &= F_0 + 19F_2 - 99F_4 + 715F_6 \\
E(^3F) &= F_0 - 10F_2 - 33F_4 - 286F_6 \\
E(^1G) &= F_0 - 30F_2 + 97F_4 + 78F_6 \\
E(^3H) &= F_0 - 25F_2 - 51F_4 - 13F_6 \\
E(^1I) &= F_0 + 25F_2 + 9F_4 + F_6
\end{aligned}$$

where $F_0 = F^{(0)}$

$$F_2 = \frac{F^{(2)}}{225}$$

$$F_4 = \frac{F^{(4)}}{1089}$$

$$F_6 = \frac{25F^{(6)}}{184041}$$

These results agree with those of Condon and Shortley.[1] If the apparently naïve assumption is made that the radial eigenfunction R_{nf} is hydrogenic, the number of effective parameters in these equations can be reduced to only one. It is found that

$$\begin{aligned}
\frac{F_4}{F_2} &= \tfrac{41}{297} = 0.138 \\
\frac{F_6}{F_2} &= \tfrac{175}{11583} = 0.0151
\end{aligned} \qquad (4\text{-}10)$$

for a $4f$ hydrogenic eigenfunction, and

$$\begin{aligned}\frac{F_4}{F_2} &= \tfrac{23255}{163559} = 0.142 \\ \frac{F_6}{F_2} &= \tfrac{102725}{6378801} = 0.0161\end{aligned} \tag{4-11}$$

for a $5f$ hydrogenic eigenfunction. A method for obtaining Eqs. (4-10) and (4-11) is described in Appendix 1. In spite of the dissimilarity between $4f$ and $5f$ radial eigenfunctions, the corresponding F_k ratios are very much the same. This insensitivity to the shape of the eigenfunction means that the relative energies of the terms of f^2 in one atom are expected to be very similar (apart from a scaling factor) to the corresponding energies in another atom. Before a direct comparison with experiment can be made, however, the spin-orbit interaction has to be examined.

4-4 THE SPIN-ORBIT INTERACTION

We now consider

$$H_2 = \sum_i \xi(r_i)\mathbf{s}_i \cdot \mathbf{l}_i$$

which is already in tensor-operator form. For a particular term $\xi(r_j)\mathbf{s}_j \cdot \mathbf{l}_j$ in the sum,

$$\begin{aligned}(\gamma SLJM_J|\mathbf{s}_j \cdot \mathbf{l}_j|\gamma' S'L'J'M'_J) &= (-1)^{S'+L+J}\delta(J,J')\delta(M_J,M'_J)\begin{Bmatrix} S & S' & 1 \\ L' & L & J \end{Bmatrix} \\ &\quad \times \sum_{\gamma''} (\gamma S\|s_j\|\gamma'' S')(\gamma'' L\|l_j\|\gamma' L')\end{aligned} \tag{4-12}$$

from Eq. (3-36). The 6-j symbol is independent of γ and γ' and recurs for every term in the sum. If it is not to vanish, the selection rules

$$\Delta S,\ \Delta L = 0,\ \pm 1$$

must be satisfied. The delta functions in Eq. (4-12) indicate that H_2 is diagonal with respect to J and M_J. When $S' = S$ and $L' = L$,

$$(-1)^{S'+L+J}\begin{Bmatrix} S & S & 1 \\ L & L & J \end{Bmatrix} = \frac{J(J+1) - S(S+1) - L(L+1)}{[S(2S+1)(2S+2)L(2L+1)(2L+2)]^{\frac{1}{2}}}$$

from Eq. (3-10). Since the matrix elements of $2\mathbf{S}\cdot\mathbf{L}$ are simply $J(J+1) - S(S+1) - L(L+1)$, we obtain the familiar result that, *within* a manifold of states of given S and L, the matrix elements of H_2 can be reproduced by $\lambda\mathbf{S}\cdot\mathbf{L}$, where λ is a suitably chosen constant. This leads to the *Landé interval rule*, namely, that in any Russell-Saunders multiplet the interval between two neighboring levels is proportional to the higher J value of the pair.

We now specialize to the configuration f^2. The quantum numbers f^2SLJM_J are sufficient to define the states, and the sum over γ'' in Eq. (4-12) reduces to a single term. From Eq. (3-37),

$$(S\|s_1\|S') = (-1)^{\frac{1}{2}+\frac{1}{2}+S'+1}\{[S][S']\}^{\frac{1}{2}} \begin{Bmatrix} S & 1 & S' \\ \frac{1}{2} & \frac{1}{2} & \frac{1}{2} \end{Bmatrix} (s\|s\|s)$$

and
$$(L\|l_1\|L') = (-1)^{3+3+L'+1}\{[L][L']\}^{\frac{1}{2}} \begin{Bmatrix} L & 1 & L' \\ 3 & 3 & 3 \end{Bmatrix} (l\|l\|l)$$

Similar expressions are obtained for the second f electron, the only difference being that S and L replace S' and L' in the phase factors. The product of the reduced matrix elements for s_1 and l_1 must be the same as the product of those for s_2 and l_2, since $S' + L'$ and $S + L$ must both be even. Using Eqs. (4-1) and (4-2), we find that Eq. (4-12) becomes

$$\begin{aligned}(f^2SLJM_J|H_2|f^2S'L'J'M'_J) &= 6(14)^{\frac{1}{2}}\zeta(-1)^{S'+L+J+1}\{[S][S'][L][L']\}^{\frac{1}{2}}\delta(J,J')\delta(M_J,M'_J) \\ &\times \begin{Bmatrix} S & S' & 1 \\ L' & L & J \end{Bmatrix} \begin{Bmatrix} S & S' & 1 \\ \frac{1}{2} & \frac{1}{2} & \frac{1}{2} \end{Bmatrix} \begin{Bmatrix} L & L' & 1 \\ 3 & 3 & 3 \end{Bmatrix} \qquad (4\text{-}13)\end{aligned}$$

where ζ is defined as in Eq. (1-34). For $J = J' = 4$, the following matrix is obtained:

	$^3H_4$	$^1G_4$	$^3F_4$
$^3H_4$	-3ζ	$-(\frac{10}{3})^{\frac{1}{2}}\zeta$	0
$^1G_4$	$-(\frac{10}{3})^{\frac{1}{2}}\zeta$	0	$(\frac{11}{3})^{\frac{1}{2}}\zeta$
$^3F_4$	0	$(\frac{11}{3})^{\frac{1}{2}}\zeta$	$3\zeta/2$

(4-14)

Before matrices of this sort are diagonalized, the energies of the terms, $E(^{2S+1}L)$, must be included on the diagonal. Calculations have been carried out for F_k ratios given by Eq. (4-11) and for a number of values of ζ/F_2. The results are given in Fig. 4-1 in the usual coordinate scheme. Precisely the same properties of f^2 are exhibited in this diagram as are shown in Fig. 1-1 for sd; the two figures are essentially

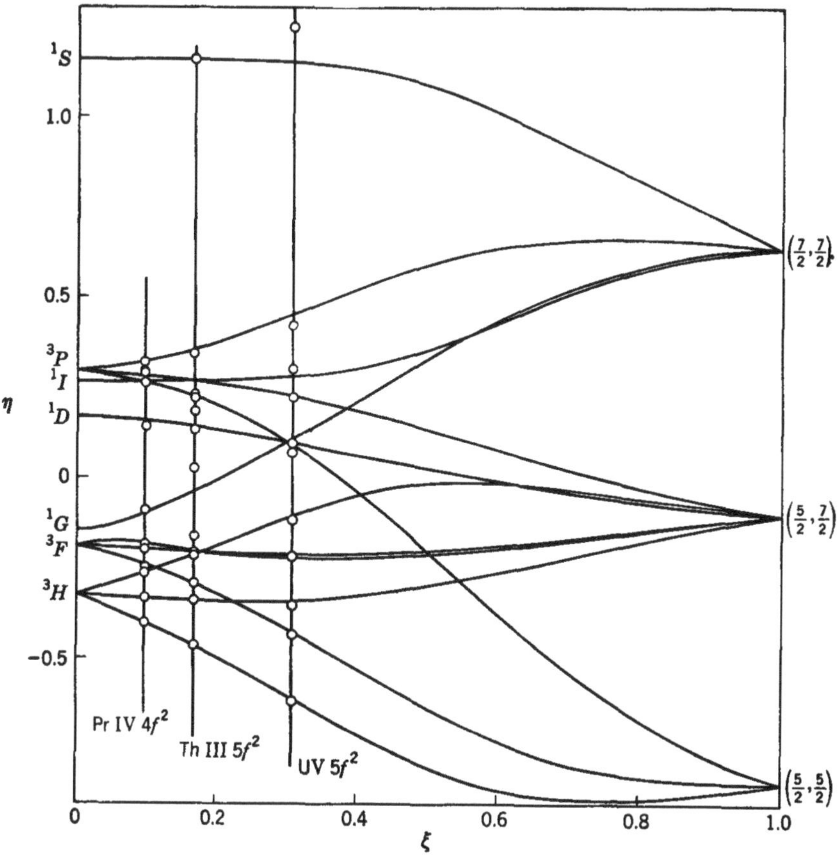

FIG. 4-1 The levels of the configuration f^2 in intermediate coupling are compared with experiment. For PrIV and ThIII, the *LSJ* designations of the observed levels are known and agree with the theoretical scheme. The energy-level scheme for UV, however, has been obtained from the spectrum of the ion U^{4+} in crystals of CaF_2; the experimental points in the diagram are the approximate centers of groups of energy levels, and it remains to be seen whether the *LSJ* designations indicated by the theoretical plot are confirmed by further experiments. The minor discrepancies between experiment and theory for PrIV and ThIII can be reduced by choosing different F_k ratios, thus altering the arrangement of levels on the extreme left of the diagram. Only the ratios of the spacings between the levels and not their absolute values are represented in the figure; indeed, for PrIV and UV, F_2 is approximately 310 and 200 cm^{-1}, respectively. (I am indebted to R. D. McLaughlin for supplying the data for the curves.)

the same in character and differ only in complexity. The agreement with the observed levels of PrIV, ThIII, and UV, remarkable as it is, could be further improved by choosing slightly different F_k ratios for the three atoms (see Runciman and Wybourne,[51] Racah,[44] and Conway[45]).

4-5 AN EXTERNAL MAGNETIC FIELD

To emphasize the separate roles of **S** and **L**, and to permit refinements of the simple Dirac theory to be incorporated into the analysis, the expression for H_4 given in Eq. (2-31) is modified to

$$H_4 = \beta\mathbf{H}\cdot(\mathbf{L} + g_s\mathbf{S})$$

It is obvious that all nonvanishing matrix elements must be diagonal with respect to S and L. The Landé g factor, defined in Eq. (2-32), can be easily found by comparing

$$(\gamma SLJ\|gJ\|\gamma SLJ) = g[J(J+1)(2J+1)]^{\frac{1}{2}}$$

which can be derived in a similar way to Eqs. (4-1) and (4-2), with

$$\begin{aligned}(\gamma SLJ\|L + g_sS\|\gamma SLJ) &= (-1)^{S+L+J+1}[J]\begin{Bmatrix}J & 1 & J\\ L & S & L\end{Bmatrix}(L\|L\|L)\\ &\quad + (-1)^{S+L+J+1}[J]\begin{Bmatrix}J & 1 & J\\ S & L & S\end{Bmatrix}g_s(S\|S\|S)\end{aligned}$$

Putting in the reduced matrix elements and using the last equation of the set (3-10), we get

$$g = 1 + (g_s - 1)\frac{J(J+1) + S(S+1) - L(L+1)}{2J(J+1)}$$

The matrix elements of the components of $\mathbf{L} + g_s\mathbf{S}$ that are diagonal with respect to J can thus be easily calculated by finding the matrix elements of $g\mathbf{J}$.

Off-diagonal matrix elements are required if two levels lie close together (in which case $H_1, H_2 \gg H_4$ is no longer a good approximation) or if, for an atom in a crystal field, H_3 is sufficiently large to mix together levels of different J (corresponding to $H_1, H_2 \gg H_3$ not being a good approximation). They may be easily found by tensor-operator techniques. For example,

$$(\gamma SLJM_J|L_z + g_sS_z|\gamma SL\,J+1\,M_J)$$
$$= (-1)^{J-M_J}\begin{pmatrix} J & 1 & J+1 \\ -M_J & 0 & M_J \end{pmatrix}(\gamma SLJ\|L + g_sS\|\gamma SL\,J+1)$$
$$= -\left[\frac{(J+1)^2 - M_J^2}{(J+1)(2J+1)(2J+3)}\right]^{\frac{1}{2}}(g_s - 1)(-1)^{S+L+J}$$
$$\times\begin{Bmatrix} S & J & L \\ J+1 & S & 1 \end{Bmatrix}(2J+1)^{\frac{1}{2}}(2J+3)^{\frac{1}{2}}(S\|S\|S)$$
$$= (g_s - 1)[(J+1)^2 - M_J^2]^{\frac{1}{2}}$$
$$\times\left[\frac{(S+L+J+2)(S+J+1-L)(L+J+1-S)(S+L-J)}{4(J+1)^2(2J+1)(2J+3)}\right]^{\frac{1}{2}}$$

4-6 THE MAGNETIC HYPERFINE INTERACTION

The magnetic field at the nucleus produced by an orbital electron is

$$e(\mathbf{v}\times\mathbf{r})/cr^3 - [\boldsymbol{\mu}r^2 - 3\mathbf{r}(\boldsymbol{\mu}\cdot\mathbf{r})]/r^5 = -2\beta[\mathbf{l} - \mathbf{s} + 3\mathbf{r}(\mathbf{s}\cdot\mathbf{r})/r^2]/r^3$$

where $\boldsymbol{\mu} = -2\beta\mathbf{s}$. The nuclear moment is sometimes written in the form $g_N\beta_N\mathbf{I}$, where β_N, the nuclear magneton, is equal to $e\hbar/2Mc$ (in which M is the mass of the proton), and where $\mathbf{I}$ is the total angular momentum of the nucleus. The symbol g_N is analogous to the Landé g factor, and we shall replace it by μ_N/I. The contribution to the Hamiltonian is thus

$$H_5 = (2\beta\beta_N\mu_N/I)\sum_i \mathbf{N}_i\cdot\mathbf{I}/r_i^3 \qquad (4\text{-}15)$$

where $$\mathbf{N}_i = \mathbf{l}_i - \mathbf{s}_i + 3\mathbf{r}_i(\mathbf{r}_i\cdot\mathbf{s}_i)/r_i^2$$

Fermi has shown that, when s electrons are present, $\mathbf{N}_i$ must be augmented by the term $(8\pi/3)\delta(r_i)\mathbf{s}_i$ to allow for the contact interaction between electron spin and nuclear spin.[52]

The first step is to express the interaction as a tensor operator. The only part that presents difficulty is

$$3\mathbf{r}_i(\mathbf{r}_i\cdot\mathbf{s}_i)/r_i^2 = 3(\mathbf{s}_i\cdot\mathbf{r}_i)\mathbf{r}_i/r_i^2 = 3(\mathbf{s}_i\cdot\mathbf{C}_i^{(1)})\mathbf{C}_i^{(1)}$$
$$= -3(3)^{\frac{1}{2}}(\mathbf{s}^{(1)}\mathbf{C}^{(1)})_i^{(0)}\mathbf{C}_i^{(1)}$$

A simple recoupling is now carried out,

$$-3(3)^{\frac{1}{2}}(\mathbf{s}^{(1)}\mathbf{C}^{(1)})_i^{(0)}\mathbf{C}_i^{(1)}$$
$$= -3(3)^{\frac{1}{2}}\sum_k((11)0,1,1|1,(11)k,1)(\mathbf{s}_i^{(1)}(\mathbf{C}^{(1)}\mathbf{C}^{(1)})_i^{(k)})^{(1)}$$

But $$((11)0,1,1|1,(11)k,1) = [k]^{\frac{1}{2}}\begin{Bmatrix} 1 & 1 & k \\ 1 & 1 & 0 \end{Bmatrix}$$
$$= (-1)^k[k]^{\frac{1}{2}}/3$$

Also, from Eqs. (4-6) and (4-7),

$$(\mathbf{C}^{(1)}\mathbf{C}^{(1)})_i^{(2)} = (\tfrac{2}{3})^{\frac{1}{2}}\mathbf{C}_i^{(2)} \qquad (\mathbf{C}^{(1)}\mathbf{C}^{(1)})_i^{(0)} = -(\tfrac{1}{3})^{\frac{1}{2}}$$

Thus

$$\begin{aligned} -3(3)^{\frac{1}{2}}(\mathbf{s}^{(1)}\mathbf{C}^{(1)})_i^{(0)}\mathbf{C}_i^{(1)} &= -3(3)^{\frac{1}{2}}[-(\tfrac{1}{3})^{\frac{1}{2}}\mathbf{s}_i + (\tfrac{10}{3})^{\frac{1}{2}}(\mathbf{s}\mathbf{C}^{(2)})_i^{(1)}]/3 \\ &= \mathbf{s}_i - (10)^{\frac{1}{2}}(\mathbf{s}\mathbf{C}^{(2)})_i^{(1)} \end{aligned}$$

and so

$$\mathbf{N}_i = \mathbf{l}_i - (10)^{\frac{1}{2}}(\mathbf{s}\mathbf{C}^{(2)})_i^{(1)} \tag{4-16}$$

From the Wigner-Eckart theorem, the matrix elements of this operator within the $2J + 1$ states belonging to a given level are proportional to those of **J**. Let us write

$$(2\beta\beta_N\mu_N/I)\sum_i \mathbf{N}_i \cdot \mathbf{I}/r_i^3 \equiv A\mathbf{J}\cdot\mathbf{I}$$

and determine A for the lowest level of f^2, namely, ${}^3H_4$. Now

$$(f^2\,{}^3H_4\|J\|f^2\,{}^3H_4) = [J(J+1)(2J+1)]^{\frac{1}{2}} = 6(5)^{\frac{1}{2}}$$

while $(f^2\,{}^3H_4\|N_i\|f^2\,{}^3H_4)$

$$= (f^2\,{}^3H_4\|l_i\|f^2\,{}^3H_4) - (10)^{\frac{1}{2}}(f^2\,{}^3H_4\|(\mathbf{s}\mathbf{C}^{(2)})_i^{(1)}\|f^2\,{}^3H_4)$$

From Eq. (3-38),

$$\begin{aligned}(f^2\,{}^3H_4\|\sum_i l_i\|f^2\,{}^3H_4) &= (f^2\,{}^3H_4\|L\|f^2\,{}^3H_4) \\ &= (-1)^{1+5+4+1}9\begin{Bmatrix}4 & 1 & 4\\ 5 & 1 & 5\end{Bmatrix}(H\|L\|H) \\ &= \frac{36}{(5)^{\frac{1}{2}}}\end{aligned}$$

The second matrix element involves the operator $(\mathbf{s}\mathbf{C}^{(2)})_i^{(1)}$. It must be evaluated as it stands before the sum over i is carried out. From Eq. (3-35),

$$\begin{aligned}&(f^2\,{}^3H_4\|(\mathbf{s}\mathbf{C}^{(2)})_i^{(1)}\|f^2\,{}^3H_4) \\ &\quad= 9(3)^{\frac{1}{2}}\begin{Bmatrix}1 & 1 & 1\\ 5 & 5 & 2\\ 4 & 4 & 1\end{Bmatrix}({}^3H\|C_i^{(2)}\|\,{}^3H)({}^3H\|s_i\|\,{}^3H) \\ &\quad= 297(3)^{\frac{1}{2}}\begin{Bmatrix}1 & 1 & 1\\ 5 & 5 & 2\\ 4 & 4 & 1\end{Bmatrix}\begin{Bmatrix}5 & 2 & 5\\ 3 & 3 & 3\end{Bmatrix}\begin{Bmatrix}1 & 1 & 1\\ \frac{1}{2} & \frac{1}{2} & \frac{1}{2}\end{Bmatrix} \\ &\qquad\times (l\|C_i^{(2)}\|l)(s\|s_i\|s) \\ &\quad= -\frac{13(2)^{\frac{1}{2}}}{75}\end{aligned}$$

This result must be multiplied by 2 to account for the second electron.

Thus

$$({}^3H_4\| \sum_i N_i/r_i^3 \| {}^3H_4) = \langle r^{-3}\rangle[36/(5)^{\frac{1}{2}} + 26(20)^{\frac{1}{2}}/75]$$
$$= 592\langle r^{-3}\rangle/15(5)^{\frac{1}{2}}$$

where $\langle r^{-3}\rangle$ is the mean value of r^{-3} for an f electron. It follows that

$$A = 296(2\beta\beta_N\mu_N/I)\langle r^{-3}\rangle/225$$

and the matrix elements of H_5 may be easily determined by considering those of $A\mathbf{J}\cdot\mathbf{I}$. It is to be stressed that it is only within a manifold of constant J states that $\Sigma\mathbf{N}_i$ can be replaced by a factor times $\mathbf{J}$. The selection rules on $\mathbf{J}$ are

$$\Delta S,\ \Delta L,\ \Delta J = 0$$

Those on $\Sigma\mathbf{N}_i$ are

$$\Delta L = 0,\ \pm 1,\ \pm 2 \qquad \Delta S,\ \Delta J = 0,\ \pm 1$$

as can be seen by inspecting the 9-j symbol

$$\begin{Bmatrix} S & S' & 1 \\ L & L' & 2 \\ J & J' & 1 \end{Bmatrix}$$

4-7 MAGNETIC SPIN-SPIN INTERACTION

The expression

$$H_6 = 4\beta^2 \sum_{i>j} \left[\frac{\mathbf{s}_i \cdot \mathbf{s}_j}{r_{ij}^3} - \frac{3(\mathbf{r}_{ij}\cdot\mathbf{s}_i)(\mathbf{r}_{ij}\cdot\mathbf{s}_j)}{r_{ij}^5} \right]$$

where $\mathbf{r}_{ij} = \mathbf{r}_i - \mathbf{r}_j$, will now be considered. The complete contribution to the Hamiltonian from spin-spin interaction contains in addition to H_6 a term representing the contact interaction of the spins; however, it does not affect the splitting of a multiplet and will not be considered any further here. The principal complication in putting H_6 in tensor-operator form lies in the presence of high inverse powers of r_{ij}. As a first step, the spin parts are separated out with the aid of a 9-j symbol. Thus

$$(\mathbf{s}_i\cdot\mathbf{s}_j)/r_{ij}^3 - 9\{\mathbf{r}_{ij}\mathbf{s}_i\}^{(0)}\{\mathbf{r}_{ij}\mathbf{s}_j\}^{(0)}/r_{ij}^5$$
$$= (\mathbf{s}_i\cdot\mathbf{s}_j)/r_{ij}^3 - 9r_{ij}^{-5}\sum_{k,t}((11)0,(11)0,0|(11)k,(11)t,0)$$
$$\times\ \{\{\mathbf{r}_{ij}\mathbf{r}_{ij}\}^{(k)}\{\mathbf{s}_i\mathbf{s}_j\}^{(t)}\}^{(0)}$$
$$= \frac{\mathbf{s}_i\cdot\mathbf{s}_j}{r_{ij}^3} - \frac{9}{r_{ij}^5}\sum_k [k] \begin{Bmatrix} 1 & 1 & 0 \\ 1 & 1 & 0 \\ k & k & 0 \end{Bmatrix} \{\{\mathbf{r}_{ij}\mathbf{r}_{ij}\}^{(k)}\{\mathbf{s}_i\mathbf{s}_j\}^{(k)}\}^{(0)}$$

The part for which $k = 0$ cancels with the first term, while $\{\mathbf{r}_{ij}\mathbf{r}_{ij}\}^{(1)}$, being proportional to the vector product of two identical vectors, is zero. Only the term in the sum for which $k = 2$ remains, and the spin-spin interaction becomes

$$H_6 = -12\beta^2(5)^{\frac{1}{2}} \sum_{i>j} \{\{\mathbf{r}_{ij}\mathbf{r}_{ij}\}^{(2)}\{\mathbf{s}_i\mathbf{s}_j\}^{(2)}\}^{(0)}/r_{ij}^5 \tag{4-17}$$

The next problem is to expand r_{ij}^{-5}. Differentiating

$$\frac{1}{r_{ij}} = \sum_k \frac{r_<^k}{r_>^{k+1}} P_k(\cos\omega)$$

with respect to $\cos\omega$, and using $P'_{k+1} = [k]P_k + P'_{k-1}$ repeatedly, we get

$$\frac{1}{r_{ij}^3} = \frac{1}{r_>^2 - r_<^2} \sum_k \frac{r_<^k}{r_>^{k+1}} [k]^{\frac{1}{2}}(-1)^k\{\mathbf{C}_i^{(k)}\mathbf{C}_j^{(k)}\}^{(0)}$$

An extension of this procedure gives

$$\frac{1}{r_{ij}^5} = \frac{1}{3(r_>^2 - r_<^2)^3} \sum_k \frac{r_<^k}{r_>^{k+1}} [(2k+3)r_>^2 - (2k-1)r_<^2] \times [k]^{\frac{1}{2}}(-1)^k\{\mathbf{C}_i^{(k)}\mathbf{C}_j^{(k)}\}^{(0)} \tag{4-18}$$

It is easy to show that

$$\{\mathbf{r}_{ij}\mathbf{r}_{ij}\}^{(2)} = (\tfrac{2}{3})^{\frac{1}{2}}[r_i^2\mathbf{C}_i^{(2)} + r_j^2\mathbf{C}_j^{(2)}] - 2r_ir_j\{\mathbf{C}_i^{(1)}\mathbf{C}_j^{(1)}\}^{(2)}$$

The tensors $\mathbf{C}^{(1)}$ and $\mathbf{C}^{(2)}$ in this equation are now combined with those occurring in Eq. (4-18) by means of Eq. (4-5). Equations such as

$$r_i^2 + r_j^2 = r_>^2 + r_<^2 \qquad \text{and} \qquad r_ir_j = r_>r_<$$

are used to simplify the expressions. The coefficient of $\{\mathbf{C}_i^{(k)}\mathbf{C}_j^{(k)}\}^{(2)}$ is found to be zero: the coefficient of $\{\mathbf{C}_i^{(k+2)}\mathbf{C}_j^{(k)}\}^{(2)}$ in $\{\mathbf{r}_{ij}\mathbf{r}_{ij}\}^{(2)}/r_{ij}^5$ is

$$\begin{aligned}\frac{(-1)^k}{3(r_>^2 - r_<^2)^3} &\left\{\frac{(k+1)(k+2)(2k+1)(2k+5)}{5(2k+3)}\right\}^{\frac{1}{2}} \\ &\times \frac{1}{r_>}\Bigg[r_j^2\frac{r_<^{k+2}}{r_>^{k+2}}\{(2k+7)r_>^2 - (2k+3)r_<^2\} \\ &+ r_i^2\frac{r_<^k}{r_>^k}\{(2k+3)r_>^2 - (2k-1)r_<^2\} \\ &- 2r_ir_j\frac{r_<^{k+1}}{r_>^{k+1}}\{(2k+5)r_>^2 - (2k+1)r_<^2\}\Bigg]\end{aligned} \tag{4-19}$$

Now if $r_j > r_i$, then $r_> = r_j$ and $r_< = r_i$. On making these substitutions in (4-19), the bracket is found to vanish. We may therefore put $r_> = r_i$ and $r_< = r_j$, with the understanding that the eventual radial integration is carried out subject to $r_j < r_i$. The bracket now simplifies to

$$-(2k+3)r_j^k(r_j^2 - r_i^2)^3/r_i^{k+2}$$

Tensors of the type $\{\mathbf{C}_i^{(k)}\mathbf{C}_j^{(k+2)}\}^{(2)}$ also occur in the expansion of $\{\mathbf{r}_{ij}\mathbf{r}_{ij}\}^{(2)}/r_{ij}^5$, of course. In all, we find

$$\begin{aligned} H_6 = -2(5)^{-\frac{1}{2}}\beta^2 \sum_k (-1)^k \left\{\frac{(2k+5)!}{(2k)!}\right\}^{\frac{1}{2}} \\ \times \sum_{i>j} \left[\frac{r_j^k}{r_i^{k+3}} (\{\mathbf{C}_i^{(k+2)}\mathbf{C}_j^{(k)}\}^{(2)} \cdot \{\mathbf{s}_i\mathbf{s}_j\}^{(2)}) \right. \\ \left. + \frac{r_i^k}{r_j^{k+3}} (\{\mathbf{C}_i^{(k)}\mathbf{C}_j^{(k+2)}\}^{(2)} \cdot \{\mathbf{s}_i\mathbf{s}_j\}^{(2)})\right] \end{aligned} \quad (4\text{-}20)$$

where the first and second terms in the bracket are to be integrated over r_j while fulfilling $r_j < r_i$ and $r_i < r_j$, respectively. Only one term is nonzero at a time. Equation (4-20) has been obtained by Innes[53] by slightly different methods.

The contribution of spin-spin interaction to the splitting of the ground term ^3H of f^2 will now be calculated. For a particular term in the summation of Eq. (4-20),

$$\begin{aligned} (f^2SLJM_J|(\{\mathbf{s}_i\mathbf{s}_j\}^{(2)} \cdot \{\mathbf{C}_i^{(k)}\mathbf{C}_j^{(k+2)}\}^{(2)})|f^2S'L'J'M_J') \\ = (-1)^{S'+L+J}\delta(J,J')\delta(M_J,M_J') \begin{Bmatrix} S & S' & 2 \\ L' & L & J \end{Bmatrix} \\ \times (f^2S\|\{\mathbf{s}_i\mathbf{s}_j\}^{(2)}\|f^2S')(f^2L\|\{\mathbf{C}_i^{(k)}\mathbf{C}_j^{(k+2)}\}^{(2)}\|f^2L') \end{aligned}$$

The triangular conditions for the nonvanishing of the 6-j symbol indicate that the selection rules for H_6 are

$$\Delta S, \Delta L = 0, \pm 1, \pm 2$$

The actual form of the 6-j symbol, i.e., its dependence on S, S', L, L', and J, indicates that, in the limit of LS coupling, spin-spin interaction produces deviations from the Landé interval rule while at the same time leaving S and L good quantum numbers. The notorious departures from the interval rule in the excited triplets of HeI are due to spin-spin interaction.

From Eq. (3-35),

$$(f^2S\|\{s_is_j\}^{(2)}\|f^2S') = (s\|s\|s)^2\{5[S][S']\}^{\frac{1}{2}} \begin{Bmatrix} \frac{1}{2} & \frac{1}{2} & 1 \\ \frac{1}{2} & \frac{1}{2} & 1 \\ S & S' & 2 \end{Bmatrix}$$

$$= (\tfrac{5}{4})^{\frac{1}{2}} \qquad \text{for } S = S' = 1$$

Again, $(f^2L\|\{C_i^{(k)}C_j^{(k+2)}\}^{(2)}\|f^2L')$

$$= (l\|C^{(k)}\|l)(l\|C^{(k+2)}\|l)\{5[L][L']\}^{\frac{1}{2}} \begin{Bmatrix} 3 & 3 & k \\ 3 & 3 & k+2 \\ L & L' & 2 \end{Bmatrix}$$

For $L = L' = 5$,

$$\begin{Bmatrix} 3 & 3 & k \\ 3 & 3 & k+2 \\ 5 & 5 & 2 \end{Bmatrix} = (\tfrac{1}{42})(\tfrac{13}{110})^{\frac{1}{2}},\ -(\tfrac{17}{1386})(\tfrac{13}{70})^{\frac{1}{2}}, \text{ and } -(\tfrac{53}{30492})(\tfrac{1}{13})^{\frac{1}{2}}$$

for $k = 0$, 2, and 4, respectively. Thus

$$(f^2\,{}^3H_J|H_6|f^2\,{}^3H_J)$$

$$= -(143)^{\frac{1}{2}}(-1)^J \begin{Bmatrix} 1 & 1 & 2 \\ 5 & 5 & J \end{Bmatrix} [-\tfrac{2}{3}I_0 + \tfrac{68}{33}I_2 + \tfrac{2650}{4719}I_4]$$

where $$I_k = \beta^2 \int_0^\infty \int_0^{r_i} \frac{r_j^k}{r_i^{k+3}} [R_{nf}(r_i)R_{nf}(r_j)]^2\, dr_j\, dr_i + \beta^2 \int_0^\infty \int_{r_i}^\infty \frac{r_i^k}{r_j^{k+3}} [R_{nf}(r_i)R_{nf}(r_j)]^2\, dr_j\, dr_i$$

If it is assumed that R_{nf} is hydrogenic, I_k can be expressed as multiples of $\langle r^{-3}\rangle$. Using a method similar to that described in Appendix 1, we find that for $n = 4$, corresponding to PrIV,

$$I_0 = 3473\beta^2\langle r^{-3}\rangle/8192 \qquad I_2 = 2115\beta^2\langle r^{-3}\rangle/8192$$
$$I_4 = 1485\beta^2\langle r^{-3}\rangle/8192$$

On taking Ridley's value of 29.4 Å^{-3} for $\langle r^{-3}\rangle$,[54] the levels ${}^3H_6$, ${}^3H_5$, and ${}^3H_4$ are found to be shifted -2.2, $+5.7$, and -3.8 cm^{-1}, respectively, owing to spin-spin interaction. This is much smaller than spin-orbit effects in PrIV, which are of the order of 10^3 cm^{-1}.

PROBLEMS

4-1. Prove that the ratios of the spacings between the energies $E({}^3L)$ of the triplet terms of f^2 are independent of the Slater integrals. Find the energies of the triplets of g^2 by tensor-operator techniques, and show that an analogous result does not hold for this configuration. (See Shortley and Fried.[55])

4-2. Prove that the effect of the nuclear quadrupole moment can be taken into account by including the term

$$H_7 = -\frac{e^2Q}{I(2I-1)}\sum_j \frac{1}{r_j^3}(\mathbf{C}_j^{(2)} \cdot \mathbf{K}^{(2)})$$

in the Hamiltonian, where $\mathbf{K}^{(2)}$ is a tensor of rank 2 having components

$$K_0^{(2)} = \tfrac{1}{2}(3I_z^2 - I(I+1)) \qquad K_{\pm1}^{(2)} = \mp(\tfrac{3}{8})^{\frac{1}{2}}(I_zI_\pm + I_\pm I_z)$$
$$K_{\pm2}^{(2)} = (\tfrac{3}{8})^{\frac{1}{2}}I_\pm^2$$

and where Q is defined by the equation

$$Q = (I, M_I = I|\sum_i (3Z_i^2 - R_i^2)|I, M_I = I)$$

The coordinates of the protons of the nucleus are denoted by capital letters.

Show that, within a manifold of states of constant I and J, H_7 can be replaced by

$$\frac{B}{2I(2I-1)J(2J-1)}\left[3(\mathbf{I}\cdot\mathbf{J})^2 + \frac{3}{2}(\mathbf{I}\cdot\mathbf{J}) - I(I+1)J(J+1)\right]$$

where

$$B = -2e^2Q[J(2J-1)/(J+1)(2J+1)(2J+3)]^{\frac{1}{2}}\left(J\|\sum_j C_j^{(2)}/r_j^3\|J\right)$$

(See Trees.[48])

4-3. Prove that, within a given LS multiplet, the spin-spin interaction H_6 has matrix elements proportional to

$$(\mathbf{S}\cdot\mathbf{L})^2 + \tfrac{1}{2}(\mathbf{S}\cdot\mathbf{L}) - \tfrac{1}{3}S(S+1)L(L+1)$$

4-4. Coles, Orton, and Owen[56] have used paramagnetic-resonance methods to investigate exchange interactions between neighboring Mn^{++} ions present as substitutional impurities in MgO. Show that an isotropic interaction of the form $A\mathbf{S}_1 \cdot \mathbf{S}_2$, where $\mathbf{S}_1$ and $\mathbf{S}_2$ are the spins of the two interacting magnetic centers, leads to a system of energy levels which follow the Landé interval rule with respect to the total spin T. Show also that a small additional anisotropic interaction of the form

$$D(\mathbf{S}_1\cdot\mathbf{S}_2 - 3S_{1z}S_{2z}) + E(S_{1x}S_{2x} - S_{1y}S_{2y})$$

produces splittings in the levels that can be described by

$$\left[\frac{T(T+1)+4S(S+1)}{2(2T-1)(2T+3)}\right][-D(3T_z^2 - T(T+1)) + E(T_x^2 - T_y^2)]$$

where S is the spin quantum number for an individual magnetic center.

4-5. Prove that, for the configuration l^2, the operator-equivalent factors α, β, and γ defined in Sec. 2-7 are given by the formula

$$(l^2SLJ\|\nu_k\|l^2S'L'J) = 2^{k+1}\delta(S,S')(-1)^{S+J+l}[J][l][L]^{\frac{1}{2}}[L']^{\frac{1}{2}}$$
$$\times\left[\frac{(2J-k)!}{(2J+k+1)!}\right]^{\frac{1}{2}}\begin{pmatrix} l & k & l \\ 0 & 0 & 0\end{pmatrix}\begin{Bmatrix} L & k & L' \\ l & l & l\end{Bmatrix}\begin{Bmatrix} J & k & J \\ L' & S & L\end{Bmatrix}$$

where $\qquad (l^2SLJ\|\nu_k\|l^2S'L'J) = \alpha, \beta, \gamma \qquad$ for $k = 2, 4, 6$

4-6. Evaluate the reduced matrix element

$$((s_1l_1)j_1,(s_2l_2)j_2,J\|\{\{\mathbf{C}_1^{(k_1)}\mathbf{C}_2^{(k_2)}\}^{(k)}\{\mathbf{s}_1^{(\kappa_1)}\mathbf{s}_2^{(\kappa_2)}\}^{(\kappa)}\}^{(K)}\|(s_1'l_1')j_1',(s_2'l_2')j_2',J')$$

(1) by recoupling the operator and (2) by transforming the states from jj to LS coupling. Equate the two results, and obtain the identity

$$\sum_{S,L,J,S',L',J'} [S][S'][L][L'][J][J'] \begin{Bmatrix} s_1 & l_1 & j_1 \\ s_2 & l_2 & j_2 \\ S & L & J \end{Bmatrix} \begin{Bmatrix} s_1' & l_1' & j_1' \\ s_2' & l_2' & j_2' \\ S' & L' & J' \end{Bmatrix} \begin{Bmatrix} j_1 & j_1' & \lambda \\ j_2 & j_2' & \mu \\ J & J' & K \end{Bmatrix}$$
$$\times \begin{Bmatrix} S & S' & \kappa \\ L & L' & k \\ J & J' & K \end{Bmatrix} \begin{Bmatrix} s_1 & s_1' & \kappa_1 \\ s_2 & s_2' & \kappa_2 \\ S & S' & \kappa \end{Bmatrix} \begin{Bmatrix} l_1 & l_1' & k_1 \\ l_2 & l_2' & k_2 \\ L & L' & k \end{Bmatrix}$$
$$= \begin{Bmatrix} s_1 & s_1' & \kappa_1 \\ l_1 & l_1' & k_1 \\ j_1 & j_1' & \lambda \end{Bmatrix} \begin{Bmatrix} s_2 & s_2' & \kappa_2 \\ l_2 & l_2' & k_2 \\ j_2 & j_2' & \mu \end{Bmatrix} \begin{Bmatrix} \kappa_1 & \kappa_2 & \kappa \\ k_1 & k_2 & k \\ \lambda & \mu & K \end{Bmatrix}$$

Show that the appropriate coupling diagram is given in Fig. 4-2, and verify that apart from a phase factor the sum could have been found by means of the rules described at the close of Sec. 3-4. Check that the identity contains Eq. (3-28) as a special case.

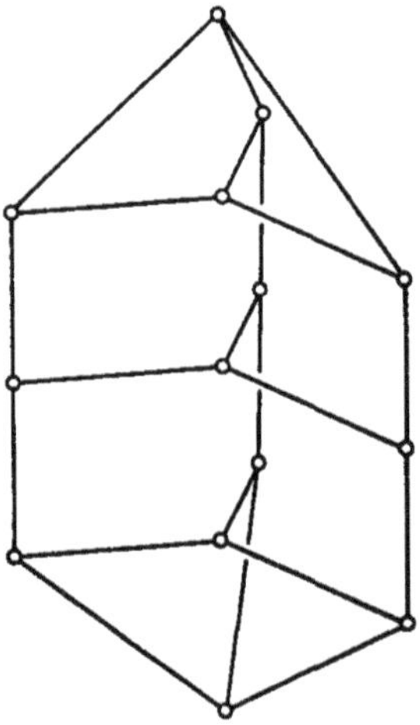

FIG. 4-2 The coupling diagram of Prob. 4-6.

4-7.‡ Prove that the familiar vector product $\mathbf{T}^{(1)} \times \mathbf{U}^{(1)}$ of two vectors can be rewritten as

$$-i(2)^{\frac{1}{2}}\{\mathbf{T}^{(1)}\mathbf{U}^{(1)}\}^{(1)}$$

and show that the linear momentum $\mathbf{p}$ can be expressed as

$$i(2)^{\frac{1}{2}}r^{-1}(\mathbf{C}^{(1)}\mathbf{l})^{(1)} - i\frac{\partial}{\partial r}\mathbf{C}^{(1)}$$

‡ Problems 4-7 to 4-9 are taken from a paper by Innes and Ufford[38]; their definition of $\mathbf{r}_{ij}$ is opposite in sign to that used above.

4-8. Derive the equations

$$(\mathbf{C}^{(k+1)}\mathbf{l})^{(k)} = \left[\frac{k(2k-1)}{(k+1)(2k+3)}\right]^{\frac{1}{2}} (\mathbf{C}^{(k-1)}\mathbf{l})^{(k)}$$

and

$$(l\|(\mathbf{C}^{(k)}\mathbf{l})^{(k)}\|l') = \frac{l(l+1) - k(k+1) - l'(l'+1)}{[2k(2k+2)]^{\frac{1}{2}}} (l\|C^{(k)}\|l')$$

4-9. Verify the expansion

$$\frac{\mathbf{r}_{ij}}{r_{ij}^3} = \left(\frac{1}{3}\right)^{\frac{1}{2}} \sum_k (-1)^k \left[\frac{r_i^{k-1}}{r_j^{k+1}} \{k(2k-1)(2k+1)\}^{\frac{1}{2}} \{\mathbf{C}_i^{(k-1)}\mathbf{C}_j^{(k)}\}^{(1)} + \frac{r_j^k}{r_i^{k+2}} \{(k+1)(2k+1)(2k+3)\}^{\frac{1}{2}} \{\mathbf{C}_i^{(k+1)}\mathbf{C}_j^{(k)}\}^{(1)}\right]$$

where the first term is to be used for $r_i < r_j$ and the second for $r_i > r_j$.

4-10. By a process of recoupling, derive the equation

$$\begin{aligned}[k(2k-1)(2k+1)]^{\frac{1}{2}}\{\{\mathbf{C}_i^{(k-1)}\mathbf{C}_j^{(k)}\}^{(1)}\mathbf{p}_i\}^{(1)} &= ir_i^{-1}(k+1)[(2k+1)/2]^{\frac{1}{2}}\{\mathbf{C}_j^{(k)}(\mathbf{C}^{(k)}\mathbf{l})_i^{(k)}\}^{(1)} \\ &- ir_i^{-1}(2k+1)[(2k-1)/2]^{\frac{1}{2}}\{\mathbf{C}_j^{(k)}(\mathbf{C}^{(k)}\mathbf{l})_i^{(k-1)}\}^{(1)} \\ &- i[(k+1)k(2k+1)/2]^{\frac{1}{2}}\{\mathbf{C}_j^{(k)}\mathbf{C}_i^{(k)}\}^{(1)}\,\partial/\partial r_i\end{aligned}$$

and obtain a similar equation for

$$[(k+1)(2k+1)(2k+3)]^{\frac{1}{2}}\{\{\mathbf{C}_i^{(k+1)}\mathbf{C}_j^{(k)}\}^{(1)}\mathbf{p}_i\}^{(1)}$$

4-11. The spin-other-orbit interaction in a system of electrons has been written as

$$H_8 = e^2\hbar^2\Xi/2m^2c^2$$

where

$$\Xi = \sum_{i\neq j}\left[\nabla_i\left(\frac{1}{r_{ij}}\right) \times \mathbf{p}_i\right] \cdot (\mathbf{s}_i + 2\mathbf{s}_j)$$

(See Condon and Shortley[1] and Marvin.[67]) Prove that, in the limit of LS coupling, H_8 does not produce deviations from the Landé interval rule.‡

Use the results of the previous problem to show that, within the configuration $(nl)(n'l')$, the direct part of the interaction between two levels, characterized by the quantum numbers SLJ and $S'L'J$, respectively, can be found by using for

‡ Although Ξ is the accepted operator to use in considering spin-other-orbit interactions, a glance at expression (1-11) shows that the terms in Ξ that involve $\mathbf{s}_i$ represent an interaction of the ordinary spin-orbit type. Aller, Ufford, and Van Vleck [*Astrophys. J.*, **109,** 42 (1949)] have pointed out that the central part of this interaction is implicitly contained in H_2 and should therefore be deleted in an exact treatment. H. Horie [*Progr. Theoret. Phys. Japan*, **10,** 296 (1953)] has studied spin-other-orbit interactions in configurations of the type l^n; S. Obi and S. Yanagawa [*Publ. Astron. Soc. Japan*, **7,** 125 (1955)] have given a detailed treatment of the configurations p^ns and p^np.

≡ the operator

$$3^{-\frac{1}{2}} \sum_{i \neq j} \sum_{k} (-1)^k \left[\frac{r_i^{k-2}}{r_j^{k+1}} (2k+1)(2k-1)^{\frac{1}{2}} \{ \mathbf{C}_j^{(k)} (\mathbf{C}^{(k)} \mathbf{1})_i^{(k-1)} \}^{(1)} - \frac{r_j^k}{r_i^{k+3}} (2k+1)(2k+3)^{\frac{1}{2}} \{ \mathbf{C}_j^{(k)} (\mathbf{C}^{(k)} \mathbf{1})_i^{(k+1)} \}^{(1)} \right] \cdot (\mathbf{s}_i + 2\mathbf{s}_j)$$

The contribution to the exchange part of the interaction from terms other than those containing the partial derivatives with respect to r involves the radial integrals

$$N^{(k)} = \frac{e^2\hbar^2}{8m^2c^2} \int_0^\infty \int_0^\infty \frac{r_<^k}{r_>^{k+3}} R_{nl}(r_i) R_{n'l'}(r_i) R_{nl}(r_j) R_{n'l'}(r_j) \, dr_i \, dr_j$$

Show that, for $(nl)(n'l') \equiv dp$, the coefficients a and c in the expression

$$aN^{(-1)} + bN^{(1)} + cN^{(3)}$$

for the contribution to the matrix element of H_8 between two triplet levels ^3L and $^3L'$ satisfy the equation

$$\frac{c}{a} = -\frac{9}{(224)^{\frac{1}{2}}} \begin{Bmatrix} L & 1 & L' \\ 1 & 3 & 2 \\ 2 & 3 & 1 \end{Bmatrix} \div \begin{Bmatrix} L & 1 & L' \\ 1 & 1 & 2 \\ 2 & 1 & 1 \end{Bmatrix}$$

4-12. Prove that

$$\begin{Bmatrix} S & S' & 1 \\ L' & L & J \end{Bmatrix}^2$$

is a quadratic function of $J(J+1)$. Use this result to show that near the LS limit a more accurate expression than that provided by the Landé interval rule for the separation of a pair of adjacent levels of a multiplet is $aJ + bJ^3$, where $a \gg b$ and J is the higher J value of the pair.

Show that, within a multiplet, slight deviations from LS coupling are indistinguishable from magnetic spin-spin interactions.

5

CONTINUOUS GROUPS

5-1 INTRODUCTION

The methods described in the previous chapter, which avoid the explicit introduction of determinantal product states, can be easily extended to any configuration comprising two electrons. However, when systems possessing more than two electrons are examined, the possibility of the occurrence of more than one term with a given S and L raises the problem of defining a state. Sometimes it is convenient to introduce a particular coupling scheme; for f^2d, for example, the state

$$|f^2SL,sd,S'L'M'_SM'_L)$$

is completely defined. A description of a state in these terms makes it easy to apply the theory of tensor operators; moreover S' and L' can assume all those values which are allowed on a simple vector-coupling scheme. This approach is not suited to configurations of equivalent electrons, since it is not obvious which values of S' and L' are allowed by the Pauli exclusion principle. We shall therefore write a state as

$$|l^n\gamma SLM_SM_L)$$

and direct our attention to finding quantum numbers γ which will serve to distinguish terms of the same kind.

It is natural to hope that the classification of the terms of the configuration by means of the quantum numbers γ, S, and L will have some special significance and that we shall be able to avoid having to expand a term as a series of determinantal product states when matrix elements of operators are required. Now the labels S, L, M_S, and M_L are essentially group-theoretical in character; for example, under rotations in three-dimensional space, the $2L + 1$ components of a term with a given S and M_S transform according to the representation $\mathfrak{D}_L$ of R_3. If a group $\mathcal{G}$ of operations could be found, one which includes the operations of R_3, then the irreducible representations of $\mathcal{G}$ could be used as the additional symbols γ to label the states. These labels would certainly have a well-defined significance, and we might hope that the terms of a configuration could be divided into sets, the eigenfunctions of each set forming the basis for an irreducible representation of $\mathcal{G}$. If, for example, the two ^2H of f^3 given in Eqs. (1-29) and (1-30) belonged to two different irreducible representations of $\mathcal{G}$, then these would be appropriate labels to distinguish the states. So great is the advantage of knowing the group-theoretical properties of the states that the question whether the symbols γ are *good* quantum numbers, e.g., whether the states of Eqs. (1-29) and (1-30) are eigenfunctions of $H_1 + H_2$, is regarded as being of secondary interest and will not be considered until later. The method of searching for appropriate groups and the use of tensor operators in the theory of continuous groups are due to Racah.[58]

5-2 LIE'S THEORY OF CONTINUOUS GROUPS

We can do no more than briefly sketch some of the relevant concepts of the theory of continuous groups, the development of which is due to Lie[59] and his contemporaries. The presentation here owes much to Racah.[58] The equations

$$x^i = f^i(x_0^1, x_0^2, \ldots, x_0^n; a^1, a^2, \ldots, a^r) \tag{5-1}$$

where $i = 1, 2, \ldots, n$, carry the point $\mathbf{x}_0 = (x_0^1, x_0^2, \ldots, x_0^n)$ into another point $\mathbf{x} = (x^1, x^2, \ldots, x^n)$ lying in the n-dimensional space. It is postulated that the set of parameters $(a^1, a^2, \ldots, a^r) = \mathbf{a}$ defines the transformation represented by Eq. (5-1) completely and uniquely; that is, no two transformations defined by distinct parameters $\mathbf{a}$ and $\mathbf{a}'$ exist such that the two points $\mathbf{x}$ and $\mathbf{y}$ into which a point $\mathbf{x}_0$ is carried by the two transformations satisfy $\mathbf{x} = \mathbf{y}$ for all $\mathbf{x}_0$. We shall suppose that, when $\mathbf{a} = 0$, the point $\mathbf{x}_0$ is transformed in itself;

$$\mathbf{x} = \mathbf{f}(\mathbf{x}, 0). \tag{5-2}$$

The transformations (5-1) for various $\mathbf{a}$ are said to form a continuous group if:

1. The result of performing in succession any two transformations $\mathbf{x} = \mathbf{f}(\mathbf{x}_0,\mathbf{a})$ and $\mathbf{x}' = \mathbf{f}(\mathbf{x},\mathbf{b})$ can be reproduced by a single transformation that belongs to the set (5-1); that is, parameters

$$c^\sigma = \varphi^\sigma(\mathbf{a},\mathbf{b}) \tag{5-3}$$

can be found for which $\mathbf{x}' = \mathbf{f}(\mathbf{x}_0,\mathbf{c})$.

2. To every transformation of the set there corresponds a unique inverse transformation, also belonging to the set.

To help fix our ideas, we shall take as an example the rotation group in three dimensions, R_3. Consider a sphere whose center is the origin, and imagine that the point (x_0,y_0,z_0) is embedded in its surface. If the sphere is rotated by an angle α about the x axis, then β about the y axis, and finally γ about the z axis, the point (x_0,y_0,z_0) is carried into the point (x,y,z), where

$$\begin{aligned}
x &= x_0 \cos\beta \cos\gamma + y_0(\sin\alpha \sin\beta \cos\gamma - \cos\alpha \sin\gamma) \\
&\qquad + z_0(\cos\alpha \cos\gamma \sin\beta + \sin\alpha \sin\gamma) \\
y &= x_0 \cos\beta \sin\gamma + y_0(\cos\alpha \cos\gamma + \sin\alpha \sin\beta \sin\gamma) \\
&\qquad + z_0(\cos\alpha \sin\beta \sin\gamma - \sin\alpha \cos\gamma) \\
z &= -x_0 \sin\beta + y_0 \sin\alpha \cos\beta + z_0 \cos\alpha \cos\beta
\end{aligned}$$

These equations correspond to Eq. (5-1). They bear a close resemblance to Eqs. (4.46) of Goldstein,[60] which relate the coordinates of a point before and after a transformation of axes corresponding to the Euler angles ϕ, θ, ψ. Euler's factorization of a three-dimensional rotation into three plane rotations is unacceptable to us here, because if $\theta = 0$, the parameters ϕ and ψ no longer uniquely define a transformation, but only their sum, $\phi + \psi$.

The correspondences $(a^1,a^2,a^3) \equiv (\alpha,\beta,\gamma)$, $(b^1,b^2,b^3) \equiv (\alpha',\beta',\gamma')$, and $(c^1,c^2,c^3) \equiv (\alpha'',\beta'',\gamma'')$ are now made, and it is assumed that the angles are suitably bounded to ensure that no two sets of parameters define the same transformation. Equations equivalent to Eq. (5-3) are not difficult to find. For example,

$$\begin{aligned}
\sin\beta'' &= \sin\beta \cos\alpha' \cos\beta' + \cos\beta \cos\gamma \sin\beta' \\
&\qquad - \cos\beta \sin\gamma \sin\alpha' \cos\beta'
\end{aligned}$$

When the parameters $\mathbf{a}$ in the general transformation $\mathbf{x} = \mathbf{f}(\mathbf{x}_0,\mathbf{a})$ are increased by infinitesimal amounts, the increments in the coordinates $\mathbf{x}$ are determined by the equations

$$dx^i = \frac{\partial f^i(\mathbf{x}_0,\mathbf{a})}{\partial a^\sigma}\, da^\sigma$$

a sum over σ being implied. However, we could equally well obtain the infinitesimal change in $\mathbf{x}$ by introducing parameters $\delta\mathbf{a}$ of infinitesimal size in Eq. (5-2). In this case,

$$dx^i = u_\sigma{}^i\,\delta a^\sigma \tag{5-4}$$

where

$$u_\sigma{}^i = \left(\frac{\partial f^i(\mathbf{x},\mathbf{a})}{\partial a^\sigma}\right)_{\mathbf{a}=0}$$

A connection between $d\mathbf{a}$ and $\delta\mathbf{a}$ can be established by writing Eq. (5-3) in the form

$$a^\sigma + da^\sigma = \varphi^\sigma(\mathbf{a},\delta\mathbf{a})$$

from which it follows that

$$da^\sigma = \mu_\rho{}^\sigma(\mathbf{a})\,\delta a^\rho$$

where

$$\mu_\rho{}^\sigma(\mathbf{a}) = \left(\frac{\partial\varphi^\sigma(\mathbf{a},\mathbf{b})}{\partial b^\rho}\right)_{\mathbf{b}=0}$$

The inverse transformation is $\delta a^\rho = \lambda_\tau{}^\rho(\mathbf{a})\,da^\tau$, in which $\lambda_\tau{}^\rho\mu_{\rho'}{}^\tau = \delta_{\rho'}{}^\rho$. Equation (5-4) can now be written as

$$\frac{\partial x^i}{\partial a^\rho} = u_\sigma{}^i(\mathbf{x})\lambda_\rho{}^\sigma(\mathbf{a}) \tag{5-5}$$

For R_3 it turns out that

$$u_\sigma{}^i(\mathbf{x}) = \begin{pmatrix} 0 & -z & y \\ z & 0 & -x \\ -y & x & 0 \end{pmatrix}$$

where i labels the columns and σ the rows; and

$$\lambda_\rho{}^\sigma(\mathbf{a}) = \begin{pmatrix} \cos\gamma\cos\beta & \cos\beta\sin\gamma & -\sin\beta \\ -\sin\gamma & \cos\gamma & 0 \\ 0 & 0 & 1 \end{pmatrix}$$

If we had adopted Euler's representation of a three-dimensional rotation, we would have found that the determinant of $\mu_\rho{}^\sigma(\mathbf{a})$ was zero and hence we would have been unable to construct $\lambda_\rho{}^\sigma(\mathbf{a})$.

If we wish to construct a finite displacement from a succession of infinitesimal displacements (and this must certainly be possible if the transformations are to form a group), the differential equations (5-5) must be integrable. The condition for this is

$$\frac{\partial^2 x^i}{\partial a^\tau\,\partial a^\rho} = \frac{\partial^2 x^i}{\partial a^\rho\,\partial a^\tau}$$

and hence

$$u_\sigma{}^i(\mathbf{x})\,\frac{\partial\lambda_\rho{}^\sigma(\mathbf{a})}{\partial a^\tau} + \frac{\partial u_\sigma{}^i(\mathbf{x})}{\partial x^j}\,\frac{\partial x^j}{\partial a^\tau}\,\lambda_\rho{}^\sigma(\mathbf{a}) = u_\sigma{}^i(\mathbf{x})\,\frac{\partial\lambda_\tau{}^\sigma(\mathbf{a})}{\partial a^\rho} + \frac{\partial u_\sigma{}^i(\mathbf{x})}{\partial x^j}\,\frac{\partial x^j}{\partial a^\rho}\,\lambda_\tau{}^\sigma(\mathbf{a})$$

or $$\left(u_\nu{}^j\,\frac{\partial u_\sigma{}^i}{\partial x^j} - u_\sigma{}^j\,\frac{\partial u_\nu{}^i}{\partial x^j}\right)\lambda_\tau{}^\nu\lambda_\rho{}^\sigma + u_\sigma{}^i\left(\frac{\partial\lambda_\rho{}^\sigma}{\partial a^\tau} - \frac{\partial\lambda_\tau{}^\sigma}{\partial a^\rho}\right) = 0$$

Multiplying by $\mu_\xi{}^\tau\mu_\eta{}^\rho$ and summing over τ and ρ, we get

$$u_\xi{}^j\,\frac{\partial u_\eta{}^i}{\partial x^j} - u_\eta{}^j\,\frac{\partial u_\xi{}^i}{\partial x^j} = c_{\xi\eta}{}^\sigma(\mathbf{a})u_\sigma{}^i \tag{5-6}$$

where $$c_{\xi\eta}{}^\sigma(\mathbf{a}) = \left(\frac{\partial\lambda_\tau{}^\sigma}{\partial a^\rho} - \frac{\partial\lambda_\rho{}^\sigma}{\partial a^\tau}\right)\mu_\xi{}^\tau\mu_\eta{}^\rho$$

Since $u_\sigma{}^i(\mathbf{x})$ is independent of $\mathbf{a}$, Eq. (5-6) gives

$$u_\sigma{}^i\,\partial c_{\xi\eta}{}^\sigma(\mathbf{a})/\partial a^\rho = 0$$

for any parameter a^ρ. The fact that the parameters δa^σ of Eq. (5-4) are the minimum number to specify a transformation uniquely guarantees that the quantities $u_\sigma{}^i$ are linearly independent; hence the partial derivative of $c_{\xi\eta}{}^\sigma(\mathbf{a})$ with respect to a^ρ vanishes and $c_{\xi\eta}{}^\sigma$ is independent of $\mathbf{a}$.

The infinitesimal transformation $\mathbf{x} \to \mathbf{x} + d\mathbf{x}$ induces in a function $F(\mathbf{x})$ the transformation $F(\mathbf{x}) \to F(\mathbf{x}) + dF(\mathbf{x})$, where

$$dF(\mathbf{x}) = \frac{\partial F}{\partial x^i}\,dx^i = u_\sigma{}^i\,\delta a^\sigma\,\frac{\partial F}{\partial x^i}$$

The operator that effects this is

$$S_a = 1 + \delta a^\sigma\,X_\sigma \tag{5-7}$$

where $$X_\sigma = u_\sigma{}^i(\mathbf{x})\,\frac{\partial}{\partial x^i}$$

The quantities X_σ are called the *infinitesimal operators* of the group. Owing to Eq. (5-6), they satisfy

$$[X_\sigma,X_\rho] = c_{\sigma\rho}{}^\tau X_\tau \tag{5-8}$$

This important equation is said to define the *structure* of the group, and the quantities $c_{\sigma\rho}{}^\tau$ are called the *structure constants*. Lie has shown that, in addition to being able to derive Eq. (5-8) from the assumption that the transformations form a group, we can deduce that the operators $1 + \delta a^\sigma\,X_\sigma$ form the elements of a group if the quantities X_σ satisfy equations of the type (5-8).

5-3 DEFINITIONS

It is shown in Sec. 2-4 that many of the concepts of finite groups can be carried over to the group R_3. The present treatment, which makes extensive use of the infinitesimal operators X_σ, confines attention to that part of the group which lies close to the identity. These transformations are carried out by means of operators of the type S_a, but it is apparent that Eq. (5-7) does not express S_a to high enough orders of smallness to evaluate usefully such simple sequences of transformations as $S_bS_aS_b^{-1}$. However, according to Racah,[58] the expressions $S_bS_aS_b^{-1}S_a^{-1}$, which contain no terms of the first order of smallness, can be written as

$$1 + \delta a^\sigma \, \delta b^\rho [X_\rho, X_\sigma]$$

to the second order, and we can overcome the difficulty.

As with finite groups, a group is said to be *Abelian* if all its elements S_a commute. The condition $S_aS_b = S_bS_a$ can be rewritten as $S_bS_aS_b^{-1}S_a^{-1} = 1$, from which it follows that $[X_\rho, X_\sigma] = 0$ for all ρ and σ. This imposes the following restriction on the structure constants:

$$c_{\rho\sigma}{}^\tau = 0 \tag{5-9}$$

In analogy with its definition for finite groups, a *subgroup* is defined as a set of transformations which are contained in the group and which by themselves satisfy the group postulates. If X_1, X_2, . . . , X_p are the infinitesimal operators of a subgroup, then the operators S_a and S_b, defined in terms of parameters δa^σ and δb^σ satisfying $\delta a^\sigma = \delta b^\sigma = 0$ when $\sigma > p$, generate transformations of the subgroup; and so does

$$S_bS_aS_b^{-1}S_a^{-1} = 1 + \delta a^\sigma \, \delta b^\rho \, c_{\rho\sigma}{}^\tau X_\tau$$

The condition that no operator X_τ appears in the sum which is not contained in the set X_1, X_2, . . . , X_p is therefore

$$c_{\rho\sigma}{}^\tau = 0 \qquad \rho, \sigma \leq p, \tau > p \tag{5-10}$$

An *invariant subgroup* contains all the conjugates of its elements. By this is meant that, for any element S_a of the subgroup, $S_bS_aS_b^{-1}$ is also contained in the subgroup, where S_b is any element of the entire group. Thus $S_bS_aS_b^{-1}S_a^{-1}$ must also be contained in the subgroup, and this gives the condition

$$c_{\rho\sigma}{}^\tau = 0 \qquad \rho \leq p, \tau > p \tag{5-11}$$

which is stronger than Eq. (5-10).

A group is said to be *simple* if it contains no invariant subgroup besides the unit element. A group is said to be *semi-simple* if it contains no Abelian invariant subgroup besides the unit element. All simple groups are necessarily semi-simple.

Suppose that the infinitesimal operators X_σ of a group $\mathcal{K}$ can be broken up into two sets characterized by $\sigma \leq p$ and $\sigma > p$, respectively, such that the sets form the operators for the respective invariant subgroups $\mathcal{G}$ and $\mathcal{H}$ of $\mathcal{K}$. Then

$$c_{\rho\sigma}{}^{\tau} = 0 \qquad \rho \leq p, \tau > p$$

and

$$c_{\rho\sigma}{}^{\tau} = 0 \qquad \sigma > p, \tau \leq p$$

It follows that

$$c_{\rho\sigma}{}^{\tau} = 0 \qquad \rho \leq p, \sigma > p \tag{5-12}$$

and the elements of $\mathcal{G}$ and $\mathcal{H}$ commute. The group $\mathcal{K}$ is called the *direct product* of $\mathcal{G}$ and $\mathcal{H}$ and is written as either $\mathcal{G} \times \mathcal{H}$ or $\mathcal{H} \times \mathcal{G}$.

5-4 THE COMMUTATORS OF TENSOR OPERATORS

The matrix of $u_\sigma{}^i$ for R_3 (see Sec. 5-2) shows that for this group

$$X_1 = y\frac{\partial}{\partial z} - z\frac{\partial}{\partial y} \qquad X_2 = z\frac{\partial}{\partial x} - x\frac{\partial}{\partial z} \qquad X_3 = x\frac{\partial}{\partial y} - y\frac{\partial}{\partial x}$$

It is apparent that these expressions are proportional to the orbital angular-momentum operators; hence Eq. (5-8) for R_3 must be equivalent to the commutation relations for the components of an angular-momentum vector. Now a vector is a tensor of rank 1, and it occurred to Racah to investigate the commutation relations of tensor operators of higher rank.[58] If it can be established that the tensor operators satisfy equations of the type (5-8), then we have at our disposal the infinitesimal operators of groups other than R_3 in a very convenient form.

Since it is intended to apply group theory to configurations of electrons that are equivalent, we restrict our attention to configurations of this kind by introducing the tensor operators $\mathbf{v}^{(k)}$ for which

$$(nl\|v^{(k)}\|n'l') = (2k+1)^{\frac{1}{2}}\delta(n,n')\delta(l,l') \tag{5-13}$$

The factor $(2k + 1)^{\frac{1}{2}}$ on the right-hand side is to be noted. These tensors must of course satisfy the usual commutation relations with respect to the total angular momentum $\mathbf{J}$ of the system. Their commutation relations with respect to one another can be obtained as follows. It is clear that

$$v_{q_2}^{(k_2)}|lm_l) = \sum_{m'_l} (-1)^{l-m'_l}[k_2]^{\frac{1}{2}} \begin{pmatrix} l & k_2 & l \\ -m'_l & q_2 & m_l \end{pmatrix} |lm'_l)$$

Hence $v_{q_1}^{(k_1)}v_{q_2}^{(k_2)}|lm_l)$

$$= \sum_{m'_l,m''_l} (-1)^{2l-m'_l-m''_l}[k_1]^{\frac{1}{2}}[k_2]^{\frac{1}{2}} \times \begin{pmatrix} l & k_1 & l \\ -m''_l & q_1 & m'_l \end{pmatrix} \begin{pmatrix} l & k_2 & l \\ -m'_l & q_2 & m_l \end{pmatrix} |lm''_l)$$

$$= \sum_{m''_l,k_3,q_3} (-1)^{3l-m''_l-q_3+k_1+k_2+k_3}[k_1]^{\frac{1}{2}}[k_2]^{\frac{1}{2}}[k_3] \times \begin{Bmatrix} k_1 & k_2 & k_3 \\ l & l & l \end{Bmatrix} \begin{pmatrix} l & k_3 & l \\ -m''_l & q_3 & m_l \end{pmatrix} \begin{pmatrix} k_1 & k_2 & k_3 \\ q_1 & q_2 & -q_3 \end{pmatrix} |lm''_l)$$

$$= \sum_{k_3,q_3} (-1)^{2l-q_3+k_1+k_2+k_3}\{[k_1][k_2][k_3]\}^{\frac{1}{2}} \times \begin{Bmatrix} k_1 & k_2 & k_3 \\ l & l & l \end{Bmatrix} \begin{pmatrix} k_1 & k_2 & k_3 \\ q_1 & q_2 & -q_3 \end{pmatrix} v_{q_3}^{(k_3)}|lm_l)$$

It is obvious that $v_{q_2}^{(k_2)}v_{q_1}^{(k_1)}|lm_l)$ differs from this result only in the phase factor $(-1)^{k_1+k_2+k_3}$, which must be introduced in reversing the first two columns of the 3-j symbol. We conclude that

$$[v_{q_1}^{(k_1)},v_{q_2}^{(k_2)}] = \sum_{k_3,q_3} (-1)^{2l-q_3}\{(-1)^{k_1+k_2+k_3} - 1\} \times \{[k_1][k_2][k_3]\}^{\frac{1}{2}} \begin{Bmatrix} k_1 & k_2 & k_3 \\ l & l & l \end{Bmatrix} \begin{pmatrix} k_1 & k_2 & k_3 \\ q_1 & q_2 & -q_3 \end{pmatrix} v_{q_3}^{(k_3)} \quad (5\text{-}14)$$

If configurations of more than one electron are being examined, it is more appropriate to use the tensors $\mathbf{V}^{(k)}$, defined by

$$V_q^{(k)} = \sum_i (v_q^{(k)})_i$$

where the sum runs over the electrons. Since the two tensors $\mathbf{v}^{(k)}$ and $\mathbf{V}^{(k)}$ become equivalent when the summation reduces to a single term, that is, when the system under examination involves the coordinates of a single electron, we shall in future use the more general forms $\mathbf{V}^{(k)}$. To obtain their commutation relations, it is necessary only to replace $v_{q_1}^{(k_1)}$, $v_{q_2}^{(k_2)}$, and $v_{q_3}^{(k_3)}$ of Eq. (5-14) by $V_{q_1}^{(k_1)}$, $V_{q_2}^{(k_2)}$, and $V_{q_3}^{(k_3)}$, respectively.

It is apparent that Eq. (5-14) is of the same form as Eq. (5-8), and hence the tensor operators $V_q^{(k)}$ can be regarded as the infinitesimal operators of a group. The triangular conditions on the 6-j symbol indicate that we can restrict our attention to the $(2l + 1)^2$ components $V_q^{(k)}$ for which $-k \leq q \leq k$ and $0 \leq k \leq 2l$. The operators S_a of Eq. (5-7) can be written as

$$1 + \sum_{k,q} \delta a_{kq} V_q^{(k)}$$

Here, as elsewhere, we make the summation explicit when the covariant or contravariant character of the indices is no longer obvious. Under the influence of an operator of this type, the eigenfunction $\psi(l,m)$ becomes

$$\psi'(l,m) = \sum_{m'} c_{mm'}\psi(l,m')$$

where $$c_{mm'} = \delta(m,m') + \sum_{k,q} \delta a_{kq}(-1)^{l-m'}[k]^{\frac{1}{2}} \begin{pmatrix} l & k & l \\ -m' & q & m \end{pmatrix}$$

Because of the linear independence of the 3-j symbols, any function $\sum_m b_m\psi(l,m)$ which differs infinitesimally from the function $\sum_m b'_m\psi(l,m)$ can be reached by a suitable choice of the parameters δa_{kq}; hence the components $V_q^{(k)}$ can be regarded as the infinitesimal operators of the *full linear group* GL_{2l+1} in $2l + 1$ dimensions. This group is, however, rather too general for many purposes, and it is convenient to limit the transformations to the subgroup which preserves the orthonormality of the eigenfunctions. The equation

$$\int \psi'^*(l,m)\psi'(l,m'')\, d\tau = \delta(m,m'')$$

is satisfied if the infinitesimal parameters obey the condition

$$\delta a_{kq}^* + (-1)^q\, \delta a_{k,-q} = 0 \tag{5-15}$$

The matrix $c_{mm'}$ is unitary, and consequently the group is called the *unitary group* in $2l + 1$ dimensions, U_{2l+1}. The infinitesimal operators are the same as those for the full linear group.

The group U_{2l+1} is not simple: the infinitesimal operator $V_0^{(0)}$ commutes with every $V_q^{(k)}$ and hence by Eq. (5-11) forms an invariant subgroup. According to Eq. (5-15), δa_{00} must be imaginary; so to the first order of smallness the operator $1 + \delta a_{00}V_0^{(0)}$ can be written as $e^{i\theta}$, where θ is real. The only significance of the operator $V_0^{(0)}$ is that it leads to a change of phase in the eigenfunctions. Since, to first order, the determinant of $c_{mm'}$ is

$$1 + \sum_{k,q,m} \delta a_{kq}(-1)^{l-m}[k]^{\frac{1}{2}} \begin{pmatrix} l & k & l \\ -m & q & m \end{pmatrix} = 1 + [l]^{\frac{1}{2}} \delta a_{00}$$

the effect of dispensing with $V_0^{(0)}$ is to restrict the transformations to those whose determinants of $c_{mm'}$ are $+1$. The group formed by the $4l(l+1)$ tensor operators $V_q^{(k)}(-k \le q \le k,\ 1 \le k \le 2l)$ is called the *unimodular* or *special unitary group* in $2l+1$ dimensions, SU_{2l+1}.

Subgroups of SU_{2l+1} are not difficult to find. If k_1 and k_2 of Eq. (5-14) are odd, then the factor $\{(-1)^{k_1+k_2+k_3} - 1\}$ ensures that only those tensor operators for which k_3 is odd appear in the summation. According to Eq. (5-10), the tensor operators of odd rank can be thought of as the infinitesimal operators of a group. Now

$$\begin{aligned} \delta a_{kq} V_q^{(k)} \sum_m (-1)^m \psi_1(l,m)\psi_2(l,-m) &= \delta a_{kq} \sum_m (-1)^{l-q}[k]^{\frac{1}{2}} \begin{pmatrix} l & k & l \\ -m-q & q & m \end{pmatrix} \psi_1(l, m+q)\psi_2(l,-m) \\ &\quad + \delta a_{kq} \sum_m (-1)^{l+2m-q}[k]^{\frac{1}{2}} \begin{pmatrix} l & k & l \\ m-q & q & -m \end{pmatrix} \psi_1(l,m)\psi_2(l, q-m) \\ &= \delta a_{kq} \sum_m (-1)^{l-q}[1 + (-1)^k][k]^{\frac{1}{2}} \begin{pmatrix} l & k & l \\ -m-q & q & m \end{pmatrix} \\ &\qquad \times \psi_1(l, m+q)\psi_2(l,-m) \end{aligned}$$

This expression vanishes if k is odd; hence the operations of the group leave invariant the expression

$$I = \sum_m (-1)^m \psi_1(l,m)\psi_2(l,-m) \tag{5-16}$$

If the functions ψ are broken up into their real and imaginary parts by writing

$$\psi_n(l,m) = (2)^{-\frac{1}{2}}(x_{nm} + iy_{nm}) \qquad m > 0$$

and

$$\psi_n(l,0) = z_n \tag{5-17}$$

then the equation $\psi(l,-m) = (-1)^m\psi^*(l,m)$, which is in keeping with our phase convention, allows I to be written as

$$I = z_1z_2 + \frac{1}{2}\sum_{m>0} [(x_{1m} - iy_{1m})(x_{2m} + iy_{2m}) + (-1)^{2m}(x_{1m} + iy_{1m})(x_{2m} - iy_{2m})]$$

The leading term z_1z_2 exists only if m can assume the value zero. Although it has been implied that l is the azimuthal quantum number of the electron, care has been taken not to assume that l is integral in the derivation of Eqs. (5-14) and (5-16). If indeed l is integral, $(-1)^{2m}$ is $+1$ and the invariant is

$$I = z_1z_2 + \sum_{m>0} (x_{1m}x_{2m} + y_{1m}y_{2m})$$

It is easy to prove from the unitarity of the matrices with elements $c_{mm'}$ that

$$z_i^2 + \sum_{m>0} (x_{im}^2 + y_{im}^2)$$

is also invariant, and hence so is the distance between two points of the type

$$(x_{i1},x_{i2}, \ . \ . \ . \ ,x_{il},y_{i1},y_{i2}, \ . \ . \ . \ ,y_{il},z_i)$$

The group whose infinitesimal operators are the $l(2l + 1)$ tensor operators $V_q^{(k)}$ of odd rank is therefore called the *rotation*, or *proper real orthogonal group*, in $2l + 1$ dimensions, R_{2l+1}.

On the other hand, if l is half-integral,

$$I = \sum_{m>0} i(x_{1m}y_{2m} - y_{1m}x_{2m})$$

Weyl[61] has given the name "symplectic" to groups that leave invariant bilinear antisymmetric forms. Thus, if l were half-integral, the $(l + 1)(2l + 1)$ components $V_q^{(k)}$ of odd rank would play the role of the infinitesimal operators of the symplectic group in $2l + 1$ dimensions, Sp_{2l+1}. This group is of importance in the jj coupling extreme, where the single-particle eigenfunctions are of the type $|j,m)$, in which j is half-integral. We shall have occasion to introduce a symplectic group in another context in Sec. 6-2.

If $k_1 = k_2 = 1$ in Eq. (5-14), then $k_3 = 1$ only. The corresponding group, which is a subgroup of R_{2l+1} or Sp_{2l+1}, has the three tensor operators $V_1^{(1)}$, $V_0^{(1)}$, and $V_{-1}^{(1)}$ as its infinitesimal operators and is nothing else but the rotation group in three dimensions, R_3.

In the case of f electrons, an apparently unique event occurs. The tensor operators of odd rank are $\mathbf{V}^{(1)}$, $\mathbf{V}^{(3)}$, and $\mathbf{V}^{(5)}$, and we should expect commutators of the type $[V_{q_1}{}^{(5)},V_{q_2}{}^{(5)}]$ to contain components of all of them. Very remarkably, however,

$$\begin{Bmatrix} 5 & 5 & 3 \\ 3 & 3 & 3 \end{Bmatrix} = 0$$

and the commutator contains components of $\mathbf{V}^{(1)}$ and $\mathbf{V}^{(5)}$ only. Moreover, $[V_{q_1}{}^{(5)},V_{q_2}{}^{(1)}]$ can be expressed as a factor times $V^{(5)}_{q_1+q_2}$, and the commutator $[V_{q_1}{}^{(1)},V_{q_2}{}^{(1)}]$ can give rise to $V^{(1)}_{q_1+q_2}$ only. Consequently the 14 components of $\mathbf{V}^{(1)}$ and $\mathbf{V}^{(5)}$ form the infinitesimal operators for a subgroup of R_7, which is called G_2. It obviously contains R_3 as a subgroup.

To indicate the succession of groups and subgroups, we write

$$R_3 \subset R_{2l+1} \subset SU_{2l+1} \subset U_{2l+1} \subset GL_{2l+1} \tag{5-18}$$

For f electrons,

$$R_3 \subset G_2 \subset R_7 \subset SU_7 \subset U_7 \subset GL_7 \tag{5-19}$$

These groups are available for studying the transformation properties of the many-electron eigenfunctions.

5-5 THE METRIC TENSOR

The symmetric tensor defined by

$$g_{\sigma\lambda} = c_{\sigma\rho}{}^{\tau}c_{\lambda\tau}{}^{\rho} \tag{5-20}$$

is called the *metric* tensor. The structure constants that appear in Eq. (5-14) may be directly substituted in this equation and the sum over the product of the 3-j symbols carried out. On replacing the symbols σ and λ with the pairs (kq) and $(k'q')$, Eq. (5-20) becomes

$$g_{(kq)(k'q')} = 2\delta(k,k')\delta(q,-q')\Sigma(-1)^q\{1 - (-1)^{k+k_2+k_3}\} \times [k_2][k_3] \begin{Bmatrix} k & k_2 & k_3 \\ l & l & l \end{Bmatrix}^2 \tag{5-21}$$

The sum runs over those values of k_2 and k_3 appropriate to the group in question. For U_{2l+1}, all integral values from zero up to $2l$ are permitted, and Eq. (5-21) reduces to

$$g_{(kq)(k'q')}(U_{2l+1}) = 2(-1)^q\delta(k,k')\delta(q,-q')[l]\{1 - \delta(k,0)\} \tag{5-22}$$

The formula for SU_{2l+1} can be easily obtained by deleting $\delta(k,0)$. For R_{2l+1}, only odd values of k_2 and k_3 are contained in the sum; the result is

$$g_{(kq)(k'q')}(R_{2l+1}) = (-1)^q\delta(k,k')\delta(q,-q')(2l-1) \tag{5-23}$$

For G_2,

$$g_{(kq)(k'q')}(G_2) = 4(-1)^q\delta(k,k')\delta(q,-q') \tag{5-24}$$

We are able to form the reciprocal tensor $g^{\sigma\mu}$ for any group for which the determinant $|g_{\sigma\lambda}|$ does not vanish. This modest requirement is, however, satisfied only by semi-simple groups. For if the indices of an Abelian invariant subgroup are indicated by primes, then, following the argument of Racah,[58]

$$\begin{aligned} g_{\sigma\lambda'} &= c_{\sigma\rho}{}^{\tau}c_{\lambda'\tau}{}^{\rho} = c_{\sigma\rho'}{}^{\tau}c_{\lambda'\tau}{}^{\rho'} && \text{by (5-11)} \\ &= -c_{\rho'\sigma}{}^{\tau}c_{\lambda'\tau}{}^{\rho'} = -c_{\rho'\sigma}{}^{\tau'}c_{\lambda'\tau'}{}^{\rho'} && \text{by (5-11)} \\ &= 0 && \text{by (5-9)} \end{aligned}$$

Hence all the entries in the column λ' of $|g_{\sigma\lambda}|$ vanish, and the determinant is zero. The group U_{2l+1} is not semi-simple, and it is easy to see from Eq. (5-22) that $g_{(kq)(00)}(U_{2l+1}) = 0$ for all k and q. The presence of delta functions in Eqs. (5-22) to (5-24) makes it easy to find $g^{\sigma\mu}$. For R_{2l+1},

$$g^{(kq)(k'q')}(R_{2l+1}) = (-1)^q\delta(k,k')\delta(q,-q')(2l-1)^{-1} \tag{5-25}$$

5-6 THE ROOT FIGURES

It can be seen from Eq. (5-14) that the tensor operators $V_0^{(k)}$ commute with one another. For if $q_1 = q_2 = 0$, then $q_3 = 0$, in which case the 3-j symbol vanishes if $k_1 + k_2 + k_3$ is odd and the factor $\{(-1)^{k_1+k_2+k_3} - 1\}$ vanishes if the sum is even. Consequently we can search for those linear combinations

$$E_\alpha = \sum_{k,q} b_\alpha{}^{(kq)}V_q{}^{(k)} \qquad q \neq 0 \tag{5-26}$$

which *simultaneously* satisfy equations of the type

$$[V_0{}^{(k')},E_\alpha] = \alpha_{k'}E_\alpha \tag{5-27}$$

for all values of k' permitted for a specific group.

The infinitesimal operators have now been broken up into two sets. The first comprises operators (or linear combinations of them) for which $q = 0$; they commute with one another and, following Weyl,[62] we denote them by H_i ($i = 1, 2, \ldots$). The second set comprises the operators E_α, each of which is a simultaneous eigenvector of the operators H_i. If we rewrite Eq. (5-27) in the form

$$[H_i,E_\alpha] = \alpha_i E_\alpha \tag{5-28}$$

it is clear that the eigenvalues α_i for a given α can be regarded as the covariant components of a vector $\boldsymbol{\alpha}$ in a certain space. This is called the *weight* space; the vectors are called *roots*. The new metric tensor corresponding to the transformation (5-26) is given by

$$g_{\alpha\beta} = \sum_{k,k',q,q'} b_\alpha{}^{(kq)} b_\beta{}^{(k'q')} g_{(kq)(k'q')}$$

where terms in the sum for which $q = 0$ or $q' = 0$ are excluded. If we suppose that $V_0{}^{(k')}$ corresponds to H_i,

$$g_{\alpha i} = \sum_{k;q\neq 0} b_\alpha{}^{(kq)} g_{(kq)(k'0)} = 0$$

for all α and i, owing to Eq. (5-21). The components $g_{(k0)(k'0)}$ are untouched by the transformation (5-26) and are simply relabeled as g_{ij}. By putting $q = q' = 0$ in Eq. (5-21) we see that g_{ij} is diagonal; in particular, the diagonal elements for the groups SU_{2l+1}, R_{2l+1}, and G_2 are independent of k. The scalar product of two covariant vectors in the weight space, say, $\boldsymbol{\alpha}$ and $\boldsymbol{\beta}$, can now be defined in terms of the inverse tensor g^{ij},

$$\boldsymbol{\alpha} \cdot \boldsymbol{\beta} = g^{ij}\alpha_i\beta_j$$

The tensor g^{ij}, like g_{ij}, is diagonal, with all elements equal; consequently we may plot out the components α_i in an ordinary Euclidean space with an orthogonal system of coordinates. Scalar products contain a numerical factor but in other respects are formed in the usual way. This simplification is a direct result of including the factor $[k]^{\frac{1}{2}}$ in Eq. (5-13).

As an example we take the group R_5. The tensors $\mathbf{V}^{(1)}$ and $\mathbf{V}^{(3)}$ play the role of the infinitesimal operators. The following relations may be readily verified:

$$\begin{aligned}
[V_0{}^{(1)},V_3{}^{(3)}] &= (\tfrac{9}{10})^{\frac{1}{2}} V_3{}^{(3)} \\
[V_0{}^{(3)},V_3{}^{(3)}] &= -(\tfrac{1}{10})^{\frac{1}{2}} V_3{}^{(3)} \\
[V_0{}^{(1)},V_2{}^{(3)}] &= (\tfrac{2}{5})^{\frac{1}{2}} V_2{}^{(3)} \\
[V_0{}^{(3)},V_2{}^{(3)}] &= (\tfrac{1}{10})^{\frac{1}{2}} V_2{}^{(3)} \\
[V_0{}^{(1)}, (\tfrac{2}{5})^{\frac{1}{2}}V_1{}^{(1)} + (\tfrac{3}{5})^{\frac{1}{2}}V_1{}^{(3)}] &= (\tfrac{1}{10})^{\frac{1}{2}}\{(\tfrac{2}{5})^{\frac{1}{2}}V_1{}^{(1)} + (\tfrac{3}{5})^{\frac{1}{2}}V_1{}^{(3)}\} \\
[V_0{}^{(3)}, (\tfrac{2}{5})^{\frac{1}{2}}V_1{}^{(1)} + (\tfrac{3}{5})^{\frac{1}{2}}V_1{}^{(3)}] &= (\tfrac{9}{10})^{\frac{1}{2}}\{(\tfrac{2}{5})^{\frac{1}{2}}V_1{}^{(1)} + (\tfrac{3}{5})^{\frac{1}{2}}V_1{}^{(3)}\} \\
[V_0{}^{(1)}, (\tfrac{3}{5})^{\frac{1}{2}}V_1{}^{(1)} - (\tfrac{2}{5})^{\frac{1}{2}}V_1{}^{(3)}] &= (\tfrac{1}{10})^{\frac{1}{2}}\{(\tfrac{3}{5})^{\frac{1}{2}}V_1{}^{(1)} - (\tfrac{2}{5})^{\frac{1}{2}}V_1{}^{(3)}\} \\
[V_0{}^{(3)}, (\tfrac{3}{5})^{\frac{1}{2}}V_1{}^{(1)} - (\tfrac{2}{5})^{\frac{1}{2}}V_1{}^{(3)}] &= -(\tfrac{2}{5})^{\frac{1}{2}}\{(\tfrac{3}{5})^{\frac{1}{2}}V_1{}^{(1)} - (\tfrac{2}{5})^{\frac{1}{2}}V_1{}^{(3)}\}
\end{aligned}$$

In addition to these equations there are eight more with values of q_2 and q_3 opposite in sign to those given above. Since $k_1 + k_2 + k_3$ is odd, the 3-j symbol of Eq. (5-14) changes sign if the signs of the arguments in its lower row are reversed and hence the eight additional values of $\alpha_{k'}$ are equal in magnitude but opposite in sign to those already obtained.

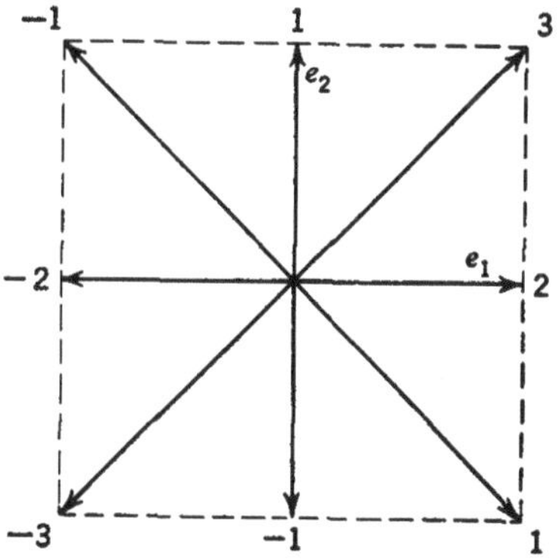

FIG. 5-1 The root figure for R_5. The numbers against the roots indicate the values of $q_2 = q_3$ to which the roots correspond.

For any eigenvector E_α, there are two eigenvalues $\alpha_{k'}$, corresponding to $k' = 1$ and 3. The eight points (α_1,α_3) can therefore be plotted in a two-dimensional weight space. It is immediately found that the array of points lies on the perimeter of a square. The square is tilted with respect to the coordinate axes; however, this defect can be easily removed by choosing appropriate linear combinations of $V_0^{(1)}$ and $V_0^{(3)}$. If we use

$$H_1 = (\tfrac{3}{5})^{\frac{1}{2}} V_0^{(1)} + (\tfrac{2}{5})^{\frac{1}{2}} V_0^{(3)}$$

and

$$H_2 = (\tfrac{2}{5})^{\frac{1}{2}} V_0^{(1)} - (\tfrac{3}{5})^{\frac{1}{2}} V_0^{(3)} \tag{5-29}$$

the coordinates of the eight points are $(1,1),(1,0),(1,-1),(0,1),(0,-1),(-1,1),(-1,0)$, and $(-1,-1)$ for $q_2 = q_3 = 3, 2, 1, 1, -1, -1, -2$, and -3, respectively. The vectors from the origin to these points are the roots, and they are shown in Fig. 5-1. Diagrams of the roots such as Fig. 5-1 are called *root figures.* The transformation (5-29) leaves g_{ij} diagonal but multiplies the two diagonal elements by 2.

This procedure can be generalized to other groups. For GL_{2l+1} or U_{2l+1} (which possess the same infinitesimal operators) we define

$$X_{\nu\mu} = \sum_{k,q} (-1)^{l-\nu}[k]^{\frac{1}{2}} \begin{pmatrix} l & k & l \\ -\nu & q & \mu \end{pmatrix} V_q^{(k)} \tag{5-30}$$

The operators $X_{\nu\mu}$ are linearly independent and for a single-electron system possess the property

$$X_{\nu\mu}|l,m) = \delta(\mu,m)|l,\nu)$$

Thus

$$X_{\xi\eta}X_{\nu\mu}|l,m) = \delta(\eta,\nu)\delta(\mu,m)|l,\xi)$$

and

$$[X_{\xi\eta},X_{\nu\mu}]|l,m) = \delta(\eta,\nu)\delta(\mu,m)|l,\xi) - \delta(\mu,\xi)\delta(\eta,m)|l,\nu)$$

Hence

$$[X_{\xi\eta},X_{\nu\mu}] = \delta(\eta,\nu)X_{\xi\mu} - \delta(\xi,\mu)X_{\nu\eta}$$

As may be easily verified, this result is equally valid for a many-electron system. The operators X_{ii} commute with one another, and for U_{2l+1} we make the identification $H_{l-i+1} = X_{ii}$. The equation

$$[X_{ii},X_{\nu\mu}] = \{\delta(\nu,i) - \delta(\mu,i)\}X_{\nu\mu} \tag{5-31}$$

shows that as i runs in integral steps from l to $-l$ the eigenvalues of X_{ii} (and of H_{l-i+1}) are given by the sequence

$$0, 0, \ldots, 0, 1, 0, \ldots, 0, -1, 0, \ldots, 0$$

The symbol 1 appears in the $(l - \nu + 1)$st position, the -1 in the $(l - \mu + 1)$st. The weight space is $(2l + 1)$-dimensional, and the roots are all of the form $\mathbf{e}_i - \mathbf{e}_j$ (with $i \neq j$), where $\mathbf{e}_1, \mathbf{e}_2, \ldots, \mathbf{e}_{2l+1}$ are a set of mutually perpendicular unit vectors. The root corresponding to $X_{\nu\mu}$ is $\mathbf{e}_{l-\nu+1} - \mathbf{e}_{l-\mu+1}$. It is interesting to observe that the $2l(2l + 1)$ vectors $\mathbf{e}_i - \mathbf{e}_j$ do not span the entire $(2l + 1)$-dimensional weight space, since they are all perpendicular to $\sum_k \mathbf{e}_k$. This is a direct consequence of the presence of the operator $V_0^{(0)}$ in the set of infinitesimal operators of the group. Its coefficient in X_{ii} is $[l]^{-\frac{1}{2}}$ for all i, and it can be removed by replacing X_{ii} by

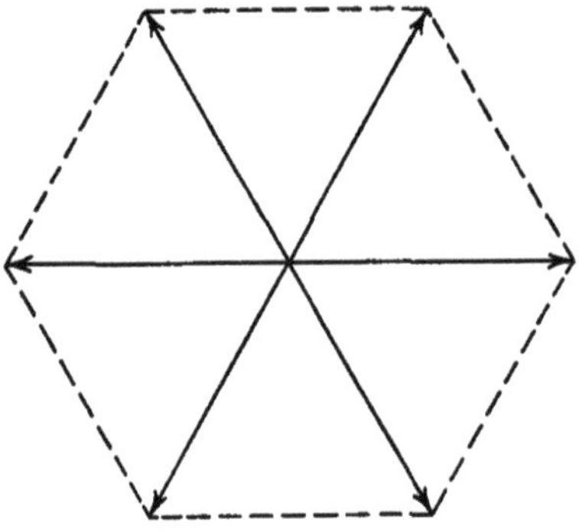

FIG. 5-2 The root figure for SU_3.

$$X'_{ii} = X_{ii} - [l]^{-1} \sum_j X_{jj}$$

These are the commuting infinitesimal operators for SU_{2l+1}; only $2l$ of them are linearly independent, and the weight space is $2l$-dimensional. For SU_{2l+1}, we take $H_{l-i+1} = X'_{ii}$. The root figure for SU_3 consists of the vectors $\pm(\mathbf{e}_2 - \mathbf{e}_3)$, $\pm(\mathbf{e}_3 - \mathbf{e}_1)$, $\pm(\mathbf{e}_1 - \mathbf{e}_2)$. It is drawn out in Fig. 5-2.

We next consider the operators

$$W_{\nu\mu} = \sum_{k,q} (-1)^{l-\nu}[1 - (-1)^k][k]^{\frac{1}{2}} \begin{pmatrix} l & k & l \\ -\nu & q & \mu \end{pmatrix} V_q^{(k)} \tag{5-32}$$

which include tensors of odd rank only. The analogue of Eq. (5-31) is

$$[W_{ii},W_{\nu\mu}] = \{\delta(\nu,i) - \delta(\mu,i) - \delta(-\nu,i) + \delta(-\mu,i)\}W_{\nu\mu} \tag{5-33}$$

It is to be noted that

$$W_{\nu\mu} = (-1)^{2l+\nu+\mu+1}W_{-\mu,-\nu} \tag{5-34}$$

and so the operators do not form a linearly independent set. If l is integral, there are l independent operators W_{ii}, where $i = 1, 2, \ldots, l$. The possibility $i = 0$ is excluded since $W_{00} = 0$. We make the identification $W_{ii} = H_{l+1-i}$ for R_{2l+1}; Eq. (5-32) reduces to Eqs. (5-29) if we put $l = 2$. If $\nu > 0$ and $\mu > 0$, the l eigenvalues of H_i are all zero except those for which $i = l - \nu + 1$ and $i = l - \mu + 1$, which are $+1$ and -1, respectively. These give roots of the type $\mathbf{e}_{l-\nu+1} - \mathbf{e}_{l-\mu+1}$ $(\nu \neq \mu)$, where $\mathbf{e}_j$ $(j = 1, 2, \ldots, l)$ stands for one of a set of l mutually perpendicular unit vectors. Operators $W_{\nu\mu}$ in Eq. (5-33) for which $\nu > 0$ and $\mu < 0$ give the roots $\mathbf{e}_{l-\nu+1} + \mathbf{e}_{l+\mu+1}$; those for which $\nu < 0$ and $\mu > 0$ give the roots $-\mathbf{e}_{l+\nu+1} - \mathbf{e}_{l-\mu+1}$. The possibility of ν being equal to $-\mu$ is excluded because in this case the 3-j symbol with odd k in Eq. (5-32) vanishes and so does $W_{\nu\mu}$. The possibility of either ν or μ (but not both) being zero gives rise to the roots $-\mathbf{e}_{l-\mu+1}$ and $\mathbf{e}_{l-\nu+1}$. All in all, the roots are given by the set of vectors $\pm\mathbf{e}_i \pm \mathbf{e}_j$, $\pm\mathbf{e}_i$ $(i, j = 1, 2, \ldots, l$, with $i \neq j)$; the root corresponding to $W_{\nu\mu}$ is $\epsilon^{\nu}\mathbf{e}_{l-|\nu|+1} - \epsilon^{\mu}\mathbf{e}_{l-|\mu|+1}$, where ϵ^{q} is $+1$ if q is positive, -1 if q is negative, and zero if q is zero. The root figure for any group R_{2l+1} can thus be easily constructed: Fig. 5-1 corresponds to the case of $l = 2$.

It is somewhat easier to find the root figures for groups of the type Sp_{2l+1}. There are $l + \frac{1}{2}$ independent operators W_{ii} corresponding to $i = \frac{1}{2}, \frac{3}{2}, \ldots, l$. The weight space is therefore $(l + \frac{1}{2})$-dimensional. Because l is half-integral, ν and μ must also be half-integral; consequently neither ν nor μ can be zero, and no vectors of the type $\pm\mathbf{e}_i$ occur. However, it is now possible to have $\mu = -\nu$, and this gives rise to vectors of the type $\pm 2\mathbf{e}_i$. The roots for Sp_{2l+1} are accordingly $\pm\mathbf{e}_i \pm \mathbf{e}_j$, $\pm 2\mathbf{e}_i$ $(i, j = 1, 2, \ldots, l + \frac{1}{2}$, with $i \neq j)$. The root figure for Sp_4 is similar to that for R_5, but rotated by $\frac{1}{4}\pi$. For higher values of l the figures are distinct.

There remains the group G_2 to consider. The operators which commute with each other are $V_0^{(1)}$ and $V_0^{(5)}$, and their commutation relations with the other components of $\mathbf{V}^{(1)}$ and $\mathbf{V}^{(5)}$ are straightforward to derive. On plotting out the eigenvalues of $V_0^{(1)}$ and $V_0^{(5)}$, the highly symmetrical figure of Fig. 5-3 is obtained. The roots can be conveniently represented by the six vectors $\mathbf{e}_i - \mathbf{e}_j$ $(i, j = 1, 2, 3;$ $i \neq j)$ together with the six $\pm 2\mathbf{e}_i \mp \mathbf{e}_j \mp \mathbf{e}_k$ $(i, j = 1, 2, 3; i \neq j \neq k)$. These vectors are all perpendicular to $\mathbf{e}_1 + \mathbf{e}_2 + \mathbf{e}_3$, and the weight space is two-dimensional. The operators

$$\begin{aligned} &4(\tfrac{1}{7})^{\frac{1}{2}}V_0^{(1)} + 6(\tfrac{1}{21})^{\frac{1}{2}}V_0^{(5)} \\ \text{and} \qquad &(\tfrac{1}{7})^{\frac{1}{2}}V_0^{(1)} - 9(\tfrac{1}{21})^{\frac{1}{2}}V_0^{(5)} \end{aligned} \tag{5-35}$$

which we take to play the roles of H_1 and H_2 for G_2, possess the eigenvalues $(1,1),(2,-1),(1,0),(0,1),(1,-1),(-1,2),(1,-2),(-1,1),(0,-1),$ $(-1,0),(-2,1),(-1,-1)$ for eigenvectors corresponding to $q = 5, 4, 3, 2, 1, 1, -1, -1, -2, -3, -4, -5$; they can be used to plot out the roots in an oblique coordinate scheme, such as that defined by the vectors $\mathbf{e}_1 - \mathbf{e}_3$ and $\mathbf{e}_1 - \mathbf{e}_2$. The metric tensor takes the form

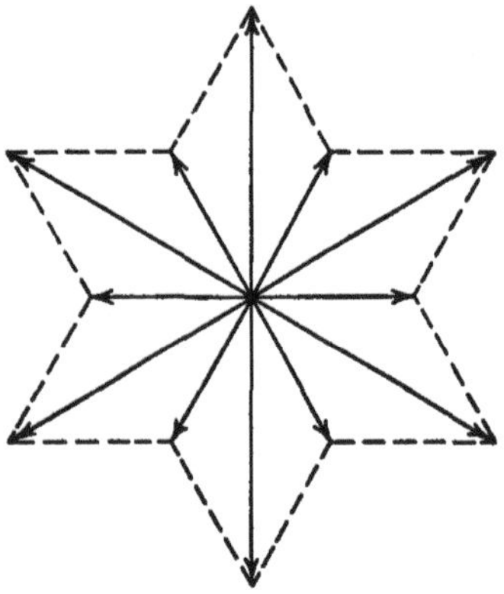

Fig. 5-3 The root figure for G_2.

$$g_{ij} = \begin{pmatrix} 16 & -8 \\ -8 & 16 \end{pmatrix} \tag{5-36}$$

Techniques of the kind described above allow the construction of the root figures for all the groups of immediate interest to us. It is natural to ask whether or not there are a great many other groups which exist but which have no relevance here. Comparison with the complete classification of all simple groups, which Cartan[63] obtained in 1894, shows that only one general class, that of the rotation groups in an even number of dimensions, and four special groups have escaped our attention. The groups SU_{2l+1}, R_{2l+1}, and Sp_{2l+1} correspond to A_{2l}, B_l, and $C_{l+\frac{1}{2}}$ in Cartan's notation. His subscripts give the dimension of the weight space. The rotation group in an even number of dimensions is labeled D_l, and the root figure is given by the vectors $\pm\mathbf{e}_i \pm \mathbf{e}_j$ $(i, j = 1, 2, \ldots, l;\ i \neq j)$. Cartan calls the five special groups, which do not fall into one of the four general classes, E_6, E_7, E_8, F_4, and G_2.

Schouten[64] has shown that all the root figures can be constructed from the simple requirement that, for any pair of roots $\boldsymbol{\alpha}$ and $\boldsymbol{\beta}$, the numbers m and n defined by the equations

$$\boldsymbol{\alpha} \cdot \boldsymbol{\beta} = \tfrac{1}{2}m\boldsymbol{\alpha} \cdot \boldsymbol{\alpha} = \tfrac{1}{2}n\boldsymbol{\beta} \cdot \boldsymbol{\beta}$$

must be integers. Van der Waerden[65] proved that to every root figure there corresponds only one simple group and thereby obtained Cartan's classification by a geometrical approach.

5-7 REPRESENTATIONS

Consider the 15 eigenfunctions $|d^2\,{}^1LM_L)$, where $L = 0$, 2, and 4. These are the singlet states of d^2. Under the operations of R_3, the $2L + 1$ components of each term transform among themselves and form a basis for the irreducible representation $\mathfrak{D}_L$ of R_3. This result can be viewed in the light of the infinitesimal transformations produced by the operator S_a of Eq. (5-7). The equation

$$\left(1 + \sum_q \delta a_{1q} V_q^{(1)}\right) |LM_L) = |LM_L) + \sum_{L',M'_L,q} \delta a_{1q} (L'M'_L|V_q^{(1)}|LM_L)|L'M'_L) \tag{5-37}$$

determines the way the 15 functions transform. Equation (5-13) shows that $\mathbf{v}^{(1)}$ is indistinguishable from $\mathbf{l}[3/l(l+1)(2l+1)]^{\frac{1}{2}}$; hence

$$\mathbf{V}^{(1)} = \mathbf{L}[3/l(l+1)(2l+1)]^{\frac{1}{2}} \tag{5-38}$$

and the only terms in the sum are those for which $L' = L$. The components of the three terms therefore transform among themselves, and the degree of freedom left in the parameters δa_{1q} ensures that in general the three separate representations whose bases are the components of the three terms are irreducible.

To examine the way the 15 eigenfunctions transform under the operations of R_5, the term $\sum_q \delta a_{3q} V_q^{(3)}$ is included in the expression for S_a. The condition $L' = L$ no longer holds, since $\mathbf{V}^{(3)}$ has nonzero matrix elements between states of ${}^1G$ and ${}^1D$. However, there is no way of coupling either ${}^1G$ or ${}^1D$ to ${}^1S$; so we conclude that ${}^1D$ and ${}^1G$ together form the basis for an irreducible representation of R_5 and that ${}^1S$ forms one by itself.

The next problem is to find labels for the representations. For R_3, the subscript to the symbol $\mathfrak{D}$ can be regarded as the maximum eigenvalue of L_z. Since L_z is related to $V_0^{(1)}$, the natural generalization is to examine the eigenvalues of $V_0^{(1)}$ and $V_0^{(3)}$; however, it turns out that it is more convenient to use the linear combinations H_1 and H_2 of Eqs. (5-29). For these,

$$\begin{aligned}
H_1|{}^1G,4) &= 2|{}^1G,4) \\
H_2|{}^1G,4) &= 0 \\
H_1|{}^1G,3) &= |{}^1G,3) \\
H_2|{}^1G,3) &= |{}^1G,3) \\
H_1[(\tfrac{3}{7})^{\frac{1}{2}}|{}^1D,2) - (\tfrac{4}{7})^{\frac{1}{2}}|{}^1G,2)] &= 0 \\
H_2[(\tfrac{3}{7})^{\frac{1}{2}}|{}^1D,2) - (\tfrac{4}{7})^{\frac{1}{2}}|{}^1G,2)] &= 2[(\tfrac{3}{7})^{\frac{1}{2}}|{}^1D,2) - (\tfrac{4}{7})^{\frac{1}{2}}|{}^1G,2)]
\end{aligned}$$

etc. In all, 28 equations of the type $H_i u = m_i u$ exist, and the 14 points (m_1,m_2) can be plotted in the weight space. Similarly, we could take the states $|d^2\,{}^3PM_L)$ and $|d^2\,{}^3FM_L)$ as the basis for a representation of R_5; these give rise to 10 points. The arrays of points for the two representations of R_5 are given in Fig. 5-4. It is to be noted that the point (0,0) occurs twice in both diagrams.

This procedure can be generalized to any group. The point $(m_1,m_2, \ldots)$ in the weight space, defined by the equations $H_i u = m_i u$,

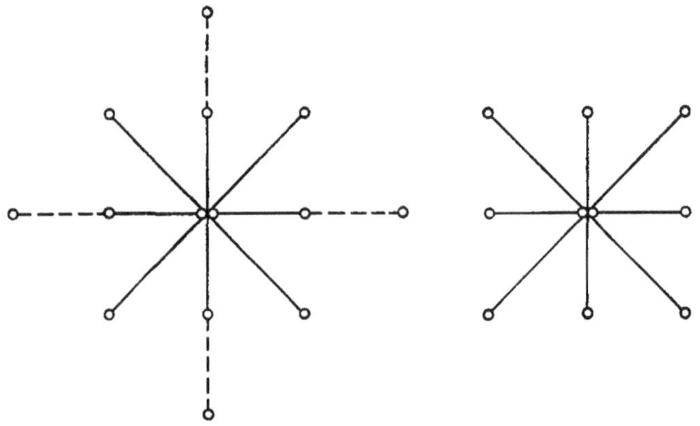

FIG. 5-4 The weights of the representations (20) and (11) of R_5, superposed on the root figure.

is called the *weight* of u. The weight $(m_1,m_2, \ldots)$ is said to be *higher* than $(m_1',m_2', \ldots)$ if the first nonvanishing quantity in the sequence $m_1 - m_1'$, $m_2 - m_2'$, . . . is positive. Thus, for R_5, $(1,-2)$ is higher than (0,2). According to the theory of semi-simple groups,[62] an irreducible representation is unambiguously specified by its highest weight, which can therefore be used as a suitable label for the representation. The highest weight of the representation whose basis is formed by the components of ^1D and ^1G is (2,0); that of the second representation of Fig. 5-4 is (1,1). These symbols (simplified by omitting the commas) are taken to label the representations. For R_3, the weights are the numbers M_L; the highest is L and is precisely the subscript to $\mathfrak{D}$ that is used to distinguish irreducible representations. As a consequence of the oblique system of coordinates introduced in Sec. 5-6 for G_2, the procedure for finding the highest weight of a representation of this group frequently yields a weight (m_1,m_2) for which m_2 is negative. By regarding (u_1,u_2) as the highest weight, where

$$(u_1,u_2) \equiv (m_1 + m_2, \; -m_2)$$

this blemish can be removed, and at the same time the conventional labeling is reproduced.

It can be seen that to each weight of the representation (20) of R_5 is attached a specific value of M_L; moreover, the weight (2,0) corresponds to $M_L = 4$. This provides an alternative method of defining the highest weight of a representation: it is the point corresponding to the maximum value of M_L. The operators H_i of the groups U_{2l+1}, SU_{2l+1}, R_{2l+1}, and G_2 have been chosen so that the two definitions of the highest weight coincide.

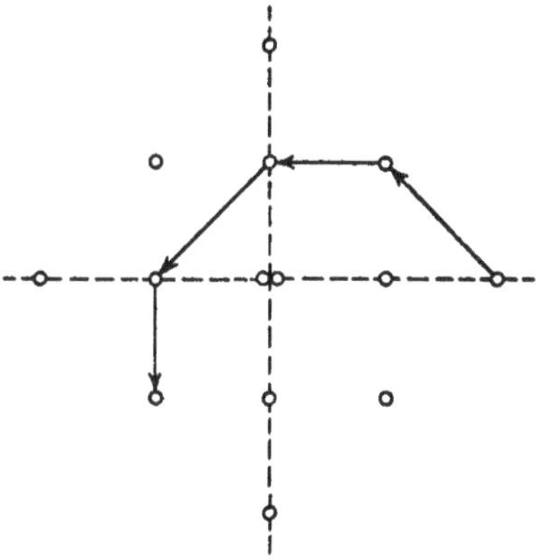

FIG. 5-5 The shift operators whose effects, shown here, correspond to the path

$$(2,0) \to (1,1) \to (0,1) \to (-1,0) \to (-1,-1)$$

produce the following changes in M_L:

$$4 \to 3 \to 1 \to -2 \to -3$$

as can be seen from the labels on the roots in Fig. 5-1.

Just as the operators H_i of a group are generalizations from R_3 of the operator L_z, so the operators E_α are the generalizations of the shift operators L_+ and L_-. The equations

$$H_i(E_\alpha u) = (E_\alpha H_i + \alpha_i E_\alpha)u = (m_i + \alpha_i)(E_\alpha u)$$

indicate that the weight of $E_\alpha u$ is $(m_1 + \alpha_1, m_2 + \alpha_2, \ldots)$. For R_3, a shift operator can act only in the one-dimensional space $M_L = L, L - 1, \ldots, -L$; in general, there are as many shift operators as there are roots. The walk within the representation (20) of R_5 produced by operating on $|^1G,4)$ in succession with the operators

$$\begin{gathered}(\tfrac{2}{5})^{\frac{1}{2}}V_{-1}{}^{(1)} + (\tfrac{3}{5})^{\frac{1}{2}}V_{-1}{}^{(3)} \\ V_{-2}{}^{(3)} \\ V_{-3}{}^{(3)} \\ (\tfrac{3}{5})^{\frac{1}{2}}V_{-1}{}^{(1)} - (\tfrac{2}{5})^{\frac{1}{2}}V_{-1}{}^{(3)} \\ \text{and} \qquad V_{-2}{}^{(3)}\end{gathered}$$

is illustrated in Fig. 5-5. The action of the last operator is to leave the array of weights, and the result is identically zero.

Suppose that u is the simultaneous eigenvector of the operators H_i of a group and that it corresponds to the highest weight $(l_1 l_2 \cdots)$ of an irreducible representation. Then for some shift operators we can be sure that $E_\alpha u = 0$. The corresponding roots α are called *posi-*

tive and are denoted by α^+. With our particular choice of operators H_i, all operators E_α comprising linear combinations of $V_q^{(k)}$ for which $q > 0$ are positive; for the highest weight corresponds to the maximum value of M_L. From Fig. 5-1, we see that the positive roots for R_5 are $\mathbf{e}_1 + \mathbf{e}_2$, $\mathbf{e}_1$, $\mathbf{e}_2$, and $\mathbf{e}_1 - \mathbf{e}_2$; if this four-vector structure is superimposed on the highest weight of any irreducible representation of R_5 (such as either of those given in Fig. 5-4), it does not lead to another point in the array.

The array of weights of a representation can be thought of as being developed from a single weight (e.g., the highest) by means of the shift operators E_α. Since these act in directions determined by the roots, it is clear that the array exhibits the symmetry of the root figure; or, to be more precise, the array is invariant with respect to reflections in the hyperplanes that pass through the origin and are perpendicular to the roots. This finite group is labeled S by Weyl.[62]

To conclude this section we consider the representations of U_{2l+1}. The infinitesimal operators for this group include all tensors $\mathbf{V}^{(k)}$ for which $0 \leq k \leq 2l$, and since they commute with $\mathbf{S}$, we anticipate that all the states in a configuration l^n with a given S and M_S form the basis for a single irreducible representation of U_{2l+1}. This expectation can be confirmed by examining the effect of the shift operators $X_{\nu\mu}$ on the determinantal product states. Furthermore, the equation

$$X_{ii}|l,m) = \delta(i,m)|l,m) \tag{5-39}$$

which the commuting operators X_{ii} obey, implies that any determinantal product state is a simultaneous eigenfunction of the l operators and hence corresponds to a definite weight. To see this, we can ignore the quantum numbers m_s and write a determinantal product state for the configuration l^n as $\{m_1m_2 \cdot\cdot\cdot m_n\}$. The equation

$$X_{ii}\{m_1m_2 \cdot\cdot\cdot m_n\} = \sum_j \delta(i,m_j)\{m_1m_2 \cdot\cdot\cdot m_n\}$$

shows that the weight corresponding to $\{m_1m_2 \cdot\cdot\cdot m_n\}$ is

$$\Big[\sum_j \delta(m_j,l), \sum_j \delta(m_j,l-1), \ . \ . \ . \ , \sum_j \delta(m_j, -l)\Big] \tag{5-40}$$

For example, $\{\overset{+}{2}\overset{+}{1}\overset{-}{1}-\overset{+}{2}-\overset{-}{2}-\overset{+}{3}\}$ of f^6 corresponds to the weight [0120021]. Strictly speaking, the appropriate linear combination of the states $\{m_1m_2 \cdot\cdot\cdot m_n\}$, with various plus and minus signs over the m_l values, should be taken to ensure that the state is an eigenfunction of S_z and $\mathbf{S}^2$.

The procedure for finding the irreducible representations of U_{2l+1} will become clear if the configuration f^6 is taken as an example. The maximum spin S is 3; if we select $M_S = 3$, there are seven determinantal product states of the type $\{\overset{+}{3}\overset{+}{2}\overset{+}{1}\overset{+}{0}-\overset{+}{2}-\overset{+}{3}\}$. Of these, $\{\overset{+}{3}\overset{+}{2}\overset{+}{1}\overset{+}{0}-\overset{+}{1}-\overset{+}{2}\}$ possesses the highest weight, namely, [1111110]. Since X_{ii} commutes with **S**, no higher weight can be found for states for which $S = 3$, $M_S < 3$. Thus a set of seven states for $S = 3$ and any value of M_S forms a basis for the irreducible representation [1111110] of U_7. If we now turn to states for which $S = 2$, $M_S = 2$, it is clear that [2111100] is the highest weight, corresponding to $\{\overset{+}{3}\overset{-}{3}\overset{+}{2}\overset{+}{1}\overset{+}{0}-\overset{+}{1}\}$. The highest weights for the states for which $S = 1$ and 0 are [2211000] and [2220000], respectively. No number greater than 2 can appear in the weights because of the Pauli exclusion principle. To summarize, all the states of f^6 belonging to a specific S and M_S form the basis for one of the irreducible representations [1111110], [2111100], [2211000], or [2220000] of U_7, according to whether $S = 3$, 2, 1, or 0. In general, $2l + 1$ integers $[\lambda_1\lambda_2 \cdots \lambda_{2l+1}]$ are required to specify the irreducible representation of U_{2l+1}; if we suppose that the first a of them are 2, the next b are 1, and all the rest are zeros, i.e., if

$$\begin{aligned} \lambda_1 &= \cdots = \lambda_a = 2 \\ \lambda_{a+1} &= \cdots = \lambda_{a+b} = 1 \\ \lambda_{a+b+1} &= \cdots = 0 \end{aligned}$$

then the corresponding values of S and n are given by

$$n = \sum_i \lambda_i \tag{5-41}$$

and

$$S = \tfrac{1}{2}b \tag{5-42}$$

It has been seen that the highest weights of the representations of R_5 and U_{2l+1} possess integral coefficients for the particular bases chosen. A general method for deciding the nature of the weights for the representations of a particular group has been given by Racah,[58] to which the reader is referred for details. Although we shall be concerned exclusively with integral representations of R_{2l+1}, we note here that half-integral representations of these groups exist and may be considered the generalizations of $\mathfrak{D}_{\frac{1}{2}}$, $\mathfrak{D}_{\frac{3}{2}}$, etc. The representation $(\frac{1}{2}\frac{1}{2})$ of R_5 is shown in Fig. 5-6; the array of weights is invariant with respect to the operations of the group S, and all the weights can be reached from a single weight by means of the shift operators. If we adopt the coordinate scheme for G_2 discussed in Sec. 5-6, it can be shown that the components u_1 and u_2 of the representations (u_1u_2) of G_2 are always integers. The representation (11) of G_2 is shown in Fig. 5-7. For SU_{2l+1} the components λ'_i of a weight $(\lambda'_1\lambda'_2 \cdots \lambda'_{2l+1})$ are fractions which differ by integers and have the denominator $2l + 1$. This follows immediately when it is recognized that Eq. (5-39) is replaced by

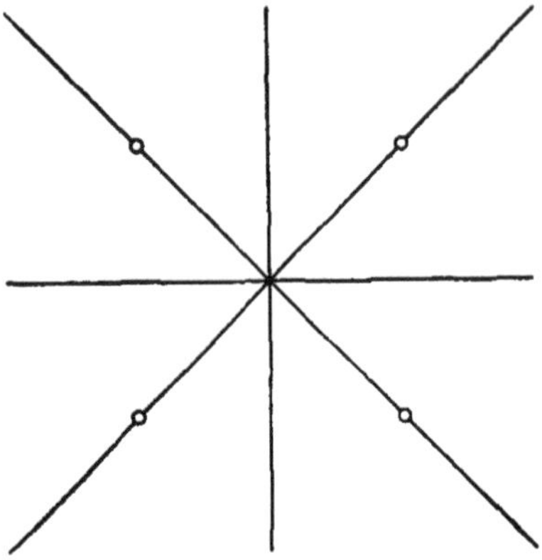

FIG. 5-6 The array of weights for the four-dimensional representation $(\frac{1}{2}\frac{1}{2})$ of R_5, superposed on the root figure.

$$X'_{ii}|l,m) = \{X_{ii} - [l]^{-\frac{1}{2}} V_0^{(0)}\}|l, m)$$
$$= \{\delta(i,m) - [l]^{-1}\}|l,m)$$

The eigenfunctions that form a basis for the irreducible representation $[\lambda_1\lambda_2 \cdots \lambda_{2l+1}]$ of U_{2l+1} also form a basis for the irreducible representation $(\lambda'_1\lambda'_2 \cdots \lambda'_{2l+1})$ of SU_{2l+1}, where

$$\lambda'_i = \lambda_i - n[l]^{-1} \tag{5-43}$$

It is clear that the sum of the components λ'_i is zero. The representation [100] of U_3 and $(\frac{2}{3},-\frac{1}{3},-\frac{1}{3})$ of SU_3 share the same basis; the weights of the latter are shown in Fig. 5-8.

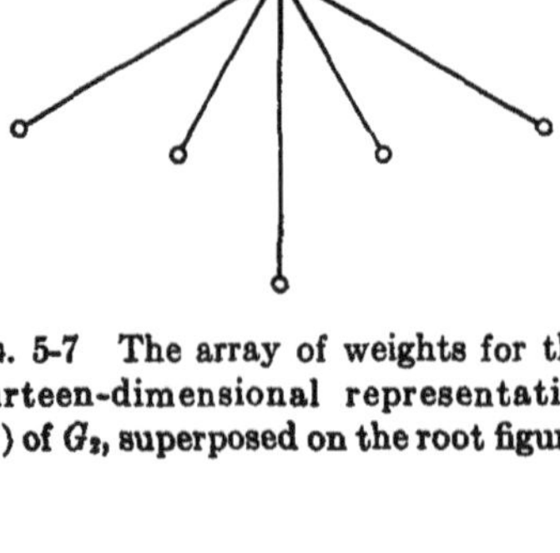

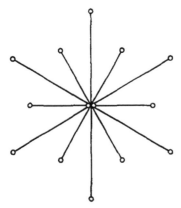

FIG. 5-7 The array of weights for the fourteen-dimensional representation (11) of G_2, superposed on the root figure.

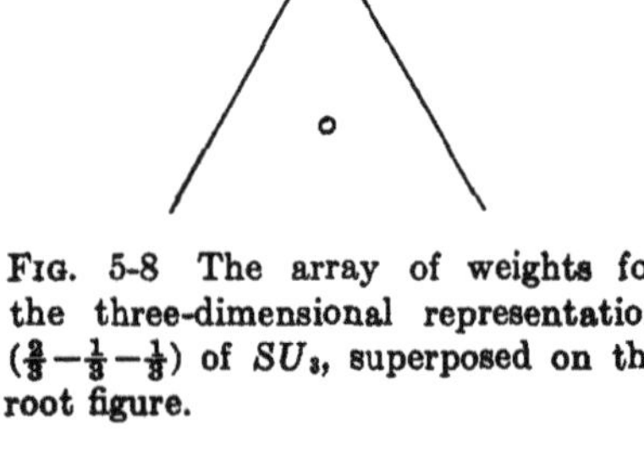

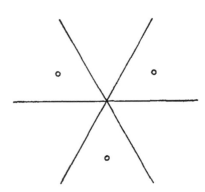

FIG. 5-8 The array of weights for the three-dimensional representation $(\frac{2}{3}-\frac{1}{3}-\frac{1}{3})$ of SU_3, superposed on the root figure.

5-8 THE YOUNG TABLEAUX

Although the many-electron eigenfunctions have been investigated with regard to their properties under the operations of such groups as R_3, R_5, U_{2l+1}, and SU_{2l+1}, yet their transformation properties with respect to an interchange of two electrons, which admit of a group-theoretical description, have not yet been considered. Of course, the Pauli exclusion principle demands that the complete eigenfunction be antisymmetric with respect to the interchange of any two electrons. There is, however, the question of the symmetry properties of the separate spin and orbital functions. For two-electron configurations, the factorization of any component $|SM_SLM_L)$ of a term into a spin and an orbital part enables the question to be easily answered. If $S = 1$, the spin function is symmetric, being $\alpha_1\alpha_2$, $\alpha_1\beta_2 + \alpha_2\beta_1$, or $\beta_1\beta_2$ according as $M_S = 1$, 0, or -1; in this case the orbital function is antisymmetric. If $S = 0$, the spin function is the antisymmetric form $\alpha_1\beta_2 - \beta_1\alpha_2$ and the orbital part is symmetric.

In general, the component of a term deriving from a configuration l^n, where $n > 2$, cannot be simply factored into a spin and an orbital part but instead can be expressed as a linear combination of spin and orbital functions which possess special symmetry properties with respect to the interchange of two electrons. To describe the construction of these functions, we first define a *partition* $[\mu_1\mu_2 \cdots \mu_r]$ as a set of integers μ_i for which

$$\mu_1 \geq \mu_2 \geq \cdots \geq \mu_r \geq 0 \tag{5-44}$$

and

$$\sum_i \mu_i = n \tag{5-45}$$

The partition can be visualized as an arrangement of n cells in r rows; each row begins from the same vertical line, and the number of cells in successive rows is μ_1, μ_2, μ_3, etc. The inequality (5-44) ensures that no row overshoots the one above it. This representation of a partition is called a *shape*. If the numbers 1, 2, . . . , n are inserted in the cells, we obtain a *tableau;* if, further, the numbers increase on going down columns and also from left to right along rows, the tableau is said to be *standard*. Thus, corresponding to the partition [211], there are just three standard tableaux, namely,

1	2
3	
4	

1	3
2	
4	

1	4
2	
3	

If we have a function $f(\mathbf{x}_1,\mathbf{x}_2, \ldots ,\mathbf{x}_n)$ of n particles, then we can construct a completely symmetric function by adding the $n!$ functions formed by permuting the n suffices. Thus from the product $\xi_1\eta_2\zeta_3$, we can construct the completely symmetric function

$$\xi_1\eta_2\zeta_3 + \xi_2\eta_1\zeta_3 + \xi_3\eta_2\zeta_1 + \xi_1\eta_3\zeta_2 + \xi_2\eta_3\zeta_1 + \xi_3\eta_1\zeta_2$$

If, however, the factor $(-1)^p$ is inserted before the terms in the sum, where p is the parity of the permutation, a completely antisymmetric function is formed. For example,

$$\xi_1\eta_2\zeta_3 - \xi_2\eta_1\zeta_3 - \xi_3\eta_2\zeta_1 - \xi_1\eta_3\zeta_2 + \xi_2\eta_3\zeta_1 + \xi_3\eta_1\zeta_2$$

is totally antisymmetric. Corresponding to any tableau, we can symmetrize with respect to the numbers in each row and *then* antisymmetrize with respect to the numbers in each column. In this way functions can be constructed that are neither completely symmetric nor completely antisymmetric. As an example we take the product $\xi_1\xi_2\eta_3\zeta_4$ and construct the functions corresponding to the three standard tableaux given above. The product is already symmetric with respect to 1 and 2; antisymmetrization with respect to 1, 3, and 4 gives the function

$$\Theta_1 = \xi_1\xi_2\eta_3\zeta_4 + \xi_3\xi_2\eta_4\zeta_1 + \xi_4\xi_2\eta_1\zeta_3 - \xi_1\xi_2\eta_4\zeta_3 - \xi_4\xi_2\eta_3\zeta_1 - \xi_3\xi_2\eta_1\zeta_4$$

This function corresponds to the first tableau. Turning now to the second tableau, the symmetrization with respect to 1 and 3 yields

$$\xi_1\xi_2\eta_3\zeta_4 + \xi_3\xi_2\eta_1\zeta_4$$

If we now antisymmetrize with respect to 1, 2, and 4, the first part is identically zero and the second gives

$$\Theta_2 = \xi_3\xi_2\eta_1\zeta_4 + \xi_3\xi_1\eta_4\zeta_2 + \xi_3\xi_4\eta_2\zeta_1 - \xi_3\xi_2\eta_4\zeta_1 - \xi_3\xi_1\eta_2\zeta_4 - \xi_3\xi_4\eta_1\zeta_2$$

Similarly, the function corresponding to the third tableau is

$$\Theta_3 = \xi_4\xi_2\eta_3\zeta_1 + \xi_4\xi_3\eta_1\zeta_2 + \xi_4\xi_1\eta_2\zeta_3 - \xi_4\xi_2\eta_1\zeta_3 - \xi_4\xi_1\eta_3\zeta_2 - \xi_4\xi_3\eta_2\zeta_1$$

As a second example, we consider the function $\alpha_1\alpha_2\alpha_3\beta_4$ and the three standard tableaux

1	3	4
2		

1	2	4
3		

1	2	3
4		

Functions possessing the symmetries imposed by these tableaux are

$$\Phi_1 = \alpha_1\alpha_4\alpha_3\beta_2 - \alpha_2\alpha_3\alpha_4\beta_1$$
$$\Phi_2 = \alpha_1\alpha_2\alpha_4\beta_3 - \alpha_2\alpha_3\alpha_4\beta_1$$

and

$$\Phi_3 = \alpha_1\alpha_2\alpha_3\beta_4 - \alpha_2\alpha_3\alpha_4\beta_1$$

It is to be noticed that the function Φ_k corresponding to a particular tableau is not simultaneously symmetric with respect to the rows and antisymmetric with respect to the columns. However, the result of permuting electrons i and j, which we may suppose is carried out by the operator P_{ij}, can be expressed as a linear combination of the other functions Φ_h. Thus

$$\begin{array}{lll} P_{12}\Phi_1 = -\Phi_1 & P_{12}\Phi_2 = \Phi_2 - \Phi_1 & P_{12}\Phi_3 = -\Phi_1 + \Phi_3 \\ P_{23}\Phi_1 = \Phi_2 & P_{23}\Phi_2 = \Phi_1 & P_{23}\Phi_3 = \Phi_3 \\ P_{34}\Phi_1 = \Phi_1 & P_{34}\Phi_2 = \Phi_3 & P_{34}\Phi_3 = \Phi_2 \end{array}$$

It is not difficult to show that any permutation on the four electrons can be expressed in terms of P_{12}, P_{23}, and P_{34}; consequently every operation of the group S_4, when applied to one of the functions Φ_i, can be expressed as a linear combination of the three of them. Thus Φ_1, Φ_2, and Φ_3 form a basis for a representation of S_4. To find the representation, we select one of the permutations from each of the five classes of S_4, which are given explicitly in Sec. 2-3. The characters turn out to be 3, 1, -1, 0, -1; from Table 2-1 we deduce that the three functions Φ_k form a basis for the irreducible representation Γ_5 of S_4. Similarly, it can be shown that the three functions Θ_k form a basis for the irreducible representation Γ_4 of S_4.

There remain three other partitions for $n = 4$, namely,

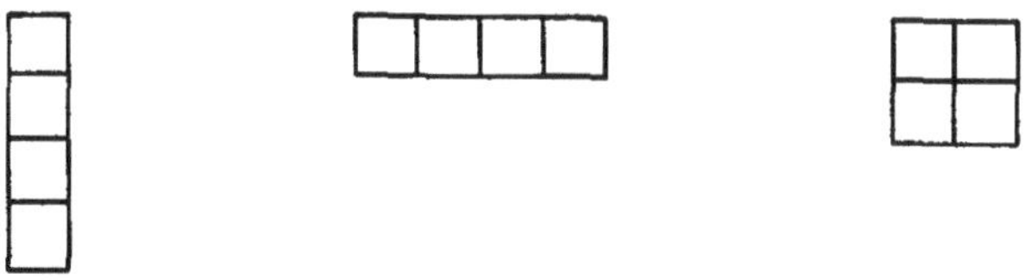

Only one standard tableau can be constructed from the first shape; the function is totally antisymmetric and forms by itself a basis for the one-dimensional irreducible representation Γ_2 of S_4. The second shape corresponds to totally symmetric functions such as $\alpha_1\alpha_2\alpha_3\alpha_4$; any one of these forms a basis for the representation Γ_1. Finally, it can be shown that the two standard tableaux corresponding to the partition [22] give rise to two functions which together form a basis for the irreducible representation Γ_3. The generalization of these results can be expressed as follows:

The partitions $[\mu_1\mu_2 \cdots \mu_r]$, where $\Sigma\mu_i = n$, each correspond to a unique irreducible representation of the group S_n. The dimension of a specific irreducible representation is equal to the number of standard tableaux that can be constructed from the corresponding partition; and a basis for the irreducible representation can be constructed by a process of symmetrization with respect to the rows and antisymmetrization with respect to the columns.

These statements will not be proved here; the reader is referred to the book by Rutherford,[66] where the methods of Young, after whom the tableaux are named, are described in detail. It should be pointed out that the procedure for deriving the basis functions is not unique; for we could equally well antisymmetrize with respect to the columns before symmetrizing with respect to the rows. Yamanouchi[67] has described a method that ensures that the matrices representing P_{ij} are orthogonal.

We are now in a position to apply the theory to many-electron eigenfunctions, and we take the configuration f^4 as an example. The components of terms of maximum multiplicity can be factorized into two parts, e.g.,

$$|{}^5I\ M_S = 2,\ M_L = 6) = \{\overset{++++}{3210}\} = \{3210\}\alpha_1\alpha_2\alpha_3\alpha_4$$

The spin part is totally symmetric and corresponds to the partition [4]; the orbital part is totally antisymmetric and corresponds to [1111]. Other components of ^5I can be obtained by applying the appropriate shift operators to $|{}^5I,2,6)$; in all cases the states can be factorized into a spin part and an orbital part. The same is true of the other quintet terms of f^4. States with $S = 1$, for example,

$$|{}^3M,1,9) = \{\overset{+-++}{3321}\}$$

cannot be broken up in so simple a way. However, if we write

$$\psi(f,\ m_l = 3) = \xi \qquad \psi(f,\ m_l = 2) = \eta \qquad \psi(f,\ m_l = 1) = \zeta$$

then it is a matter of elementary algebra to show that

$$\{\overset{+-++}{3321}\} = (4!)^{-\frac{1}{2}}(\Theta_1\Phi_1 + \Theta_2\Phi_2 + \Theta_3\Phi_3) \qquad (5\text{-}46)$$

The Kronecker product $\Gamma_4 \times \Gamma_5$ has as its basis the nine functions $\Theta_i\Phi_k$; the linear combination (5-46) is the part that transforms according to Γ_2. The two tableaux that are used to construct Θ_i and its companion Φ_i are *adjoint* in the sense that one can be obtained from the other simply by replacing rows by columns. This result, which is true in general, ensures that the resulting linear combination is totally antisymmetric. The existence of only two spin functions makes it impossible to construct from them a function that is antisymmetrical with respect to more than two electrons; for the antisymmetrization procedure gives identically zero in these cases. Consequently the tableaux for the spin functions are limited to two rows at most. The tableau given by a single row of cells always corresponds to the maximum spin S for a configuration; as cells are removed from this row and placed in a second row, the resulting tableaux define spin functions with ever-decreasing S. For f^4, the partitions [4], [31], and [22] correspond to $S = 2$, 1, and 0, respectively. The symmetries of the associated orbital functions are defined by the adjoint tableaux; thus the orbital functions derived from the partitions [1111], [211], and [22] correspond to $S = 2$, 1, and 0. In general, the partition $[\mu_1\mu_2 \cdot \cdot \cdot \mu_r]$, where

$$\mu_1 = \cdot \cdot \cdot = \mu_a = 2$$
$$\mu_{a+1} = \cdot \cdot \cdot = \mu_{a+b} = 1$$

gives rise to orbital functions that must be combined with the spin functions for which $S = \frac{1}{2}b$ in order that the complete eigenfunction may be totally antisymmetric. The association between the partition $[\mu_1\mu_2 \cdot \cdot \cdot \mu_r]$ in the orbital space and the spin S is virtually identical to the association between the irreducible representation $[\lambda_1\lambda_2 \cdot \cdot \cdot \lambda_{2l+1}]$ of U_{2l+1} and the spin S which is established in Sec. 5-7; in fact, if $2l + 1 - r$ zeros are added to the partition, we may regard the resulting symbol $[\mu_1\mu_2 \cdot \cdot \cdot \mu_r00 \cdot \cdot \cdot 0]$ as defining either the partition that determines the way the orbital functions transform under the operations of S_n or the irreducible representation of U_{2l+1} for which they serve as a basis. Just as the existence of two spin functions limits the spin tableaux to two rows, so the existence of $2l + 1$ orbital functions $\psi(l,m)$ limits the orbital tableaux to $2l + 1$ rows. This ensures that $2l + 1 - r$ is not negative, and that the connection between partitions and irreducible representations of U_{2l+1} is always possible. The value of this connection is that many results that have been derived for the tableaux can be immediately taken over and interpreted in terms of the irreducible representations of U_{2l+1}.

5-9 CASIMIR'S OPERATOR

Owing to the definition of the structure constants given in Eq. (5-8), $c_{\sigma\rho}{}^{\tau} = -c_{\rho\sigma}{}^{\tau}$. It can also be shown that the tensors $c_{\rho}{}^{\tau\lambda} = g^{\sigma\lambda}c_{\sigma\rho}{}^{\tau}$

satisfy the equation $c_\rho{}^{\tau\lambda} = -c_\rho{}^{\lambda\tau}$. Hence

$$\begin{aligned}[g^{\rho\sigma}X_\rho X_\sigma,X_\tau] &= g^{\rho\sigma}X_\rho[X_\sigma,X_\tau] + g^{\rho\sigma}[X_\rho,X_\tau]X_\sigma \\ &= g^{\rho\sigma}c_{\sigma\tau}{}^{\mu}X_\rho X_\mu + g^{\rho\sigma}c_{\rho\tau}{}^{\nu}X_\nu X_\sigma \\ &= (c_\tau{}^{\mu\rho} + c_\tau{}^{\rho\mu})X_\rho X_\mu = 0\end{aligned}$$

The operator

$$G = g^{\rho\sigma}X_\rho X_\sigma$$

commutes with all infinitesimal operators X_τ of the group and is called Casimir's operator.[68] It is the natural extension of the operator $\mathbf{L}^2$ of R_3, which commutes with L_x, L_y, and L_z.

The formulas for the metric tensor given in Sec. 5-5 permit Casimir's operator to be rapidly expressed in terms of tensor operators for any group of interest to us. Thus, for R_{2l+1}, Eq. (5-25) gives

$$\begin{aligned}G(R_{2l+1}) &= \sum_{q,\ \text{odd}\ k} (-1)^q V_q{}^{(k)} V_{-q}{}^{(k)} (2l-1)^{-1} \\ &= (2l-1)^{-1} \sum_{\text{odd}\ k} (\mathbf{V}^{(k)})^2 \qquad (5\text{-}47)\end{aligned}$$

Similarly,

$$G(SU_{2l+1}) = \tfrac{1}{2}[l]^{-1} \sum_{k>0} (\mathbf{V}^{(k)})^2 \qquad (5\text{-}48)$$

and

$$G(G_2) = \tfrac{1}{4}[(\mathbf{V}^{(1)})^2 + (\mathbf{V}^{(5)})^2] \qquad (5\text{-}49)$$

Equations (5-47) and (5-49) have been given, in a slightly different form, by Racah.[69]

Although the formulas for G given above are simple functions of the tensors $\mathbf{V}^{(k)}$, they are not suitable for finding the eigenvalues of G. For R_{2l+1}, we make the substitution

$$V_q{}^{(k)} = \frac{1}{2} \sum_{\nu,\mu} (-1)^{l-\nu}[k]^{\frac{1}{2}} \begin{pmatrix} l & k & l \\ -\nu & q & \mu \end{pmatrix} W_{\nu\mu}$$

which follows immediately from Eq. (5-32). Casimir's operator becomes

$$G(R_{2l+1}) = \tfrac{1}{4}(2l-1)^{-1} \sum_{\nu,\mu} W_{\nu\mu}W_{\mu\nu}$$

The terms $W_{\nu\mu}$ are of two kinds. Those for which $\nu = \mu$ commute with one another and are of the type H_i; those for which $\nu \neq \mu$ are the generalizations of E_α and correspond to a particular root $\boldsymbol{\alpha}$. Suppose that the operator $G(R_{2l+1})$ is applied to an eigenfunction of l^n, which, with various other eigenfunctions, forms a basis for the irreducible representation $(w_1w_2 \cdots w_l)$ of R_{2l+1}. Since G commutes with all elements of the group, it is immaterial, for the purpose of calculating its eigenvalues, which eigenfunction of the set is chosen;

however, it is greatly to our advantage to select that particular eigenfunction, say, u, which has the highest weight. Our aim is to express G as a function of operators of the type W_{ii}, for then we can make use of the equations

$$H_i u = W_{l+1-i,l+1-i} u = w_i u \qquad i = 1, 2, \ldots, l$$

To this end, we write

$$G(R_{2l+1}) = \tfrac{1}{2}(2l-1)^{-1}\{\sum_i W_{ii}^2 + \sum_{\nu\neq\mu} W_{\nu\mu}W_{\mu\nu}\}$$

Only the second sum presents any difficulty. We first observe that the root corresponding to $W_{\nu\mu}$ is positive if $\nu > \mu$. Hence

$$\sum_{\nu\neq\mu} W_{\nu\mu}W_{\mu\nu}u = \sum_{\nu>\mu} W_{\nu\mu}W_{\mu\nu}u = \sum_{\nu>\mu} [W_{\nu\mu},W_{\mu\nu}]u$$

Unless $\nu = -\mu$, in which case the commutator vanishes, we find

$$[W_{\nu\mu},W_{\mu\nu}] = W_{\nu\nu} - W_{\mu\mu}$$

and so $$\sum_{\nu\neq\mu} W_{\nu\mu}W_{\mu\nu}u = \sum_{\nu>\mu,\ \nu\neq-\mu} (W_{\nu\nu} - W_{\mu\mu})u$$
$$= 2\sum_{\nu>0} (2\nu - 1)W_{\nu\nu}u$$
$$= 2\sum_{i=1}^{l} (2l - 2i + 1)H_i u$$

Therefore $$G(R_{2l+1})u = \tfrac{1}{2}(2l-1)^{-1}\sum_{i=1}^{l} [w_i^2 + (2l-2i+1)w_i]u$$
$$= \tfrac{1}{2}(2l-1)^{-1}\sum_{i=1}^{l} w_i(w_i + 1 + 2l - 2i)u \qquad (5\text{-}50)$$

This procedure can be carried out for other groups. The eigenvalue of $G(SU_{2l+1})$ for any eigenfunction of the set that forms a basis for the irreducible representation $(\lambda_1'\lambda_2' \cdots \lambda_{2l+1}')$ of SU_{2l+1} is found to be

$$\tfrac{1}{2}[l]^{-1}\sum_{i=1}^{2l+1} \lambda_i(\lambda_i + 2 + 2l - 2i) - \tfrac{1}{2}n^2[l]^{-2}$$

where λ_i is related to λ_i' by Eq. (5-43). If the shape $[\lambda_1\lambda_2 \cdots]$ is interpreted in terms of n and S through Eqs. (5-41) and (5-42), this expression assumes the more convenient form

$$\tfrac{1}{2}[l]^{-1}\{3n + 2nl - \tfrac{1}{2}n^2 - 2S(S+1)\} - \tfrac{1}{2}n^2[l]^{-2} \qquad (5\text{-}51)$$

Racah[58] has expressed the eigenvalues of Casimir's operator for the irreducible representation $(l_1l_2 \cdots)$ of a group in the form

$\mathbf{K}^2 - \mathbf{R}^2$, where

$$\mathbf{R} = \frac{1}{2} \sum_{\alpha^+} \alpha \tag{5-52}$$

and K is a vector having the covariant components

$$K_i = R_i + l_i \tag{5-53}$$

These results can be obtained by a generalization of the techniques leading to Eq. (5-50); they circumvent the explicit construction of Casimir's operator in tensor-operator form. Take, for example, the irreducible representation (u_1u_2) of G_2. Summing the roots corresponding to $q > 0$, we obtain, in the oblique coordinate scheme associated with the operators (5-35),

$$\mathbf{R} = 2(\mathbf{e}_1 - \mathbf{e}_3) + (\mathbf{e}_1 - \mathbf{e}_2)$$

corresponding to the point (2,1). The coordinates defining $\mathbf{K}$ are $(u_1 + 2, u_2 + 1)$. From Eq. (5-36),

$$g^{ij} = \begin{pmatrix} \frac{1}{12} & \frac{1}{24} \\ \frac{1}{24} & \frac{1}{12} \end{pmatrix}$$

and the eigenvalues of Casimir's operator are given by

$$\begin{aligned} \mathbf{K}^2 - \mathbf{R}^2 &= g^{ij}(K_iK_j - R_iR_j) \\ &= (u_1^2 + u_2^2 + u_1u_2 + 5u_1 + 4u_2)/12 \end{aligned} \tag{5-54}$$

5-10 THE CHARACTERS OF REPRESENTATIONS

Two important topics in the theory of continuous groups have still to be considered: these are the Kronecker product of two representations and the branching rules for determining how an irreducible representation of a group breaks up when the transformations are limited to a subgroup. For finite groups, these problems are most readily solved by introducing the character of a representation, and an analogous procedure is available to us here. As basis functions for a representation of a continuous group, we may choose the quantities $u(\gamma,\mathbf{m})$, each of which corresponds to a definite weight $\mathbf{m} = (m_1,m_2, \ldots)$. The symbol γ is required to distinguish functions possessing the same weight [such as $|d^2\,{}^1D,0)$ and $|d^2\,{}^1G,0)$ for the irreducible representation (20) of R_5]. The equations

$$S_a u(\gamma,\mathbf{m}) = (1 + \delta a^\sigma X_\sigma)u(\gamma,\mathbf{m}) = \sum_{\gamma',\mathbf{m}'} C(\gamma,\mathbf{m};\gamma',\mathbf{m}')u(\gamma',\mathbf{m}')$$

define matrices C that form the natural extension of a matrix representation for the elements of a finite group. In the calculation of the characters of a representation of R_3, we are able to simplify matters by choosing the axis of rotation to be the z axis, so that the matrices become diagonal. Since the operators E_α are shift operators, the assumption that C is diagonal restricts our attention to infinitesimal operators of the type H_i, which commute with all elements of the group. The equations

$$H_i u(\gamma,\mathbf{m}) = m_i u(\gamma,\mathbf{m}) \tag{5-55}$$

enable the diagonal elements of C to be rapidly found; however, it is at once apparent that a symmetry of the weights about the origin, such as is displayed in Fig. 5-4, produces entries on the diagonal which, when the trace is formed, cancel in pairs. The character of an irreducible representation of a group for which this occurs is equal to the dimension of the representation and is useless for the purposes we have in mind.

This difficulty can be overcome by considering finite rather than infinitesimal transformations. The effect of n successive operations can be reproduced by the operator

$$(1 + \delta a^j H_j)^n$$

Equation (5-15), combined with the fact that H_j is a linear combination of tensor operators for which $q = 0$, implies that all quantities δa^j are purely imaginary. In the simultaneous limits $n \to \infty$, $\delta a^j \to 0$, we may suppose that the product $n\delta a^j$ becomes $i\varphi^j$, where φ^j is real. At the same time,

$$(1 + \delta a^j H_j)^n \to \exp\,(i\varphi^j H_j)$$

The sum over j is implicit. To make the algebra easier, it is sometimes convenient to relabel the parameters φ^j. For R_{2l+1}, we write $\varphi^j = \theta^{(l+1-j)}$; for SU_{2l+1}, $\varphi^j = \omega^{(l+1-j)}$. We note that the operator $\exp\,(i\varphi^j H_j)$ is the generalization of $\exp\,(i\alpha J_z)$ of Sec. 2-5. Thus, for R_{2l+1}, we find

$$\exp\left(-i \sum_{j=1}^{l} \theta^{(j)} W_{jj}\right)(x_m \pm iy_m) \exp\left(i \sum_{j=1}^{l} \theta^{(j)} W_{jj}\right) = (x_m \pm iy_m) \exp\,(\mp i\theta^{(m)})$$

$$\exp\left(-i \sum_{j=1}^{l} \theta^{(j)} W_{jj}\right) z \exp\left(i \sum_{j=1}^{l} \theta^{(j)} W_{jj}\right) = z$$

where x_m, y_m, and z are defined as in Eqs. (5-17). These are the generalizations to R_{2l+1} of Eqs. (2-15).

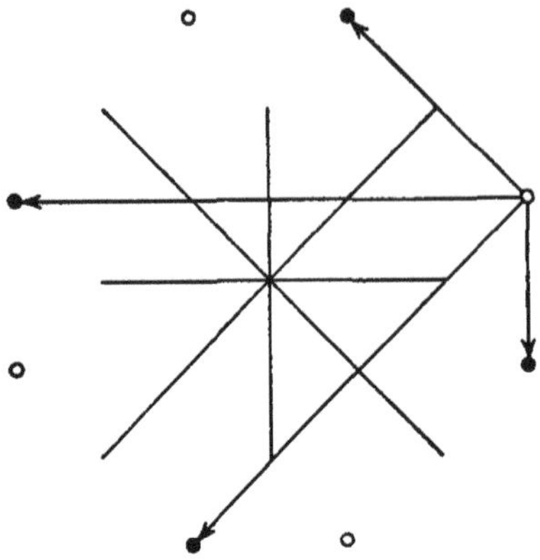

FIG. 5-9 Under single operations of Weyl's group S for R_5, the weight $(\frac{3}{2},\frac{1}{2})$ can be carried into the weights $(\frac{1}{2},\frac{3}{2})$, $(\frac{3}{2},-\frac{1}{2})$, $(-\frac{3}{2},\frac{1}{2})$, and $(-\frac{1}{2},-\frac{3}{2})$ Weights that require an even number of operations to be reached are represented by open circles.

It is now a simple matter to compute the character χ. For the irreducible representation $(l_1l_2 \cdot \cdot \cdot)$,

$$\chi(l_1l_2 \cdot \cdot \cdot) = \Sigma \exp (i\varphi^j m_j)$$

where the sum runs over all weights of the representation $(l_1l_2 \cdot \cdot \cdot)$. Thus, for the irreducible representation (11) of R_5,

$$\begin{aligned}\chi(11) &= e^{i\theta'+i\theta''} + e^{i\theta'-i\theta''} \\ &\quad + e^{-i\theta'+i\theta''} + e^{-i\theta'-i\theta''} + e^{i\theta'} \\ &\quad + e^{-i\theta'} + e^{i\theta''} + e^{-i\theta''} + 2 \end{aligned} \qquad (5\text{-}56)$$

where $\theta' = \theta^{(1)}$ and $\theta'' = \theta^{(2)}$.

The expressions for the characters of R_3 corresponding to Eq. (5-56) are geometric series and can be easily summed [see Eq. (2-12)]. Weyl[62] has shown that analogous simplifications can be carried out for the characters of any semi-simple group. His formula is

$$\chi(l_1l_2 \cdot \cdot \cdot) = \frac{\xi(\mathbf{K})}{\xi(\mathbf{R})} \qquad (5\text{-}57)$$

where $\mathbf{K}$ and $\mathbf{R}$ are defined as in Eqs. (5-53) and (5-52), and where

$$\xi(\mathbf{M}) = \sum_\sigma (-1)^p \exp [i(\mathcal{R}_\sigma \mathbf{M})_j \varphi^j]$$

The operators $\mathcal{R}_\sigma$ are elements of the group S, mentioned in Sec. 5-7, and they carry the point $\mathbf{M} = (M_1, M_2, \ldots)$ to the positions that can be reached by any number of reflections in the hyperplanes through the origin perpendicular to the roots. The parity p is $+1$ or -1 according as the number of reflections is even or odd. For R_5, $\mathbf{R} = \frac{1}{2}(3\mathbf{e}_1 + \mathbf{e}_2)$, corresponding to the point $(\frac{3}{2},\frac{1}{2})$ in the weight space. The points that are linked by the operations of S are $(\pm\frac{3}{2},\pm\frac{1}{2})$, $(\pm\frac{1}{2},\pm\frac{3}{2})$, with any combination of signs; they are shown in Fig. 5-9. For the group R_{2l+1}, it is obvious that the effect of reflecting any point

$$\mathbf{M} = \sum_j a_j \mathbf{e}_j$$

in the hyperplane perpendicular to the root $\mathbf{e}_k$ is simply to reverse the sign of the coefficient a_k. Reflection in the hyperplane perpendicular to either $\mathbf{e}_i + \mathbf{e}_k$ or $\mathbf{e}_i - \mathbf{e}_k$ can be found by writing

$$a_i\mathbf{e}_i + a_k\mathbf{e}_k = \tfrac{1}{2}(a_i + a_k)(\mathbf{e}_i + \mathbf{e}_k) + \tfrac{1}{2}(a_i - a_k)(\mathbf{e}_i - \mathbf{e}_k)$$

In the first case, we note that

$$\begin{aligned}-\tfrac{1}{2}(a_i + a_k)(\mathbf{e}_i + \mathbf{e}_k) &+ \tfrac{1}{2}(a_i - a_k)(\mathbf{e}_i - \mathbf{e}_k) \\ &= \tfrac{1}{2}(-a_k - a_i)(\mathbf{e}_i + \mathbf{e}_k) + \tfrac{1}{2}(-a_k + a_i)(\mathbf{e}_i - \mathbf{e}_k)\end{aligned}$$

and so the effect of the reflection is to interchange the coefficients a_i and a_k and to reverse their signs. In the second case, a_i and a_k are interchanged, but their signs are not reversed. Thus the operations of the group S for R_{2l+1} carry $\mathbf{M}$ into those positions which are given by permuting the coefficients a_j, with all possible changes of sign.

With this preliminary point disposed of, we can proceed to calculate $\xi(\mathbf{R})$ for R_{2l+1}. The positive roots are all of the type $\mathbf{e}_k$, $\mathbf{e}_i \pm \mathbf{e}_k$ $(i < k)$, and

$$\mathbf{R} = \frac{1}{2}\sum_{j=1}^{l} (2l + 1 - 2j)\mathbf{e}_j$$

The first member of $\xi(\mathbf{R})$, corresponding to $\mathcal{R}_\sigma = \mathcal{I}$, is

$$\exp\left[\tfrac{1}{2}i \sum_{j=1}^{l} (2l + 1 - 2j)\theta^{(l+1-j)}\right]$$

If we denote by $\det|a_{jk}|$ the determinant whose element in the jth row and kth column is a_{jk}, the entire sum of $2^l l!$ terms can be engineered into the form

$$\begin{aligned}\xi(\mathbf{R}) = \det|&\exp\left[\tfrac{1}{2}i(2l + 1 - 2k)\theta^{(l+1-j)}\right] \\ &- \exp\left[-\tfrac{1}{2}i(2l + 1 - 2k)\theta^{(l+1-j)}\right]|\end{aligned} \qquad (5\text{-}58)$$

The function $\xi(\mathbf{K})$, corresponding to the irreducible representation $(w_1w_2 \cdots w_l)$ of R_{2l+1}, is very similar,

$$\begin{aligned}\xi(\mathbf{K}) = \det|&\exp\left[i(w_k + l - k + \tfrac{1}{2})\theta^{(l+1-j)}\right] \\ &- \exp\left[-i(w_k + l - k + \tfrac{1}{2})\theta^{(l+1-j)}\right]|\end{aligned} \qquad (5\text{-}59)$$

The character of $(w_1w_2 \cdots w_l)$ is simply $\xi(\mathbf{K})/\xi(\mathbf{R})$. If, for example, $l = 2$ and $(w_1w_2) = (11)$, then

$$\chi(11) = \frac{\begin{vmatrix} e^{5i\theta''/2} - e^{-5i\theta''/2} & e^{3i\theta''/2} - e^{-3i\theta''/2} \\ e^{5i\theta'/2} - e^{-5i\theta'/2} & e^{3i\theta'/2} - e^{-3i\theta'/2} \end{vmatrix}}{\begin{vmatrix} e^{3i\theta''/2} - e^{-3i\theta''/2} & e^{i\theta''/2} - e^{-i\theta''/2} \\ e^{3i\theta'/2} - e^{-3i\theta'/2} & e^{i\theta'/2} - e^{-i\theta'/2} \end{vmatrix}} \tag{5-60}$$

It is straightforward to show that this equation is identical to Eq. (5-56).

The group SU_{2l+1} is somewhat easier to deal with. We find

$$\mathbf{R} = \sum_{j=1}^{2l+1} (l + 1 - j)\mathbf{e}_j$$

and the leading term of $\xi(\mathbf{R})$ is

$$\exp\left[i \sum_{j=1}^{2l+1} (l + 1 - j)\omega^{(l+1-j)}\right]$$

The effect of reflecting a point $\mathbf{M} = (M_1, M_2, \ldots)$ in the hyperplane perpendicular to the root $\mathbf{e}_i - \mathbf{e}_k$ is simply to interchange the components M_i and M_k; consequently

$$\xi(\mathbf{R}) = \det|\exp[i(l + 1 - k)\omega^{(l+1-j)}]|$$

and

$$\chi(\lambda_1'\lambda_2' \cdots \lambda_{2l+1}') = \frac{\det|\exp[i(l + 1 - k + \lambda_k')\omega^{(l+1-j)}]|}{\det|\exp[i(l + 1 - k)\omega^{(l+1-j)}]|} \tag{5-61}$$

The character of the matrix representing the identity element is nothing else but the dimension of the representation. Equations such as (5-61) may be used to compute the dimension $D(l_1l_2 \cdots)$ as a function of the components of the highest weight $(l_1l_2 \cdots)$; however, the procedure for doing this requires some care, since the direct substitution $\omega^{(l+1-j)} = 0$ gives the indeterminate expression 0/0. In Eq. (5-61) we may first set $\omega^{(l+1-j)} = (j - 1)\omega$, and both determinants become alternants, e.g.,

$$\det|\exp[i(j - 1)(l + 1 - k + \lambda_k')\omega]| = \prod_{h>k=1}^{2l+1} \{\exp[i(l + 1 - k + \lambda_k')\omega] - \exp[i(l + 1 - h + \lambda_h')\omega]\}$$

In the limit $\omega \to 0$, this expression goes over into

$$(i\omega)^{l(2l+1)} \prod_{h>k=1}^{2l+1} (h - k + \lambda'_k - \lambda'_h)$$

Hence
$$D(\lambda'_1\lambda'_2 \cdots \lambda'_{2l+1}) = \prod_{h>k=1}^{2l+1} (h - k + \lambda'_k - \lambda'_h)/(h - k) \quad (5\text{-}62)$$

By taking the general formula (5-57) as his starting point, Weyl[62] obtained the expression

$$D(l_1l_2 \cdots) = \prod_{\alpha^+} (\alpha \cdot \mathbf{K})/(\alpha \cdot \mathbf{R}) \quad (5\text{-}63)$$

for the dimension of the irreducible representation $(l_1l_2 \cdots)$ of any semi-simple group. For the irreducible representation (w_1w_2) of R_5, for example,

$$\mathbf{R} = \tfrac{3}{2}\mathbf{e}_1 + \tfrac{1}{2}\mathbf{e}_2,$$
$$\mathbf{K} = (w_1 + \tfrac{3}{2})\mathbf{e}_1 + (w_2 + \tfrac{1}{2})\mathbf{e}_2$$

and
$$\alpha^+ = \mathbf{e}_1 \pm \mathbf{e}_2,\ \mathbf{e}_1,\ \mathbf{e}_2$$

Thus

$$D(w_1w_2) = (w_1 + w_2 + 2)(w_1 - w_2 + 1)(2w_1 + 3)(2w_2 + 1)/6 \quad (5\text{-}64)$$

Similarly, the dimension $D(w_1w_2w_3)$ of the irreducible representation $(w_1w_2w_3)$ of R_7 is found to be

$$D(w_1w_2w_3) = (w_1 + w_2 + 4)(w_1 + w_3 + 3)(w_2 + w_3 + 2) \times (w_1 - w_2 + 1)(w_1 - w_3 + 2)(w_2 - w_3 + 1)(2w_1 + 5) \times (2w_2 + 3)(2w_3 + 1)/720 \quad (5\text{-}65)$$

For G_2, we get

$$D(u_1u_2) = (u_1 + u_2 + 3)(u_1 + 2)(2u_1 + u_2 + 5)(u_1 + 2u_2 + 4) \times (u_1 - u_2 + 1)(u_2 + 1)/120 \quad (5\text{-}66)$$

5-11 THE KRONECKER PRODUCT OF TWO REPRESENTATIONS

Suppose that the two sets of functions $u(\gamma,\mathbf{m})$ and $u'(\gamma',\mathbf{m}')$ form bases for the respective irreducible representations $(l_1l_2 \cdots)$ and $(l'_1l'_2 \cdots)$ of a continuous group. Since the operators H_i must behave like differential operators,

$$\begin{aligned} H_i[u(\gamma,\mathbf{m})u'(\gamma',\mathbf{m}')] &= u'(\gamma',\mathbf{m}')[H_iu(\gamma,\mathbf{m})] + u(\gamma,\mathbf{m})[H_iu'(\gamma',\mathbf{m}')] \\ &= (m_i + m'_i)[u(\gamma,\mathbf{m})u'(\gamma',\mathbf{m}')] \end{aligned} \quad (5\text{-}67)$$

from Eq. (5-55). The products $u(\gamma,\mathbf{m})u'(\gamma',\mathbf{m}')$ form a representation of the group of dimension $D(l_1l_2 \cdots)D(l'_1l'_2 \cdots)$; the weights of the representation consist of all possible sums of the weights of the representations $(l_1l_2 \cdots)$ and $(l'_1l'_2 \cdots)$. If diagrams such as Fig. 5-4 are available for many irreducible representations, it is a simple matter to form the Kronecker product $(l_1l_2 \cdots) \times (l'_1l'_2 \cdots)$. Suppose, for example, that the decomposition of the Kronecker product (10) × (10) for the group R_5 is required. The weights of (10) are (1,0), (0,1), (−1,0), (0,−1), and (0,0); they form a five-point array in the weight space. If we take an identical array of points [corresponding to the second representation (10)] and, by translating the array, superpose its center successively on each of the points of the first array, we obtain the arrangement of points shown in Fig. 5-10. Since this can be produced by the superposition of the arrays for (20), (11), and (00), we conclude that the representation (10) × (10) decomposes into (20), (11), and (00):

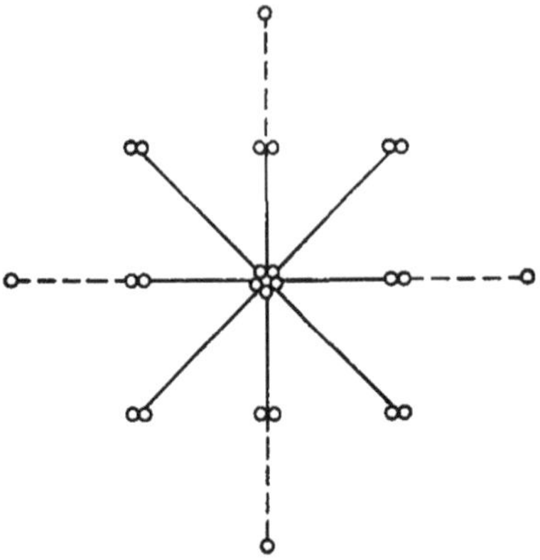

FIG. 5-10 By superposing the five-point array (1,0), (0,1), (−1,0), (0,−1), and (0,0) on every point of itself, we obtain the weights of the representation (10) × (10), shown here. The decomposition

$$(10) \times (10) = (20) + (11) + (00)$$

follows from Fig. 5-4.

$$(10) \times (10) = (20) + (11) + (00)$$

Equation (5-64) furnishes a dimension check:

$$5 \times 5 = 14 + 10 + 1$$

Kronecker products of representations of R_3 can be dealt with very simply by this procedure. The weights of the representation $\mathfrak{D}_L$ are $(L), (L-1), \ldots, (-L)$; those of $\mathfrak{D}_S$ are $(S), (S-1), \ldots, (-S)$. If the first array of weights is translated in the one-dimensional weight space so that its center lies in turn on each weight of the second set, we obtain a linear array extending from $(S+L)$ to $(-S-L)$. It is not difficult to see that this array can be formed by adding the weights of the representations $\mathfrak{D}_{S+L}, \mathfrak{D}_{S+L-1}, \ldots, \mathfrak{D}_{|S-L|}$; consequently these are the irreducible representations that occur in the decomposition of $\mathfrak{D}_S \times \mathfrak{D}_L$. This result is equivalent to Eq. (2-13). For the representations $\mathfrak{D}_S$ and $\mathfrak{D}_L$ of R_3, Eq. (5-67) reduces to

$$J_z|SM_SLM_L) = (M_S + M_L)|SM_SLM_L)$$

In general, the number of times the irreducible representation $(l_1''l_2'' \cdots)$ occurs in the decomposition of $(l_1l_2 \cdots) \times (l_1'l_2' \cdots)$ is given by the coefficient

$$c((l_1l_2 \cdots)(l_1'l_2' \cdots)(l_1''l_2'' \cdots))$$

in the equation

$$\chi(l_1l_2 \cdots)\chi(l_1'l_2' \cdots) = \sum_{(l_1'''l_2''' \cdots)} c((l_1l_2 \cdots)(l_1'l_2' \cdots)(l_1'''l_2''' \cdots))\chi(l_1'''l_2''' \cdots) \quad (5\text{-}68)$$

To solve for the coefficients, we adapt a method due to Weyl.[61] First, the substitution (5-57) is made for $\chi(l_1l_2 \cdots)$. Second, $\chi(l_1'l_2' \cdots)$ is replaced by

$$\sum_{\mathbf{m}'} \exp\,[im_j'\varphi^j]$$

The sum runs over all the weights $\mathbf{m}' = (m_1', m_2', \ldots)$ of the irreducible representation $(l_1'l_2' \cdots)$; a weight that occurs more than once [for example, (0,0) of the representation (20) of R_5] gives rise to as many equal terms in the sum as its multiplicity. The left-hand side of Eq. (5-68) becomes

$$[\xi(\mathbf{R})]^{-1} \sum_{\sigma,\mathbf{m}'} (-1)^p \exp\,[i\varphi^j\{m_j' + (\mathcal{R}_\sigma\mathbf{K})_j\}]$$

Since the array of weights $\mathbf{m}'$ is invariant under the operations of the group S,

$$\sum_{\mathbf{m}'} f(\mathbf{m}') = \sum_{\mathbf{m}'} f(\mathcal{R}_\sigma\mathbf{m}')$$

where $f(\mathbf{m}')$ is any function of $\mathbf{m}'$. Hence

$$\begin{aligned}\chi(l_1l_2 \cdots)\chi(l_1'l_2' \cdots) &= [\xi(\mathbf{R})]^{-1} \sum_{\sigma,\mathbf{m}'} (-1)^p \exp\,[i\varphi^j\mathcal{R}_\sigma(\mathbf{K} + \mathbf{m}')_j] \\ &= \sum_{\mathbf{m}'} \xi(\mathbf{K} + \mathbf{m}')/\xi(\mathbf{R}) \\ &= \sum_{\mathbf{m}'} \chi(l_1 + m_1', l_2 + m_2', \ldots) \qquad (5\text{-}69)\end{aligned}$$

Because of the range of values over which the components m_j' can run, the symbol $(l_1 + m_1', l_2 + m_2', \ldots)$ is not necessarily admissible as an irreducible representation of the group. It might turn out, for example, that, for a particular weight $\mathbf{m}'$, the component $l_1 + m'_1$ is negative. In cases such as this, the explicit expression for the character must be consulted to reexpress $\chi(l_1 + m'_1, l_2 + m'_2, \ldots)$ in an acceptable form. Usually it is necessary only to reorder the columns of a determinant, though occasionally the signs of a column need to be reversed. For R_5, for example, the equations

$$\chi(w_1w_2) = -\chi(w_2 - 1, w_1 + 1) = -\chi(w_1, -w_2 - 1) = -\chi(-w_1 - 3, w_2)$$

are readily derived. Their generalization to R_{2l+1} can be obtained without difficulty from Eq. (5-59). If columns h and k of the determinant are interchanged, we get

$$\chi(w_1w_2 \cdots w_h \cdots w_k \cdots w_l) = -\chi(w_1w_2 \cdots w_k - k + h \cdots w_h - h + k \cdots w_l) \quad (5\text{-}70)$$

As an illustration we consider the Kronecker product of the representations (11) and (10) of R_5. The five weights of (10) are (1,0), (0,1), (−1,0), (0,−1), and (0,0). Using Eq. (5-69), we obtain

$$\chi(11)\chi(10) = \chi(21) + \chi(12) + \chi(01) + \chi(10) + \chi(11)$$

But $\chi(12) = -\chi(12)$ and $\chi(01) = -\chi(01)$

Thus $\chi(12) = \chi(01) = 0$ and $(11) \times (10) = (21) + (10) + (11)$

The dimension check is

$$10 \times 5 = 35 + 5 + 10$$

Equation (5-69) can be applied to other groups with equal ease. The $2l + 1$ orbital states of a single electron with azimuthal quantum number l form the basis for the irreducible representation $[100 \cdots 0]$ of U_{2l+1} and also for the irreducible representation

$$(2l[l]^{-1}, -[l]^{-1}, \ldots, -[l]^{-1})$$

of SU_{2l+1}. The $2l + 1$ possible arrangements of the symbols in the parentheses give the weights of the representation. Consequently,

$$\chi(\lambda'_1\lambda'_2 \cdots \lambda'_{2l+1})\chi(2l[l]^{-1}, -[l]^{-1}, \ldots, -[l]^{-1}) = \sum_{j=1}^{2l+1} \chi(\lambda'_1 - [l]^{-1} + \delta(j,1), \lambda'_2 - [l]^{-1} + \delta(j,2), \ldots, \lambda'_{2l+1} - [l]^{-1} + \delta(j, 2l + 1))$$

If $\lambda'_k = \lambda'_{k+1} = \cdots = \lambda'_r$, then the terms in the sum for which $k < j \leq r$ are zero. This is because the corresponding determinants in the numerator on the right-hand side of Eq. (5-61) possess two equal columns. Hence

$$
\begin{aligned}
(\lambda_1'\lambda_2' \cdots \lambda_{2l+1}') &\times (2l[l]^{-l}, -[l]^{-l}, \ldots, -[l]^{-l}) \\
&= (\lambda_1'' + 1, \lambda_2'', \ldots, \lambda_{2l+1}'') + \cdots \\
&\quad + (\lambda_1'', \lambda_2'', \ldots, \lambda_{k-1}'', \lambda_k'' + 1, \ldots, \lambda_{2l+1}'') + \cdots \qquad (5\text{-}71)
\end{aligned}
$$

where $\lambda_j'' = \lambda_j' - [l]^{-1}$ and $\lambda_k'' < \lambda_{k-1}''$.

If we suppose that the basis functions for $(\lambda_1'\lambda_2' \cdots \lambda_{2l+1}')$ involve the coordinates of n electrons, the bases for the representations on the right of Eq. (5-71) involve the coordinates of $n + 1$ electrons. With the aid of Eq. (5-43), we find that Eq. (5-71) is equivalent to

$$
\begin{aligned}
[\lambda_1\lambda_2 \cdots \lambda_{2l+1}] \times [10 \cdots 0] \\
&= [\lambda_1 + 1, \lambda_2, \ldots, \lambda_{2l+1}] + \cdots \\
&\quad + [\lambda_1, \lambda_2, \ldots, \lambda_{k-1}, \lambda_k + 1, \ldots, \lambda_{2l+1}] + \cdots
\end{aligned}
$$

Only those representations occur for which $\lambda_k < \lambda_{k-1}$. Thus

$$
\begin{aligned}
[2210 \cdots 0] \times [10 \cdots 0] \\
&= [3210 \cdots 0] + [2220 \cdots 0] + [22110 \cdots 0]
\end{aligned}
$$

An alternative way of representing this decomposition is as follows:

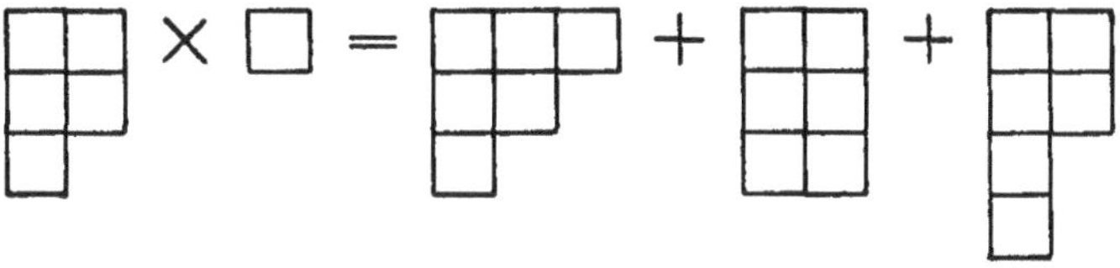

(5-72)

The shapes on the right can be obtained simply by adding a single cell to the shape [221] in all permissible ways; the condition $\lambda_k < \lambda_{k-1}$ eliminates the irregular shape

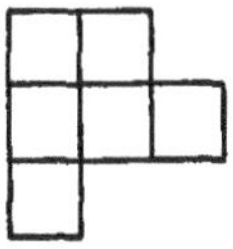

As might be anticipated from the extreme simplicity of Eq. (5-72), a set of rules can be constructed for finding the Kronecker product of two irreducible representations of U_{2l+1}. No derivation of the prescription will be given here. Apart from a few changes in notation, we shall simply quote Littlewood's[70] formulation, which, although derived in connection with the properties of the symmetric groups S_n, can be immediately taken over to the group U_{2l+1} because of the correspondence established in Sec. 5-8. We begin with a definition. Consider the partition $[\lambda_1\lambda_2 \cdots \lambda_p]$, where $\sum_i \lambda_i = n$. If among the first r terms of any permutation of the n factors of the product

$$x_1^{\lambda_1}x_2^{\lambda_2} \cdots x_p^{\lambda_p}$$

the number of times x_1 occurs $\geq$ the number of times x_2 occurs $\geq$ the number of times x_3 occurs, etc., for all values of r, this permutation is called a *lattice permutation*. Littlewood's recipe can now be stated in the following form: The shapes appearing in

$$[\lambda_1\lambda_2 \cdots \lambda_p] \times [\mu_1\mu_2 \cdots \mu_p]$$

are those which can be built by adding to the shape $[\lambda_1\lambda_2 \cdots \lambda_p]$ μ_1 cells containing the same symbol α, then μ_2 cells containing the same symbol β, etc., subject to the two conditions:

1. After the addition of each set of cells labeled by a common symbol we must have a permissible shape with no two identical symbols in the same column.

2. If the total set of added symbols is read from right to left in the consecutive rows of the final shape, we obtain a lattice permutation of

$$\alpha^{\mu_1}\beta^{\mu_2}\gamma^{\mu_3} \cdots$$

For example,

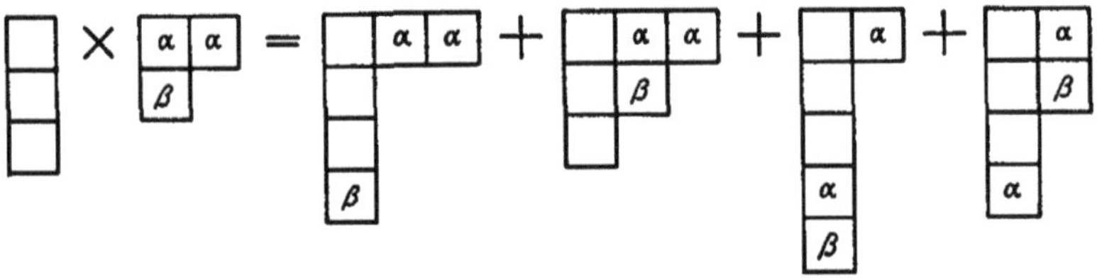

(5-73)

To interpret an equation such as this in terms of irreducible representations of U_{2l+1}, we have merely to replace the shape $[\lambda_1\lambda_2 \cdots \lambda_p]$ by the representation $[\lambda_1\lambda_2 \cdots \lambda_p 00 \cdots 0]$. The dimension of the latter depends on l; from Eqs. (5-43) and (5-62), we find that the dimension check on Eq. (5-73) for $l = 3$ is

$$35 \times 112 = 840 + 2352 + 140 + 588$$

5-12 BRANCHING RULES

On restricting a group of transformations to a subgroup, irreducible representations of the group decompose into irreducible representations of the subgroup. Tables 2-2 and 2-4 are examples of such decompositions. Our present aim is to find analogous tables for the groups that occur in the sequences (5-18) and (5-19). The most direct method is to make use of the formulas for the characters of the representations, in complete analogy with the procedure for decomposing the irreducible representations of finite groups. This involves finding the minimum number of restrictions on the parameters φ^j that ensure that the operator $\varphi^j H_j$ of the group reduces to an operator of the subgroup: under these restrictions the character $\chi(l_1 l_2 \cdots)$ of the irreducible representation $(l_1 l_2 \cdots)$ breaks up into a linear combination of characters of the subgroup and thereby determines the decomposition of $(l_1 l_2 \cdots)$. However, if tables of Kronecker products of irreducible representations of both the group and the subgroup are available, this rather tedious procedure can be avoided. As is shown in Sec. 5-13, a chain calculation can be set up to determine the branching rules from the known decompositions of the Kronecker products.

Before illustrating these two methods in detail, it is convenient to describe the somewhat atypical reduction $U_{2l+1} \to SU_{2l+1}$. The branching rules are particularly simple: each irreducible representation $[\lambda_1 \lambda_2 \cdots \lambda_{2l+1}]$ of U_{2l+1} becomes the irreducible representation $(\lambda_1' \lambda_2' \cdots \lambda_{2l+1}')$ of SU_{2l+1}, where λ_i' is related to λ_i by Eq. (5-43). We observe here that all irreducible representations of U_{2l+1} of the type

$$[\lambda_1 - a, \lambda_2 - a, \ldots, \lambda_{2l+1} - a]$$

where a is any integer such that $\lambda_{2l+1} - a \geq 0$ become the same irreducible representation of SU_{2l+1}. This implies that when U_5 is limited to SU_5, for example, the following equivalence holds:

(5-74)

To illustrate the direct method for finding branching rules, we consider first the reduction $R_{2l+1} \to R_3$. The operator $\exp(i\varphi^j H_j)$, the sum of whose eigenvalues determines the character of a representation of a group, is given by $\exp\left(i \sum_{j=1}^{l} W_{jj}\theta^{(j)}\right)$ for R_{2l+1} and $\exp(iL_z\theta)$,

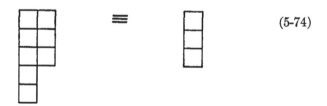

say, for R_3. If we set $\theta^{(j)} = j\theta$, the two become equivalent; for

$$\sum_{j=1}^{l} W_{jj}j\theta = \tfrac{1}{2}\theta \sum_{j} W_{jj}(-1)^{l-j}[l(l+1)(2l+1)]^{\frac{1}{2}} \begin{pmatrix} l & 1 & l \\ -j & 0 & j \end{pmatrix}$$

$$= \tfrac{1}{2}\theta[l(l+1)(2l+1)]^{\frac{1}{2}} \sum_{k,j} \{1-(-1)^k\}[k]^{\frac{1}{2}} \begin{pmatrix} l & 1 & l \\ -j & 0 & j \end{pmatrix} \times \begin{pmatrix} l & k & l \\ -j & 0 & j \end{pmatrix} V_0^{(k)}$$

$$= \theta[l(l+1)(2l+1)]^{\frac{1}{2}} \sum_{k} \delta(k,1) V_0^{(k)}(3)^{-\frac{1}{2}}$$

$$= V_0^{(1)}[l(l+1)(2l+1)/3]^{\frac{1}{2}}\theta = L_z\theta$$

The last step follows from Eq. (5-38). To find how the irreducible representation $(w_1w_2 \cdots w_l)$ of R_{2l+1} breaks up when it is considered as a representation of R_3, we have merely to replace $\theta^{(j)}$ by $j\theta$ in the expression for $\chi(w_1w_2 \cdots w_l)$ and reduce the resultant function to the form

$$\sum_{L} \frac{e^{i(L+\frac{1}{2})\theta} - e^{-i(L+\frac{1}{2})\theta}}{e^{\frac{1}{2}i\theta} - e^{-\frac{1}{2}i\theta}} \tag{5-75}$$

Every term in the sum is the character of an irreducible representation of R_3, and therefore the values of L that appear immediately determine the representations $\mathfrak{D}_L$ of R_3. For example, on putting $\theta' = \theta$ and $\theta'' = 2\theta$ in either Eq. (5-56) or Eq. (5-60), we get

$$\chi(11) = \frac{e^{7i\theta/2} - e^{-7i\theta/2}}{e^{i\theta/2} - e^{-i\theta/2}} + \frac{e^{3i\theta/2} - e^{-3i\theta/2}}{e^{i\theta/2} - e^{-i\theta/2}}$$

Thus, under the reduction $R_5 \rightarrow R_3$, the irreducible representation (11) of R_5 breaks up into $\mathfrak{D}_3$ and $\mathfrak{D}_1$. This can be concisely expressed by $(11) \rightarrow P + F$. We could, of course, have anticipated this result, since the terms ^3P and ^3F of d^2 are used in Sec. 5-7 to form basis functions for the representation (11). The convention adopted here, which replaces $\mathfrak{D}_L$ by the spectroscopic symbol for L, will be frequently used in subsequent work.

The reduction $SU_{2l+1} \rightarrow R_3$ can be dealt with in an analogous fashion. The methods of the previous paragraph can be used to show that the operator

$$\sum_{j=-l}^{l} X'_{jj}\omega^{(j)}$$

reduces to $L_z\theta$ if the substitutions $\omega^{(j)} = j\theta$ are made; hence, by substituting $\omega^{(l+1-j)} = (l+1-j)\theta$ in $\chi(\lambda_1'\lambda_2' \cdots \lambda_{2l+1}')$ and expressing the resultant function as a sum of the form (5-75), the decomposition of any irreducible representation of SU_{2l+1} can be obtained (see Prob. 5-4).

The simplicity of the expression for the character of an irreducible representation of R_3 makes it comparatively easy to decompose a particular representation of SU_{2l+1} or R_{2l+1} into irreducible representations of R_3. The method is unsuited to study the reduction $SU_{2l+1} \rightarrow R_{2l+1}$ because of the complexity of $\chi(\lambda_1'\lambda_2' \cdots \lambda_{2l+1}')$ and $\chi(w_1w_2 \cdots w_l)$. Fortunately, the general solution to the problem of decomposing irreducible representations $[\lambda_1\lambda_2 \cdots \lambda_{2l+1}]$ of U_{2l+1} into irreducible representations of R_{2l+1} can be described by a number of simple rules, and we can thereby avoid having to construct the characters explicitly. All the irreducible representations of U_{2l+1} for which determinantal product states serve as bases satisfy the inequality $\lambda_i \leq 2$; in these cases the rules, as stated by Littlewood,[70] can be adapted as follows: We draw the shape $[\lambda_1\lambda_2 \cdots]$ corresponding to the representation $[\lambda_1\lambda_2 \cdots \lambda_{2l+1}]$ and perform the successive operations:

1. Leave the shape untouched.
2. Delete the two cells at the feet of the two columns, if this is possible.
3. Take the new shape, and delete the two cells at the feet of its two columns.
4. Continue in this way until a shape possessing a single column is obtained or, if the two columns of the original shape are of equal length, until no cells remain.

Parentheses rather than brackets are used to describe all the shapes that can be obtained by this process, and the resulting symbols are interpreted directly as the irreducible representations of R_{2l+1} into which $[\lambda_1\lambda_2 \cdots \lambda_{2l+1}]$ decomposes. Thus the representation [2210000] of U_7 corresponds to the shape [221], which, under the above operations, becomes successively

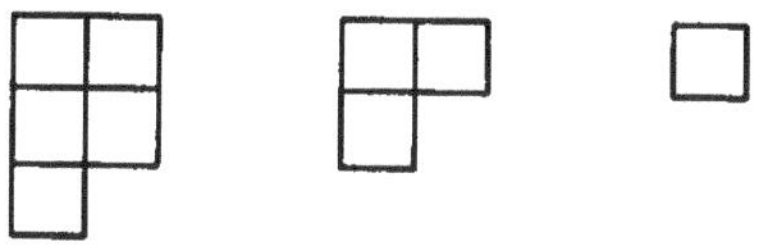

Accordingly,

$$[2210000] \rightarrow (221) + (210) + (100)$$

The shape [111] already possesses but one column; hence

$$[1110000] \rightarrow (111)$$

Since irreducible representations of R_{2l+1} are described by only l numbers, it is not possible to carry out the procedure if more than l symbols λ_j are nonzero. We notice, however, that if we take each of the determinants of Eq. (5-61) in turn, interchange rows 1 and $2l+1$, rows 2 and $2l$, rows 3 and $2l-1$, etc., and then interchange columns 1 and $2l+1$, columns 2 and $2l$, etc., we get

$$\chi(\lambda_1'\lambda_2' \cdots \lambda_{2l+1}') = \frac{\det|\exp[i(-l-1+k+\lambda_{2l+2-k}')\omega^{(-l-1+j)}]|}{\det|\exp[i(-l-1+k)\omega^{(-l-1+j)}]|} \tag{5-76}$$

The reduction $SU_{2l+1} \rightarrow R_{2l+1}$ is accomplished by setting

$$\omega^{(j)} = -\omega^{(-j)} = \theta^{(j)}$$

under which

$$\sum_{j=-l}^{l} X_{jj}'\omega^{(j)} \rightarrow \sum_{j=1}^{l} W_{jj}\theta^{(j)}$$

On replacing $\omega^{(-l-1+j)}$ by $-\omega^{(l+1-j)}$ in Eq. (5-76), we obtain

$$\chi(\lambda_1'\lambda_2' \cdots \lambda_{2l+1}') = \frac{\det|\exp[i(l+1-k-\lambda_{2l+2-k}')\omega^{(l+1-j)}]|}{\det|\exp[i(l+1-k)\omega^{(l+1-j)}]|}$$

Comparison with Eq. (5-61) shows that this is nothing else but $\chi(-\lambda_{2l+1}', -\lambda_{2l}', \ldots, -\lambda_1')$; consequently the two irreducible representations $(\lambda_1'\lambda_2' \cdots \lambda_{2l+1}')$ and $(-\lambda_{2l+1}', -\lambda_{2l}', \ldots, -\lambda_1')$ break up into the same set of irreducible representations of R_{2l+1}. This in turn implies the equivalence

$$[\lambda_1' + a, \lambda_2' + a, \ldots, \lambda_{2l+1}' + a] \equiv [-\lambda_{2l+1}' + b, -\lambda_{2l}' + b, \ldots, -\lambda_1' + b]$$

where a and b are constants which ensure that the entries in the brackets are integers. If we set $a = n[l]^{-1}$ and $b = \lambda_1'$, we get, using Eq. (5-43),

$$[\lambda_1\lambda_2 \cdots \lambda_{2l+1}] \equiv [\lambda_1 - \lambda_{2l+1}, \lambda_1 - \lambda_{2l}, \ldots, \lambda_1 - \lambda_2, 0] \tag{5-77}$$

It is interesting to note that the two shapes corresponding to these irreducible representations can be fitted together to form a rectangle

$\lambda_1 \times (2l + 1)$. For d electrons, for example,

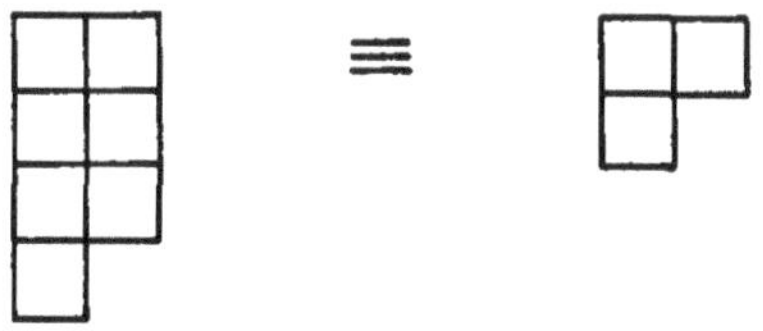

(5-78)

If we use Eq. (5-42), this equivalence tells us that the states of d^7 with $S = \frac{1}{2}$ form bases for the same irreducible representations of R_5 as do the states of d^3 with $S = \frac{1}{2}$; in fact, it establishes the familiar correspondence between electrons and holes. For U_7, Eq. (5-77) gives $[2222100] \equiv [2210000]$; consequently

$$[2222100] \rightarrow (221) + (210) + (100)$$

Similarly,

$$[1111000] \rightarrow (111)$$

There remain a number of representations of U_{2l+1} that cannot be decomposed by these methods. An example is [2211100] of U_7. In cases such as this, we perform the operations 1 to 4 above, using parentheses to describe the resulting shapes. Thus

$$[2211100] \rightarrow (22111) + (2111) + (111)$$

Murnaghan[71] has given modification rules for converting nonstandard symbols such as (22111) to admissible representations of the appropriate rotation group. When the number of 2's is not greater than l, the rules can be concisely summarized by the formula

$$(\underbrace{22 \cdots 2}_{c}\underbrace{11 \cdots 1}_{d}) = (\underbrace{22 \cdots 2}_{c}\underbrace{11 \cdots 1}_{x}\underbrace{00 \cdots 0}_{y}) \qquad 0 \leq c \leq l \quad (5\text{-}79)$$

where $x = 2l + 1 - 2c - d$ and $y = l - c - x$. For example, $(22111) = (220)$. Equation (5-79) can be applied only to those cases for which $x \geq 0$; all other nonstandard symbols can be avoided by a judicious use of Eq. (5-77).

For f electrons, the reductions $R_7 \rightarrow G_2$ and $G_2 \rightarrow R_3$ remain to be considered. The operators H_i for R_7 are as follows:

$$\begin{aligned}
H_1 &= W_{33} = 3(\tfrac{1}{7})^{\frac{1}{2}}V_0^{(1)} + (\tfrac{2}{3})^{\frac{1}{2}}V_0^{(3)} + (\tfrac{1}{21})^{\frac{1}{2}}V_0^{(5)} \\
H_2 &= W_{22} = 2(\tfrac{1}{7})^{\frac{1}{2}}V_0^{(1)} - (\tfrac{2}{3})^{\frac{1}{2}}V_0^{(3)} - 4(\tfrac{1}{21})^{\frac{1}{2}}V_0^{(5)} \\
H_3 &= W_{11} = (\tfrac{1}{7})^{\frac{1}{2}}V_0^{(1)} - (\tfrac{2}{3})^{\frac{1}{2}}V_0^{(3)} + 5(\tfrac{1}{21})^{\frac{1}{2}}V_0^{(5)}
\end{aligned}$$

On writing $\varphi^3 = \varphi^1 - \varphi^2$, the operator $H_j\varphi^j$ reduces to a linear combination of the operators (5-35). The characters $\chi(w_1w_2w_3)$ for R_7

decompose into sums of the characters $\chi(u_1u_2)$ for G_2. With a considerable amount of algebraic manipulation, we arrive at the result

$$\chi(w_1w_2w_3) = \Sigma[\chi(i-k, j+k) + \chi(j-k-1, i-j)] \quad (5\text{-}80)$$

where the sum runs over all integral values of i, j, and k satisfying

$$w_1 \geq i \geq w_2 \qquad w_2 \geq j \geq w_3 \qquad w_3 \geq k \geq -w_3$$

The equations

$$\chi(u_1u_2) = -\chi(u_2 - 1, u_1 + 1) \qquad \chi(u_1, -1) = 0$$

and

$$\chi(u_1, -2) = -\chi(u_1 - 1, 0)$$

are used to remove characters $\chi(u_1u_2)$ whose arguments do not give admissible irreducible representations of G_2; they can be obtained by methods similar to those that led to Eq. (5-70). The reduction $G_2 \to R_3$ can be accomplished by writing $\varphi^1 = 3\theta$ and $\varphi^2 = 2\theta$.

5-13 CLASSIFICATION OF THE TERMS OF f^n

It is stated in Sec. 5-7 that the totality of states of the configuration l^n for a given S and M_S form a basis for a single irreducible representation of U_{2l+1}. Since the representation does not depend on M_S, the labeling of states by $[\lambda_1\lambda_2 \cdots \lambda_{2l+1}]$ is equivalent to specifying S. The actual correspondence is given in Eq. (5-42). The terms that occur in the configuration l^n are very easy to find when $n < 3$, and hence the branching rules for the decompositions of a few representations of U_{2l+1} to irreducible representations of R_3 can be rapidly obtained. It will now be shown how a chain calculation can be set up to determine the branching rules for other representations of U_{2l+1}. The general method will become clear if we take the special case of $l = 3$.

From the configurations f^0, f^1, and f^2, the following branching rules for the reduction $U_7 \to R_3$ are readily derived:

$$\begin{aligned} &[0000000] \to S \\ &[1000000] \to F \\ &[2000000] \to S + D + G + I \\ &[1100000] \to P + F + H \end{aligned}$$

For conciseness, we shall frequently omit plus signs and arrows in tabulations such as this. By using the corresponding partitions to label the irreducible representations of U_{2l+1}, the necessity of writing out many zeros in the brackets can also be avoided. Suppose that we select any two of the four representations of U_7 listed above and form their Kronecker product. The irreducible representations of R_3 contained in this product can be found by making use of the known branching rules for the two representations of U_7 and then applying Eq. (2-13) repeatedly. The representations of R_3 obtained by this procedure give the structure of the collection of irreducible representations of U_7 into which the Kronecker product decomposes and hence provide information about branching rules. For example, [11] × [1] must decompose to those irreducible representations of R_3 contained in the Kronecker product $(P + F + H) \times F$. However,

$$P \times F = D + F + G$$
$$F \times F = S + P + D + F + G + H + I$$
$$H \times F = D + F + G + H + I + K + L$$

and

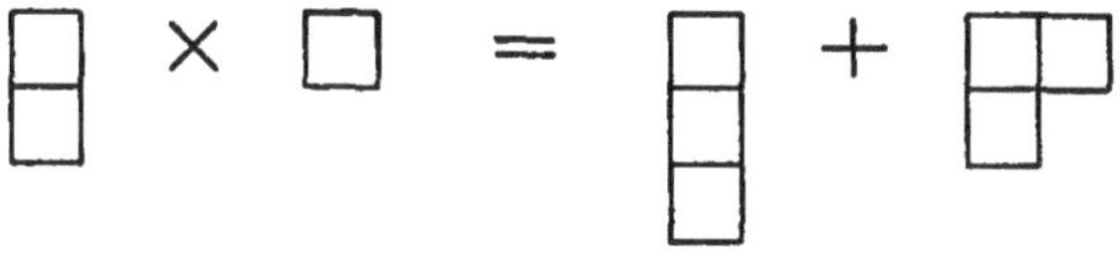

Hence $$[111] + [21] \rightarrow SPD^3F^3G^3H^2I^2KL$$

A superscript to a symbol indicates the number of times the corresponding irreducible representation occurs. It is clear that, if the decomposition of either [111] or [21] were known, then that of the other would follow immediately by subtraction. The problem of finding the irreducible representations of R_3 into which [111] decomposes is equivalent to that of finding the terms of f^3 for which $S = \frac{3}{2}$. The latter can be solved by using the elementary techniques of Sec. 1-6, and the required terms are ^4S, ^4D, ^4F, ^4G, and ^4I. Thus

$$[111] \rightarrow SDFGI$$

and so $$[21] \rightarrow PD^2F^2G^2H^2IKL$$

The decomposition of [21] gives the terms of f^3 for which $S = \frac{1}{2}$; we notice that there are two ^2H terms, in agreement with the conclusion arrived at in Sec. 1-6.

The method used above to find the decomposition of [111] is essentially alien to the present approach. Fortunately, the branching rules for all the remaining irreducible representations $[\lambda_1\lambda_2 \cdots \lambda_7]$ of U_7 for which $\lambda_1 \leq 2$ can be calculated without having to resort to such a device again. As an illustration, we calculate the terms of f^4:

The Kronecker product

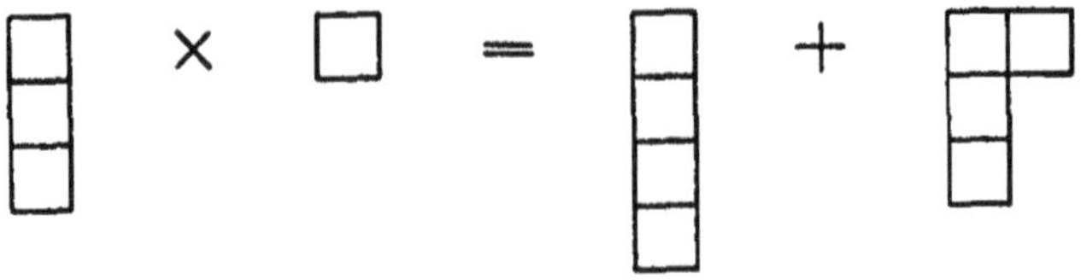

corresponds to

$$(SDFGI) \times F = SP^3D^3F^5G^4H^4I^3K^2LM$$

According to Eq. (5-77), the decompositions of [1111] and [111] are identical; hence

$$[1111] \rightarrow SDFGI$$

and so

$$[211] \rightarrow P^3D^2F^4G^3H^4I^2K^2LM$$

These decompositions give the terms of f^4 for which $S = 2$ and $S = 1$, respectively. To find the singlets, we use

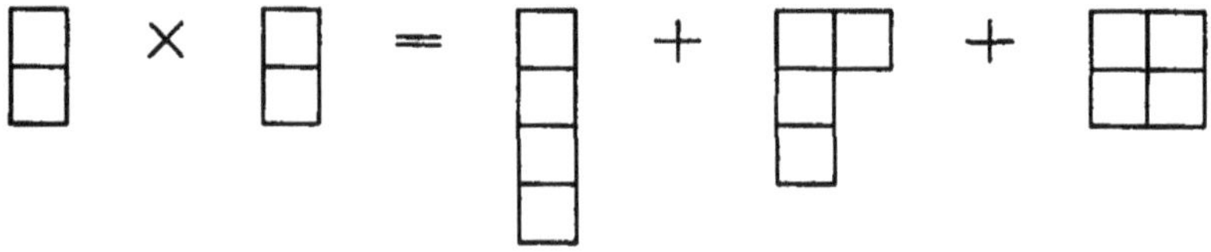

to which corresponds

$$(PFH) \times (PFH) = S^3P^3D^7F^6G^8H^6I^6K^3L^3MN$$

On subtracting the parts that arise from [1111] and [211], we get

$$[22] \rightarrow S^2D^4FG^4H^2I^3KL^2N$$

There is no difficulty in extending this procedure to other configurations of the type f^n.

The method for finding branching rules that is described in Sec. 5-12 treats each particular reduction as a self-contained problem, to be solved by using the characters of the group and the subgroup. Once the branching rules for a few reductions have been found, however, they can often be used with great effect to find others. To illustrate this approach, we show how the branching rules for $U_7 \rightarrow R_7$ and $U_7 \rightarrow R_3$ can be combined to find those for $R_7 \rightarrow R_3$. Suppose, for example, that the irreducible representations of R_3 into which (220) decomposes are required. It is necessary only to compare the decompositions

$$[22] \to (220) + (200) + (000)$$

and

$$[22] \to S^2D^4FG^4H^2I^3KL^2N$$

with

$$[2] \to (200) + (000)$$

and

$$[2] \to SDGI$$

to obtain the result

$$(220) \to SD^3FG^3H^2I^2KL^2N$$

We can sometimes use a knowledge of the dimensions of irreducible representations to obtain branching rules. Suppose that we are interested in the reduction $G_2 \to R_3$. We draw up a table for $R_7 \to R_3$ and list the dimensions of the representations of R_7 in order of increasing magnitude.

$D(w_1w_2w_3)$	$(w_1w_2w_3)$	L
1	(000)	S
7	(100)	F
21	(110)	PFH
27	(200)	DGI
35	(111)	$SDFGI$
...	.....	......

The dimensions of the representations of G_2 are calculated from Eq. (5-66) and tabulated in order of increasing magnitude.

$D(u_1u_2)$	(u_1u_2)
1	(00)
7	(10)
14	(11)
27	(20)
64	(21)
..	....

The only representation of R_3 with dimension 1 is $\mathfrak{D}_0$, corresponding to an S state; hence $(00) \to S$. The representation (100) has a dimension of 7 and could decompose either into seven representations of the type (00) or simply into (10). The former case can be excluded, because it implies that, under the reduction $R_7 \to R_3$, $(100) \to S^7$. Therefore $(10) \to F$. Similarly, (110) is of dimension 21; since an S state is not included in its decomposition, it can break down into $(10)^3$ or into $(11) + (10)$. The first possibility can be discarded, since we have already shown that $(10) \to F$, and (110) does not decompose

TABLE 5-1 BRANCHING RULES FOR THE REDUCTION $U_7 \to R_7$

Configuration	Dimension	$[\lambda_1\lambda_2 \cdots]$	$(w_1w_2w_3)$
f^0	1	[0]	(000)
f^1	7	[1]	(100)
f^2	21	[11]	(110)
	28	[2]	(000)(200)
f^3	35	[111]	(111)
	112	[21]	(100)(210)
f^4	35	[1111]	(111)
	210	[211]	(110)(211)
	196	[22]	(000)(200)(220)
f^5	21	[11111]	(110)
	224	[2111]	(111)(211)
	490	[221]	(100)(210)(221)
f^6	7	[111111]	(100)
	140	[21111]	(111)(210)
	588	[2211]	(110)(211)(221)
	490	[222]	(000)(200)(220)(222)
f^7	1	[1111111]	(000)
	48	[211111]	(110)(200)
	392	[22111]	(111)(211)(220)
	784	[2221]	(100)(210)(221)(222)

TABLE 5-2 BRANCHING RULES FOR THE REDUCTION $R_7 \to G_2$

$D(w_1w_2w_3)$	$(w_1w_2w_3)$	(u_1u_2)
1	(000)	(00)
7	(100)	(10)
21	(110)	(10)(11)
27	(200)	(20)
35	(111)	(00)(10)(20)
105	(210)	(11)(20)(21)
189	(211)	(10)(11)(20)(21)(30)
168	(220)	(20)(21)(22)
378	(221)	(10)(11)(20)(21)(30)(31)
294	(222)	(00)(10)(20)(30)(40)

into three F states. We can continue in this manner and unambiguously obtain the decompositions of (11), (20), and (21).

By means of the various techniques described above, the branching rules for all the groups of interest to us can be found. The results for the reductions $U_7 \to R_7$, $R_7 \to G_2$, and $G_2 \to R_3$ are given in Tables 5-1 to 5-3. The original derivation of these decompositions is due to Racah.[69] Branching rules for the reductions $U_7 \to G_2$,

TABLE 5-3 BRANCHING RULES FOR THE REDUCTION $G_2 \rightarrow R_3$

$D(u_1u_2)$	(u_1u_2)	L
1	(00)	S
7	(10)	F
14	(11)	PH
27	(20)	DGI
64	(21)	$DFGHKL$
77	(30)	$PFGHIKM$
77	(22)	$SDGHILN$
189	(31)	$PDF^2GH^2I^2K^2LMNO$
182	(40)	$SDFG^2HI^2KL^2MNQ$

$U_7 \rightarrow R_3$, and $R_7 \rightarrow R_3$ can be easily obtained from these tables and are therefore not written out explicitly. In Table 5-1, only those representations of U_7 corresponding to the first half of the f shell are given; the decomposition of the remainder can be obtained by means of the equivalence (5-77).

5-14 IRREDUCIBLE REPRESENTATIONS AS QUANTUM NUMBERS

In Sec. 5-1 we initiated a search for symbols γ to distinguish terms of the same kind in configurations of the type l^n. The aim of the group-theoretical approach is to replace γ by the irreducible representations of various groups, and we are now in a position to appraise this method. For configurations of fewer than three equivalent electrons, the symbols SM_SLM_L are sufficient to define the states; as a modest test of the group-theoretical approach, we investigate the classification of the 17 terms of f^3. From Tables 5-1 to 5-3, the following scheme is obtained:

$[\lambda_1\lambda_2 \cdots]$	$(w_1w_2w_3)$	(u_1u_2)	L
[111]	(111)	(00)	S
		(10)	F
		(20)	DGI
[21]	(100)	(10)	F
	(210)	(11)	PH
		(20)	DGI
		(21)	$DFGHKL$

It is at once apparent that no two terms share the same set of quantum numbers; hence for f^3 the states can be completely defined by writing

$$|f^3(w_1w_2w_3)(u_1u_2)SM_SLM_L)$$

The irreducible representations $[\lambda_1\lambda_2 \cdots]$ of U_7 have been dropped because they are equivalent to specifying S. It can be seen that no two identical irreducible representations of R_7 occur in the decomposition of any of the representations of U_7 listed in Table 5-1; similarly, no two identical irreducible representations of G_2 occur in the decomposition of any of the representations of R_7 listed in Table 5-2. The group-theoretical classification of the terms of f^n would be complete if no two identical irreducible representations of R_3 occurred in the decomposition of any of the representations of G_2; however, it can be seen from Table 5-3 that this is not true for (31) and (40). In these cases a symbol τ must be included in addition to the irreducible representations of R_7 and G_2 in order completely to define the states. Thus, on making the abbreviations

$$(w_1w_2w_3) = W \qquad (u_1u_2) = U$$

the states of f^n can be unambiguously written as

$$|f^nW\,U\tau SM_SLM_L)$$

For the configuration p^n, the quantum numbers SM_SLM_L are sufficient to define the states completely; for d^n, the classification according to the irreducible representations of R_5 separates all terms of the same kind (see Prob. 5-6). For g^n, however, the only useful group-theoretical labels are the irreducible representations of R_9, and these are quite inadequate for such complex configurations. We shall not consider this shortcoming here, since only a few very simple configurations involving g electrons have so far been reported.

The explicit construction of a state defined by group-theoretical labels is usually a straightforward, though sometimes a tedious, matter. Suppose, for example, that

$$|f^3(210)(11)\,{}^2H,\ M_S = \tfrac{1}{2},\ M_L = 5)$$

is required. Since $\mathbf{V}^{(5)}$ is an infinitesimal operator of the group G_2, it must be diagonal with respect to U; consequently

$$(f^3(210)(21)\,{}^2L,\tfrac{1}{2},8|V_3^{(5)}|f^3(210)(11)\,{}^2H,\tfrac{1}{2},5) = 0 \qquad (5\text{-}81)$$

Since $$(f^3(210)(21)\,{}^2L,\tfrac{1}{2},8| = \{\overset{+\!-\!+}{332}\}^*$$

it is a simple matter to express the matrix element in terms of the parameter x of Sec. 1-6. The condition (5-81) gives $x = \infty$, and the required state is just the linear combination of determinantal product states given in Eq. (1-29). In fact,

$$\gamma_1 = (210)(11) \qquad \gamma_2 = (210)(21)$$

PROBLEMS

5-1. Prove the Jacobi identity

$$[[X_\rho,X_\sigma],X_\tau] + [[X_\sigma,X_\tau],X_\rho] + [[X_\tau,X_\rho],X_\sigma] = 0$$

and show that it leads to the following equation for the structure constants:

$$c_{\rho\sigma}{}^{\mu}c_{\mu\tau}{}^{\nu} + c_{\sigma\tau}{}^{\mu}c_{\mu\rho}{}^{\nu} + c_{\tau\rho}{}^{\mu}c_{\mu\sigma}{}^{\nu} = 0$$

5-2. Prove that the operations of G_2 leave invariant the trilinear antisymmetric form

$$\sum_{m,m',m''} \begin{pmatrix} 3 & 3 & 3 \\ m & m' & m'' \end{pmatrix} \psi_1(3,m)\psi_2(3,m')\psi_3(3,m'')$$

in the notation of Sec. 5-4. (See Racah.[58])

5-3. It can be seen from Fig. 5-11 that the projection of the weights of the representation (20) of R_5 on a line making an angle $\tan^{-1} 2$ with an axis gives a

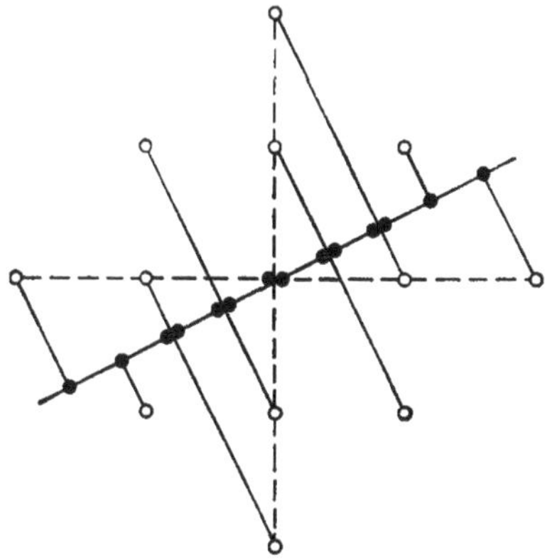

FIG. 5-11 The diagram for Prob. 5-3.

pattern of points that is identical to the combined system of weights of $\mathfrak{D}_2$ and $\mathfrak{D}_4$ of R_3. Justify this method of obtaining branching rules, and use it to prove that for the reduction $R_7 \to G_2$,

$$(100) \to (10)$$

and

$$(\tfrac{1}{2}\tfrac{1}{2}\tfrac{1}{2}) \to (00) + (10)$$

5-4. Prove that the number of terms with a given S and L in the configuration l^N is given by the coefficient n_{LS} in the equation

$$-\frac{(1-q)P_1(q)P_2(q)}{q^\gamma D_1(q)D_2(q)} = \sum_L n_{LS}(q^L - q^{-L-1})$$

where

$$\gamma = Nl + 1 - \tfrac{1}{4}[N(N-2) + 4S^2]$$

and where the functions P and D are defined in terms of the quantities

$$\alpha = \tfrac{1}{2}N - S \qquad \beta = 2S \qquad \nu = 2l + 1$$

as follows:

$$\begin{aligned}
P_1(q) &= (1 - q^{\nu+1})(1 - q^{\nu})^2(1 - q^{\nu-1})^2 \cdots (1 - q^{\nu+2-\alpha})^2(1 - q^{\nu+1-\alpha}) \\
P_2(q) &= (1 - q^{\nu-\alpha})(1 - q^{\nu-\alpha-1}) \cdots (1 - q^{\nu+1-\alpha-\beta}) \\
D_1(q) &= (1 - q)(1 - q^2)(1 - q^3) \cdots (1 - q^{\alpha}) \\
D_2(q) &= (1 - q)(1 - q^2)(1 - q^3) \cdots (1 - q^{\alpha+\beta+1})/(1 - q^{\beta+1})
\end{aligned}$$

except when $\alpha = 0$, in which case $P_1(q) = D_1(q) = 1$; or when $\beta = 0$, in which case $P_2(q) = 1$. (This result is due to Curl and Kilpatrick.[72])

5-5. Prove that, for the reduction $R_7 \rightarrow G_2$, the irreducible representation $(ww0)$ decomposes as follows:

$$(ww0) \rightarrow (ww) + (w, w - 1) + \cdots + (w0)$$

5-6. Derive Table 5-4 for the reduction $U_5 \rightarrow R_5$ and Table 5-5 for the reduction $R_5 \rightarrow R_3$. (See Jahn.[73])

TABLE 5-4 BRANCHING RULES FOR THE REDUCTION $U_5 \rightarrow R_5$

Configuration	Dimension of $[\lambda_1\lambda_2 \cdots]$	$[\lambda_1\lambda_2 \cdots]$	(w_1w_2)
d^0	1	[0]	(00)
d^1	5	[1]	(10)
d^2	10	[11]	(11)
	15	[2]	(00)(20)
d^3	10	[111]	(11)
	40	[21]	(10)(21)
d^4	5	[1111]	(10)
	45	[211]	(11)(21)
	50	[22]	(00)(20)(22)
d^5	1	[11111]	(00)
	24	[2111]	(11)(20)
	75	[221]	(10)(21)(22)

TABLE 5-5 BRANCHING RULES FOR THE REDUCTION $R_5 \rightarrow R_3$

$D(w_1w_2)$	(w_1w_2)	L
1	(00)	S
5	(10)	D
10	(11)	PF
14	(20)	DG
35	(21)	$PDFGH$
35	(22)	$SDFGI$

5-7. The tensors $\mathbf{v}^{(k)}(ll')$ satisfy the equation

$$(l'''\|v^{(k)}(ll')\|l'') = \delta(l,l''')\delta(l',l'')[k]^{\frac{1}{2}}$$

Prove

$$[v_{q_1}{}^{(k_1)}(ll'),v_{q_2}{}^{(k_2)}(l''l''')] = \sum_{k_3,q_3} \{[k_1][k_2][k_3]\}^{\frac{1}{2}} \begin{pmatrix} k_1 & k_2 & k_3 \\ q_1 & q_2 & -q_3 \end{pmatrix} (-1)^{2l'''+l''-l'-q_3}$$
$$\times \left[\delta(l',l'')(-1)^{k_1+k_2+k_3+l+l'+l''+l'''} \begin{Bmatrix} k_1 & k_2 & k_3 \\ l''' & l & l'' \end{Bmatrix} v_{q_3}{}^{(k_3)}(ll''') - \delta(l,l''') \begin{Bmatrix} k_1 & k_2 & k_3 \\ l'' & l' & l \end{Bmatrix} v_{q_3}{}^{(k_3)}(l''l') \right]$$

Show:

1. That the $(2l+1)^2$ operators $v_q{}^{(k)}(ll)$, the $(2l'+1)^2$ operators $v_q{}^{(k)}(l'l')$, together with the $2(2l+1)(2l'+1)$ operators $v_q{}^{(k)}(ll')$ and $v_q{}^{(k)}(l'l)$, can be regarded as the infinitesimal operators of the group U_s, where $s = 2(l+l'+1)$.

2. That the $l(2l+1)$ operators $v_q{}^{(k)}(ll)$ with odd k, the $l'(2l'+1)$ operators $v_q{}^{(k)}(l'l')$ with odd k, together with the $(2l+1)(2l'+1)$ operators

$$v_q{}^{(k)}(ll') + v_q{}^{(k)}(l'l)(-1)^k$$

can be regarded as the infinitesimal operators of the group R_s.

5-8. Suppose that the configurations d^n, $d^{n-1}s$, and $d^{n-2}s^2$ are separated by energies that are sufficiently small compared with the separations of the terms to make it necessary to consider the configurations as a single entity. Show that the states may be classified by using the irreducible representations of the groups in the sequence

$$R_3 \subset R_5 \subset R_6 \subset U_6$$

Investigate the properties of the irreducible representations $(\nu_1\nu_2\nu_3)$ of R_6, and verify the formula

$$D(\nu_1\nu_2\nu_3) = (\nu_1 - \nu_2 + 1)(\nu_1 - \nu_3 + 2)(\nu_2 - \nu_3 + 1)(\nu_1 + \nu_2 + 3)(\nu_1 + \nu_3 + 2) \times (\nu_2 + \nu_3 + 1)/12$$

Obtain the scheme given in Table 5-6 for the reduction $U_6 \to R_6$. Prove that, under the reduction $R_6 \to R_5$, the decompositions of $(\nu_1\nu_2\nu_3)$ and $(\nu_1\nu_2\,-\nu_3)$ are identical. Derive the branching rules of Table 5-7 for the reduction $R_6 \to R_5$. Deduce that states of the configurations d^n, $d^{n-1}s$, and $d^{n-2}s^2$ are uniquely defined by writing

$$|n,(\nu_1\nu_2\nu_3)(w_1w_2)SM_SLM_L)$$

(This method of classifying the states of mixed configurations is due to Elliott,[74] who obtained many of the results quoted here and in the preceding problem.)

5-9. Prove that R_4 is a semi-simple group, but not a simple group. Show that, under the reduction $R_4 \to R_3$, the irreducible representation $(\nu_1\nu_2)$ of R_4

TABLE 5-6 BRANCHING RULES FOR THE REDUCTION $U_6 \to R_6$

Dimension	$[\lambda_1\lambda_2 \cdots]$	$(\nu_1\nu_2\nu_3)$
1	[0]	(000)
6	[1]	(100)
15	[11]	(110)
21	[2]	(000)(200)
20	[111]	(111)(11 − 1)
70	[21]	(100)(210)
105	[211]	(110)(211)(21 − 1)
105	[22]	(000)(200)(220)
84	[2111]	(111)(11 − 1)(210)
210	[221]	(100)(210)(221)(22 − 1)
35	[21111]	(110)(200)
189	[2211]	(110)(211)(21 − 1)(220)
175	[222]	(000)(200)(220)(222)(22 − 2)

TABLE 5-7 BRANCHING RULES FOR THE REDUCTION $R_6 \to R_5$

$D(\nu_1\nu_2\nu_3)$	$(\nu_1\nu_2\nu_3)$	(w_1w_2)
1	(000)	(00)
6	(100)	(00)(10)
15	(110)	(10)(11)
10	(111)	(11)
20	(200)	(00)(10)(20)
64	(210)	(10)(11)(20)(21)
45	(211)	(11)(21)
84	(220)	(20)(21)(22)
70	(221)	(21)(22)
35	(222)	(22)

decomposes according to the rule

$$(\nu_1\nu_2) \to (\nu_1)(\nu_1 - 1) \cdots (|\nu_2|)$$

where (ν) denotes the irreducible representation $\mathfrak{D}_\nu$ of R_3.‡

‡ The weights ν_1 and ν_2 used to define a representation of R_4 are the natural analogs of those of Prob. 5-8 and correspond exactly to the scheme used, for example, by L. C. Biedenharn [J. Math. Phys. **2**, 433 (1961)]. The weights p and q of G. Ya. Lyubarskii ["The Application of Group Theory in Physics," Pergamon Press, New York (1960)] are related to ν_1 and ν_2 by the equations

$$p = \tfrac{1}{2}(\nu_1 + \nu_2) \qquad q = \tfrac{1}{2}(\nu_1 - \nu_2)$$

5-10. Show that the dimension of the irreducible representation $(\sigma_1\sigma_2 \cdots \sigma_\nu)$ of $Sp_{2\nu}$ is given by

$$D(\sigma_1\sigma_2 \ldots \sigma_\nu) = \prod_{i=1}^{\nu} \frac{\sigma_i + \nu - i + 1}{\nu - i + 1} \prod_{k>i}^{\nu} \frac{(\sigma_i - \sigma_k + k - i)(\sigma_i + \sigma_k + 2\nu - i - k + 2)}{(k - i)(2\nu + 2 - i - k)}$$

Derive the branching rules of Table 5-8 for the reduction $Sp_4 \to R_3$. (These results have been given by Flowers.[75])

TABLE 5-8 BRANCHING RULES FOR THE REDUCTION $Sp_4 \to R_3$

$D(\sigma_1\sigma_2)$	$(\sigma_1\sigma_2)$	$\mathfrak{D}_J$
1	(00)	$\mathfrak{D}_0$
4	(10)	$\mathfrak{D}_{\frac{3}{2}}$
10	(20)	$\mathfrak{D}_1 + \mathfrak{D}_3$
5	(11)	$\mathfrak{D}_2$
16	(21)	$\mathfrak{D}_{\frac{3}{2}} + \mathfrak{D}_{\frac{5}{2}} + \mathfrak{D}_{\frac{9}{2}}$
14	(22)	$\mathfrak{D}_2 + \mathfrak{D}_4$

5-11. Prove that the symbols γ_1, γ_2, and γ_3 of Prob. 1-9 can be replaced by (111)(20), (210)(20), and (210)(21), respectively.

5-12. Prove that the dimension of the irreducible representation $[\lambda_1\lambda_2 \cdots \lambda_{2l+1}]$ of U_{2l+1} that labels the terms of l^n with spin S is

$$(2S + 1) \frac{(2l + 2)!(2l + 1)!}{(2l + 2 - b)!(2l + 1 - a)!(a + 1)!b!}$$

where $\qquad a = \frac{1}{2}n + S \qquad b = \frac{1}{2}n - S$

6

SENIORITY

6-1 DOUBLE-TENSOR OPERATORS

In the preceding chapter, the spin and orbital parts of a many-electron eigenfunction have been treated as far as possible as two separate entities, impinging on each other only at the point where their respective tableaux are constructed to be adjoint. As a preliminary step toward studying the relationship between spin and orbit in greater detail, we introduce a type of operator having well-defined properties with respect to these quantities.

The operator $\mathbf{T}^{(\kappa k)}$, comprising the $(2\kappa + 1)(2k + 1)$ components $T_{\pi q}{}^{(\kappa k)}$, is said to be a double tensor if it behaves as a tensor of rank κ with respect to the total spin angular momentum and as a tensor of rank k with respect to the total orbital angular momentum. In detail,

$$\begin{aligned}
[L_z,T_{\pi q}{}^{(\kappa k)}] &= qT_{\pi q}{}^{(\kappa k)} \\
[L_\pm,T_{\pi q}{}^{(\kappa k)}] &= \{k(k+1) - q(q \pm 1)\}^{\frac{1}{2}}T_{\pi q\pm 1}{}^{(\kappa k)} \\
[S_z,T_{\pi q}{}^{(\kappa k)}] &= \pi T_{\pi q}{}^{(\kappa k)} \\
[S_\pm,T_{\pi q}{}^{(\kappa k)}] &= \{\kappa(\kappa+1) - \pi(\pi \pm 1)\}^{\frac{1}{2}}T_{\pi\pm 1,q}{}^{(\kappa k)}
\end{aligned}$$

These equations serve as a basis for the development of the theory of double tensors; however, their resemblance to Eq. (2-26) is so marked that it is easy to see how the theory of ordinary tensor operators is to be generalized. For example, the Wigner-Eckart theorem for double tensors is

$$\begin{aligned}
(\gamma SM_SLM_L|T_{\pi q}{}^{(\kappa k)}|\gamma' S'M'_SL'M'_L) \\
= (-1)^{S+L-M_S-M_L}\begin{pmatrix} S & \kappa & S' \\ -M_S & \pi & M'_S \end{pmatrix}\begin{pmatrix} L & k & L \\ -M_L & q & M_L \end{pmatrix} \\
\times (\gamma SL\|T^{(\kappa k)}\|\gamma' S'L') \quad (6\text{-}1)
\end{aligned}$$

As a second illustration, we give the reduced matrix element of a double tensor $\mathbf{T}^{(\kappa k)}$ acting only on part 1 of a system,

$$\begin{aligned}(\gamma s_1 l_1, s_2 l_2, SL\|T^{(\kappa k)}\|\gamma' s_1' l_1', s_2' l_2', S'L') \\ = \delta(s_2,s_2')\delta(l_2,l_2')(-1)^{s_1+s_2+l_1+l_2+S'+L'+\kappa+k} \\ \times \{[S][S'][L][L']\}^{\frac{1}{2}} \begin{Bmatrix} S & \kappa & S' \\ s_1' & s_2 & s_1 \end{Bmatrix} \begin{Bmatrix} L & k & L' \\ l_1' & l_2 & l_1 \end{Bmatrix} \\ \times (\gamma s_1 l_1\|T^{(\kappa k)}\|\gamma' s_1' l_1')\end{aligned} \qquad (6\text{-}2)$$

This equation can be considered to be a generalization of Eq. (3-37).

Suppose that $\mathbf{t}^{(\kappa)}$ is a tensor operator which acts in the spin space of a single particle and for which

$$(s\|t^{(\kappa)}\|s') = (2\kappa + 1)^{\frac{1}{2}}\delta(s,s') \qquad (6\text{-}3)$$

We see that $\mathbf{t}^{(\kappa)}$ is the analogue of $\mathbf{v}^{(k)}$, defined in Eq. (5-13). The quantities

$$w_{\pi q}{}^{(\kappa k)} = t_\pi{}^{(\kappa)} v_q{}^{(k)} \qquad (6\text{-}4)$$

clearly form the components of a double tensor, and so do the many-electron operators

$$W_{\pi q}{}^{(\kappa k)} = \sum_i (w_{\pi q}{}^{(\kappa k)})_i \qquad (6\text{-}5)$$

Owing to Eqs. (5-13), (6-3), and (6-4), the reduced matrix elements for $\mathbf{w}^{(\kappa k)}$ are given by

$$(sl\|w^{(\kappa k)}\|s'l') = \{[\kappa][k]\}^{\frac{1}{2}}\delta(s,s')\delta(l,l') \qquad (6\text{-}6)$$

An application of the Wigner-Eckart theorem to matrix elements of $w_{0q}{}^{(0k)}$ between single-electron states yields

$$\mathbf{w}^{(0k)} = [s]^{-\frac{1}{2}}\mathbf{v}^{(k)} \qquad (6\text{-}7)$$

For all applications of the formulas in this paragraph, $s = s' = \frac{1}{2}$, and κ is either 0 or 1; however, it is convenient to retain the general notation to emphasize the analogy with the corresponding orbital quantities. It is to be noticed that $W_{\pi q}{}^{(\kappa k)}$ is not in general the same as $T_\pi{}^{(\kappa)} V_q{}^{(k)}$, since the latter includes terms of the type

$$(t_\pi{}^{(\kappa)})_i (v_q{}^{(k)})_j$$

with $i \neq j$. However,

$$\mathbf{W}^{(0k)} = \sum_i [s]^{-\frac{1}{2}}\mathbf{v}_i{}^{(k)} = [s]^{-\frac{1}{2}}\mathbf{V}^{(k)} \qquad (6\text{-}8)$$

The commutation relations satisfied by the operators $\mathbf{V}^{(k)}$ serve to determine which groups are relevant for classifying the orbital parts of the n-electron eigenfunctions. In our attempt to remove the somewhat arbitrary barrier between spin and orbit, it is natural to examine the commutators of the double tensors $\mathbf{W}^{(\kappa k)}$. However, it should be pointed out in passing that these tensors enjoy an importance in their own right. The spin-orbit interaction and part of the hyperfine interaction are treated later (Secs. 7-9 and 8-9) as components of double tensors; that this is possible is evident from a glance at Eqs. (6-4) and (6-5). The double-tensor formalism also allows us to treat two-particle operators involving both spin and orbital variables. By a process of recoupling, the matrix elements of an operator of this kind can be expressed as a sum over weighted products of the matrix elements of double tensors. This method is illustrated in Prob. 6-2 at the end of the chapter. For extending the procedure to evaluate matrix elements of two-particle operators between coupled groups of equivalent electrons, the reader is referred to the appendixes of a paper by Innes and Ufford.[38]

6-2 THE COMMUTATORS OF DOUBLE-TENSOR OPERATORS

By methods analogous to those of Sec. 5-4, it is straightforward to derive the commutation relations

$$[w_{\pi_1 q_1}{}^{(\kappa_1 k_1)}, w_{\pi_2 q_2}{}^{(\kappa_2 k_2)}] = \sum_{\kappa_3, k_3, \pi_3, q_3} (-1)^{2s+2l-\pi_3-q_3}\{(-1)^{\kappa_1+\kappa_2+\kappa_3+k_1+k_2+k_3} - 1\} \times \begin{Bmatrix} \kappa_1 & \kappa_2 & \kappa_3 \\ s & s & s \end{Bmatrix} \begin{Bmatrix} k_1 & k_2 & k_3 \\ l & l & l \end{Bmatrix} \begin{pmatrix} \kappa_1 & \kappa_2 & \kappa_3 \\ \pi_1 & \pi_2 & -\pi_3 \end{pmatrix} \begin{pmatrix} k_1 & k_2 & k_3 \\ q_1 & q_2 & -q_3 \end{pmatrix} \times \{[\kappa_1][\kappa_2][\kappa_3][k_1][k_2][k_3]\}^{\frac{1}{2}} w_{\pi_3 q_3}{}^{(\kappa_3 k_3)} \quad (6\text{-}9)$$

This equation, which is the generalization of Eq. (5-14), is equally valid if the substitutions

$$w_{\pi_i q_i}{}^{(\kappa_i k_i)} \rightarrow W_{\pi_i q_i}{}^{(\kappa_i k_i)}$$

are made. The number of components of this kind is

$$(2l + 1)^2(2s + 1)^2 = (4l + 2)^2$$

and they can be taken to be the infinitesimal operators of the group U_{4l+2}, corresponding to the unitary transformation

$$\psi'(slm_s m_l) = \sum_{m_s', m_l'} c(m_s m_l, m_s' m_l')\psi(slm_s' m_l') \quad (6\text{-}10)$$

The n-electron eigenfunctions $|l^n\gamma SM_S LM_L)$ are totally antisymmetric with respect to the interchange of any two electrons, and they form the basis for the irreducible representation

$$\mathcal{a}_n = [\underbrace{111 \cdots 1}_{n}]$$

of U_{4l+2}. Equation (5-62) can be readily adapted to find the dimension of this representation; the result is

$$\prod_{h>k=1}^{4l+2} (h - k + \lambda_k - \lambda_h)/(h - k)$$

where $\lambda_i = 1$ for $1 \leq i \leq n$ and zero otherwise. This expression evaluates to

$$^{4l+2}C_n = (4l + 2)!/n!(4l + 2 - n)!$$

which is precisely the number of totally antisymmetric functions that can be constructed from n electrons and $4l + 2$ functions $\psi(slm_sm_l)$.

The investigation of the subgroups of U_{4l+2} proceeds in a similar manner to the study of the subgroups of U_{2l+1}, described in Sec. 5-4. On discarding $\mathbf{W}^{(00)}$ from the set of infinitesimal operators for U_{4l+2}, the group is limited to the special unitary group SU_{4l+2}. From the $(4l + 3)(4l + 1)$ operators for SU_{4l+2}, we select the set comprising $\mathbf{W}^{(0k)}(1 \leq k \leq 2l)$ and $\mathbf{W}^{(10)}$. It is clear from Eq. (6-9) that the commutator of any two components of this set involves members of the set only. Moreover, the components of $\mathbf{W}^{(0k)}$ form the infinitesimal operators of a subgroup, and so do the three components of $\mathbf{W}^{(10)}$. That both these subgroups are invariant can be seen by setting $\kappa_1 = 0$, $k_1 = k$, $\kappa_2 = 1$, and $k_2 = 0$ in Eq. (6-9); for the triangular conditions on the 6-j symbols restrict κ_3 to 1 and k_3 to k, and hence the factor

$$(-1)^{\kappa_1+\kappa_2+\kappa_3+k_1+k_2+k_3} - 1 \tag{6-11}$$

is zero. It follows from Sec. 5-3 that the group with infinitesimal operators $\mathbf{W}^{(0k)}(1 \leq k \leq 2l)$ and $\mathbf{W}^{(10)}$ is the direct product $SU_2 \times SU_{2l+1}$, and by limiting SU_{4l+2} to this subgroup, we render independent the transformations in the orbital and spin spaces. Of course, we could have separated these two kinds of transformations without imposing the unimodular condition on U_{4l+2}; but if the operator $\mathbf{W}^{(00)}$ is included in the infinitesimal operators of both invariant subgroups, the group $U_2 \times U_{2l+1}$ is not strictly a direct product. Under the reduction $U_{4l+2} \rightarrow U_2 \times U_{2l+1}$, the coefficients in Eq. (6-10) factorize according to

$$c(m_sm_l,m_s'm_l') = c'(m_s,m_s')c''(m_l,m_l')$$

and the irreducible representation $\mathcal{a}_n$ of U_{4l+2} decomposes into a number of product representations of the type $[\lambda]\dagger \times [\lambda]$, where $[\lambda] = [\lambda_1\lambda_2 \cdots \lambda_{2l+1}]$ denotes an irreducible representation of U_{2l+1}, and $[\lambda]\dagger$, the shape adjoint to $[\lambda]$, denotes an irreducible representation of U_2. This method of combining spin and orbital functions is described in detail in Sec. 5-8, and we have reached the point where we can use the results of Sec. 5-4 to decompose U_{2l+1} further.

So far the introduction of groups whose infinitesimal operators are double tensors has added very little to our appreciation of the group-theoretical classification of states. However, there exists an alternative and more interesting way of decomposing $\mathfrak{A}_n$. It can be seen from Eq. (6-9) that the double tensors $\mathbf{W}^{(\kappa k)}$ for which $\kappa + k$ is odd form the infinitesimal operators for a subgroup of SU_{4l+2}; for if both $\kappa_1 + k_1$ and $\kappa_2 + k_2$ are odd, then $\kappa_3 + k_3$ must be odd or else the factor (6-11) vanishes. From the Wigner-Eckart theorem,

$$(l^2SM_SLM_L|W_{\pi q}^{(\kappa k)}|l^20000) = (-1)^{S+L-M_S-M_L}\begin{pmatrix} S & \kappa & 0 \\ -M_S & \pi & 0 \end{pmatrix}\begin{pmatrix} L & k & 0 \\ -M_L & q & 0 \end{pmatrix}(l^2SL\|W^{(\kappa k)}\|l^200)$$

The sum $S + L$ for terms of l^2 is even, and hence if $\kappa + k$ is odd, the triangular conditions for the nonvanishing of the 3-j symbols cannot be satisfied. Consequently the operations of the group leave invariant the ^1S state of l^2, namely, the bilinear antisymmetric form

$$\sum_m [\alpha_1\psi_1(l,m)\beta_2\psi_2(l,-m) - \beta_1\psi_1(l,-m)\alpha_2\psi_2(l,m)](-1)^m$$

The group is therefore the symplectic group in $4l + 2$ dimensions, Sp_{4l+2}.

From the collection of $(2l + 1)(4l + 3)$ components of $\mathbf{W}^{(\kappa k)}$ for which $\kappa + k$ is odd, we select the set $\mathbf{W}^{(0k)}(k = 1, 3, \ldots, 2l - 1)$ and $\mathbf{W}^{(10)}$. In a manner similar to that used in reducing SU_{4l+2} to $SU_2 \times SU_{2l+1}$, it can be shown that $SU_2 \times R_{2l+1}$ is a subgroup of Sp_{4l+2}. It is therefore apparent that the sequence

$$SU_2 \times R_{2l+1} \subset Sp_{4l+2} \subset SU_{4l+2}$$

is an alternative to

$$SU_2 \times R_{2l+1} \subset SU_2 \times SU_{2l+1} \subset SU_{4l+2}$$

6-3 THE SENIORITY NUMBER

Rather than repeat the analyses of Chap. 5 for the particular case of the group Sp_{4l+2}, we shall simply state most of the results that are presently required. The representations of Sp_{4l+2} are defined by $2l + 1$ integers $(\sigma_1\sigma_2 \cdots \sigma_{2l+1})$ satisfying

$$\sigma_1 \geq \sigma_2 \geq \cdots \geq \sigma_{2l+1} \geq 0$$

The branching rules for the reduction $U_{4l+2} \to Sp_{4l+2}$ can be obtained by considering the characters of the representations of these two groups; however, like the rules for $U_{2l+1} \to R_{2l+1}$, they can be expressed by means of a simple prescription. The decomposition of $\mathcal{a}_n$ of U_{4l+2} is particularly easy to describe; we merely write down n ones followed by $2l + 1 - n$ zeros, $n - 2$ ones followed by $2l + 3 - n$ zeros, etc., and enclose the sets of symbols in parentheses,

$$[\underbrace{111 \cdots 1}_{n}] \to (\underbrace{111 \cdots}_{n} 100 \cdots 0) + (\underbrace{1 \cdots 1}_{n-2}0000 \cdots 0) + \cdots \qquad (6\text{-}12)$$

The last representation in the collection is $(100 \cdots 0)$ if n is odd and $(000 \cdots 0)$ if n is even. If n exceeds $2l + 1$, the equivalence $\mathcal{a}_n \equiv \mathcal{a}_{4l+2-n}$, which follows from Eq. (5-77), must first be used. The rules for decomposing any representation of $U_{2\nu}$ into representations of $Sp_{2\nu}$ can be obtained from the book by Littlewood[70]; they have been given explicitly by Flowers.[75]

Tables for $U_{2l+1} \to R_{2l+1}$ such as Table 5-1 allow the branching rules for the reduction $U_{4l+2} \to SU_2 \times R_{2l+1}$ to be rapidly written down. For f electrons, for example,

$$\begin{aligned}
[0] &\to {}^1(000) \\
[1] &\to {}^2(100) \\
[11] &\to {}^3(110) + {}^1(200) + {}^1(000) \\
[111] &\to {}^4(111) + {}^2(210) + {}^2(100)
\end{aligned}$$

etc. The representations of SU_2 are defined by the total spin S, and the corresponding multiplicity, $2S + 1$, is indicated by the prefixes to the irreducible representations $W = (w_1w_2w_3)$ of R_7. With the aid of the decomposition (6-12), the branching rules for the reduction $Sp_{4l+2} \to SU_2 \times R_{2l+1}$ can be easily found by a process of subtraction. They are given in Table 6-1 for $l = 3$.

All the representations $(\sigma_1\sigma_2 \cdots \sigma_{2l+1})$ of Sp_{4l+2} that are used to classify the states of l^n are of the type $(11 \cdots 10 \cdots 0)$. To specify a representation of this kind it is necessary only to give the number of ones it contains; this number, defined formally by

$$v = \sum_i \sigma_i$$

is the *seniority number* and is listed in Table 6-1. Since we may in general decompose any representation $W = (w_1 w_2 \cdots w_l)$ of R_{2l+1} into irreducible representations of R_3, we can speak of the seniority number, or simply the seniority, of a given term of l^n.

The decomposition represented by (6-12) shows that the terms of l^n break up into sets, each set labeled by a value of v. For example, for l^2, there are two sets, corresponding to $v = 2$ and 0; for l^3, there

TABLE 6-1 BRANCHING RULES FOR THE REDUCTION $Sp_{14} \rightarrow SU_2 \times R_7$

v	$(\sigma_1\sigma_2 \cdots \sigma_7)$	${}^{2S+1}W$
0	(0000000)	${}^1(000)$
1	(1000000)	${}^2(100)$
2	(1100000)	${}^3(110)$ ${}^1(200)$
3	(1110000)	${}^4(111)$ ${}^2(210)$
4	(1111000)	${}^5(111)$ ${}^3(211)$ ${}^1(220)$
5	(1111100)	${}^6(110)$ ${}^4(211)$ ${}^2(221)$
6	(1111110)	${}^7(100)$ ${}^5(210)$ ${}^3(221)$ ${}^1(222)$
7	(1111111)	${}^8(000)$ ${}^6(200)$ ${}^4(220)$ ${}^2(222)$

are also two, corresponding to $v = 3$ and 1; for l^4, there are three, and so on. It is clear that every set of terms labeled by the same value of v must have an identical structure, since the reduction of Sp_{4l+2} to its subgroups is independent of n. Identical sets of this kind, characterized by a fixed seniority v, occur in the configurations $l^v, l^{v+2}, \ldots, l^{4l+2-v}$; hence the seniority of every term belonging to a given set is equal to the number of electrons comprising the configuration in which the set first makes its appearance. It was in this sense that the concept of seniority was first introduced by Racah.[76] Now W and S uniquely specify v; that this is so for the configurations f^n can be seen from Table 6-1. Hence, if a collection of terms characterized by a given W and S occur in the configurations $l^n, l^{n'}, l^{n''}, \ldots$, the seniority of every term in the collection is equal to the smallest integer of the set $n, n', n'', \ldots$. For example, the seniority of the term $(100)(10)\ {}^2F$ of f^3 is 1, because $(100)(10)\ {}^2F$ occurs in f^1.

It is not difficult to establish the connection between v and the weights w_i. A given irreducible representation $W = (w_1 w_2 \cdots w_l)$ of R_{2l+1} occurs in the decomposition of various representations $[\lambda_1\lambda_2 \cdots]$

of U_{2l+1}; of these, the one corresponding to the shape with the smallest number of cells is $[w_1 w_2 \cdots w_l]$. If numbers c and d are defined by the equations

$$\begin{aligned} w_1 &= \cdots = w_c = 2 \\ w_{c+1} &= \cdots = w_{c+d} = 1 \\ w_{c+d+1} &= \cdots = 0 \end{aligned}$$

then it follows from Eq. (5-42) that the spin associated with the shape is $\frac{1}{2}d$. Hence the seniority of the terms labeled by W and $S = \frac{1}{2}d$ is given by the number of cells in $[w_1 w_2 \cdots w_l]$, namely,

$$v = 2c + d$$

Since $c + d \leq l$, it follows that $2S < 2l + 1 - v$. Of course, the spin of the terms labeled by a given W need not be $\frac{1}{2}d$, since the modification rule (5-79) may have been used in the process of decomposing representations of U_{2l+1} to those of R_{2l+1}. In this case,

$$\begin{aligned} S &= \tfrac{1}{2}(2l + 1 - 2c - d) \\ v &= 2c + (2l + 1 - 2c - d) \end{aligned}$$

and

$$2S > 2l + 1 - v$$

Both possibilities are included in the formulas

$$\begin{aligned} v &= 2c + 2S \\ d &= \min\,(2S,\, 2l + 1 - v) \end{aligned} \tag{6-13}$$

which not only determine v for a given W and S but also enable the irreducible representations $W = (w_1 w_2 \cdots w_l)$ to be constructed from the numbers v and S. [Equations (6-13) are due to Racah.[69]]

The group-theoretical classification of states described in Sec. 5-14 provides the set of quantum numbers

$$l^n W \xi S M_S L M_L \tag{6-14}$$

to define a state of l^n. The symbol ξ is required when n exceeds 2 to distinguish terms that possess the same WSL classification; for f electrons it may be replaced by U_T. The introduction of seniority gives the alternative set

$$l^n v \xi S M_S L M_L \tag{6-15}$$

Both schemes have their advantages: W displays the group-theoretical properties of the representation; on the other hand, v enters more naturally into quantitative calculations.‡ For example, Eq. (5-50) for the eigenvalues of Casimir's operator for R_{2l+1} simplifies considerably if v and S are used rather than the weights w_i. We obtain

$$G(R_{2l+1})u = \tfrac{1}{2}(2l - 1)^{-1}[\tfrac{1}{2}v(4 + 4l - v) - 2S(S + 1)]u \quad (6\text{-}16)$$

for either of the two possibilities for d given in Eqs. (6-13).

6-4 THE LAPORTE-PLATT DEGENERACIES

In Sec. 4-3 are given the energies $E(^{2S+1}L)$ of the terms of f^2 as linear combinations of the integrals F_k. If these integrals are regarded as variable parameters completely at our disposal, then we can evidently choose them to make the energies of any three terms coincide. If, for example, we take the ratios

$$\begin{aligned} F_4/F_2 &= \tfrac{6}{11} \\ F_6/F_2 &= \tfrac{1}{11} \end{aligned} \quad (6\text{-}17)$$

we find

$$E(^3P) = E(^3F) = E(^3H) = F_0 - 54F_2$$

However, on calculating the energies of the remaining terms of the configuration, using the ratios (6-17), we find, surprisingly, that three of the singlets coincide,

$$E(^1D) = E(^1G) = E(^1I) = F_0 + 30F_2$$

The energy of the remaining term is given by

$$E(^1S) = F_0 + 324F_2$$

Unexpected degeneracies of this kind were first noticed by Laporte and Platt,[77] who showed that a large number of terms in configurations of the type l^n coalesce for a single choice of F_k ratios.

The energies of the terms of f^2 can be obtained from the effective Hamiltonian

$$F^{(0)} + F^{(2)}(\mathbf{C}_i^{(2)} \cdot \mathbf{C}_j^{(2)}) + F^{(4)}(\mathbf{C}_i^{(4)} \cdot \mathbf{C}_j^{(4)}) + F^{(6)}(\mathbf{C}_i^{(6)} \cdot \mathbf{C}_j^{(6)})$$

To study the Laporte-Platt degeneracies, we drop $F^{(0)}$, replace the remaining integrals $F^{(k)}$ by their counterparts F_k, and use Eqs. (6-17)

‡ The schemes (6-14) and (6-15) are not the only ones available. If we are prepared to abandon the quantum number S, for example, then we may use the subgroup R_4 of U_{4l+2} whose infinitesimal operators are $W_{0q}^{(\kappa 1)}$ ($\kappa = 0, 1$; $q = -1, 0, 1$). The states of l^n can now be classified by the set of quantum numbers $l^n M_S \eta(\nu_1\nu_2)LM_L$, where $(\nu_1\nu_2)$ has the same significance here as in Prob. 5-9, and η is an additional classificatory symbol.

to eliminate F_4 and F_6. The resulting expression is

$$225F_2(\mathbf{C}_i^{(2)} \cdot \mathbf{C}_j^{(2)}) + 594F_2(\mathbf{C}_i^{(4)} \cdot \mathbf{C}_j^{(4)}) + (\tfrac{16731}{25})F_2(\mathbf{C}_i^{(6)} \cdot \mathbf{C}_j^{(6)}) \quad (6\text{-}18)$$

Provided that all calculations are performed within the configurations l^n, the tensors $\mathbf{C}^{(k)}$ and $\mathbf{v}^{(k)}$ for even k differ only through their reduced matrix elements; hence we can replace $\mathbf{C}^{(k)}$ by $\mathbf{v}^{(k)}$ in (6-18) provided that the factor

$$\frac{(l\|C^{(k)}\|l)^2}{(l\|v^{(k)}\|l)^2} = \frac{[l]^2}{[k]}\begin{pmatrix} l & k & l \\ 0 & 0 & 0 \end{pmatrix}^2$$

is included. The expression (6-18) now assumes the strikingly simple form

$$84F_2[(\mathbf{v}_i^{(2)} \cdot \mathbf{v}_j^{(2)}) + (\mathbf{v}_i^{(4)} \cdot \mathbf{v}_j^{(4)}) + (\mathbf{v}_i^{(6)} \cdot \mathbf{v}_j^{(6)})]$$

This suggests that an investigation be made of the operator

$$\Xi = \sum_{i>j} \sum_{\text{even } k} (\mathbf{v}_i^{(k)} \cdot \mathbf{v}_j^{(k)}) \qquad (k > 0)$$

The form of Ξ prompts us to examine Casimir's operator for SU_{2l+1}, since

$$\begin{aligned} G(SU_{2l+1}) &= \tfrac{1}{2}[l]^{-1} \sum_{k>0} (\mathbf{V}^{(k)})^2 \\ &= [l]^{-1} \sum_{i>j} \sum_{k>0} (\mathbf{v}_i^{(k)} \cdot \mathbf{v}_j^{(k)}) + \tfrac{1}{2}[l]^{-1} \sum_{i} \sum_{k>0} (\mathbf{v}_i^{(k)})^2 \end{aligned}$$

There are $2ln$ terms in the double sum on the extreme right, every one of which satisfies equations of the type

$$\begin{aligned} (lm|(\mathbf{v}^{(k)})^2|lm) &= \sum_{m',q} (-1)^q (lm|v_q^{(k)}|lm')(lm'|v_{-q}^{(k)}|lm) \\ &= (2k+1)/(2l+1) \qquad (6\text{-}19) \end{aligned}$$

The eigenvalues of $G(SU_{2l+1})$ are given by Eq. (5-51), and it is straightforward to deduce that the eigenvalues of

$$\Xi_1 = \sum_{i>j} \sum_{k>0} (\mathbf{v}_i^{(k)} \cdot \mathbf{v}_j^{(k)}) \qquad (6\text{-}20)$$

are

$$n - \tfrac{1}{4}n^2 - S(S+1) - \tfrac{1}{2}n(n-1)/(2l+1) \qquad (6\text{-}21)$$

Of course, Ξ_1 contains terms in the sum for which k is odd. To exclude them, we use Eq. (5-47) and write

$$\begin{aligned} \sum_{i>j} \sum_{\text{odd } k} (\mathbf{v}_i^{(k)} \cdot \mathbf{v}_j^{(k)}) &= \Xi_2 \\ &= \tfrac{1}{2}(2l-1)G(R_{2l+1}) - \frac{1}{2} \sum_{i} \sum_{\text{odd } k} (\mathbf{v}_i^{(k)})^2 \end{aligned}$$

The eigenvalues of $G(R_{2l+1})$ are given in Eq. (6-16), and we use Eq. (6-19) again to evaluate the eigenvalues of the double sum on the right. The eigenvalues of Ξ_2 are found to be

$$\tfrac{1}{8}v(4 + 4l - v) - \tfrac{1}{2}S(S + 1) - \tfrac{1}{2}nl \tag{6-22}$$

Since $\Xi = \Xi_1 - \Xi_2$, the eigenvalues of Ξ are obtained by subtracting (6-22) from (6-21). The result is

$$-\tfrac{1}{2}S(S + 1) + \tfrac{1}{8}v(v + 2) + \tfrac{1}{4}(n - v)(2l + 3) - \tfrac{1}{4}n(n - 1)(2l + 3)/(2l + 1) \tag{6-23}$$

The striking feature of the expression for the eigenvalues of Ξ is that L does not make its appearance, in spite of the fact that the tensors $\mathbf{v}^{(k)}$ from which Ξ is built act on the orbital variables. The degeneracies in f^2 may now be readily understood, since each group of degenerate terms corresponds to a particular pair of values for S and v. Thus, on making the substitutions $S = 0$, $v = 2$, $n = 2$, $l = 3$ in (6-23) and multiplying the result by $84F_2$, we obtain an energy of $30F_2$ for the terms ^1D, ^1G, and ^1I of f^2, in agreement with the initial calculation. The general form of (6-23) allows us to extend the results to other configurations; for f^3, the terms fall into three sets, with relative energies as follows:

^4SDFGI:	$-162F_2$	$(v = 3, S = \tfrac{3}{2})$
$^2PDDFGGHHIKL$:	$-36F_2$	$(v = 3, S = \tfrac{1}{2})$
^2F:	$216F_2$	$(v = 1, S = \tfrac{1}{2})$

Much of the preceding discussion may appear somewhat hypothetical, since the ratios (6-17) that are necessary for the degeneracies are so far removed from the ratios (4-11) that give good agreement with the experimental data for f^2. However, for d electrons there exists a physical interpretation for the degeneracies. When $l = 2$,

$$\sum_{i>j} [(\mathbf{v}_i^{(2)} \cdot \mathbf{v}_j^{(2)}) + (\mathbf{v}_i^{(4)} \cdot \mathbf{v}_j^{(4)})] = (\tfrac{7}{10}) \sum_{i>j} [5(\mathbf{C}_i^{(2)} \cdot \mathbf{C}_j^{(2)}) + 9(\mathbf{C}_i^{(4)} \cdot \mathbf{C}_j^{(4)})]$$

The matrix elements of the tensors $\mathbf{C}^{(k)}$ for which k is odd vanish if taken between states of the configuration l^n; moreover, the matrix elements of all tensors of rank greater than 4 vanish for d electrons. Hence the matrix elements of Ξ are identical, apart from a constant additive term, to those of

$$(\tfrac{7}{10}) \sum_{i>j} \sum_{k} (2k + 1)(\mathbf{C}_i^{(k)} \cdot \mathbf{C}_j^{(k)}) = (\tfrac{7}{10}) \sum_{i>j} \sum_{k} (2k + 1)P_k(\cos \omega_{ij})$$

where, in analogy to the notation of Sec. 4-3, ω_{ij} is the angle between the vectors $\mathbf{r}_i$ and $\mathbf{r}_j$. The sum

$$\frac{1}{2}\sum_k (2k+1)P_k(\cos\omega_{ij})$$

although divergent,[78] behaves under integration like $\delta(\cos\omega_{ij} - 1)$, where $\delta(x)$ is the delta function of Dirac;[3] it can therefore be regarded as the angular part of the function $\delta(\mathbf{r}_i - \mathbf{r}_j)$. Hence the Laporte-Platt degeneracies correspond to interelectron forces of infinitesimally short range. This is not true in general; for example, the numerical coefficients in the effective Hamiltonian (6-18), corresponding to f electrons, are not proportional to $2k+1$.

PROBLEMS

6-1. Verify the relation

$$(\gamma SL\|W^{(0k)}\|\gamma' SL') = [S]^{\frac{1}{2}}[s]^{-\frac{1}{2}}(\gamma SL\|V^{(k)}\|\gamma' SL')$$

6-2. Show that

$$\Big(\gamma SL\Big\|\sum_{i,j}\{\mathbf{s}_i^{(\kappa_1)}\mathbf{s}_j^{(\kappa_2)}\}^{(\kappa)}\{\mathbf{T}_i^{(k_1)}\mathbf{Y}_j^{(k_2)}\}^{(k)}\Big\|\gamma' S'L'\Big)$$
$$= \{[\kappa][k]\}^{\frac{1}{2}}(-1)^{S+S'+\kappa+L+L'+k}\sum_{\gamma'',S'',L''}(\gamma SL\|U^{(\kappa_1 k_1)}\|\gamma'' S''L'')$$
$$\times(\gamma'' S''L''\|Z^{(\kappa_2 k_2)}\|\gamma' S'L')\begin{Bmatrix} S & \kappa & S' \\ \kappa_2 & S'' & \kappa_1 \end{Bmatrix}\begin{Bmatrix} L & k & L' \\ k_2 & L'' & k_1 \end{Bmatrix}$$

where
$$\mathbf{U}^{(\kappa k)} = \sum_i \mathbf{s}_i^{(\kappa)}\mathbf{T}_i^{(k)}$$
and
$$\mathbf{Z}^{(\kappa k)} = \sum_i \mathbf{s}_i^{(\kappa)}\mathbf{Y}_i^{(k)}$$

(See Innes.[53])

6-3. Use the formula for $D(\sigma_1\sigma_2\cdots\sigma_\nu)$ given in Prob. 5-10 to prove that the dimension $D(v)$ of the irreducible representation $(11\cdots10\cdots0)$ of Sp_{4l+2}, corresponding to a seniority number v, is given by

$$D(v) = 2(2l+2-v)\frac{(4l+3)!}{v!(4l+4-v)!}$$

Check some of the decompositions of Table 6-1 by means of this equation.

7

FRACTIONAL PARENTAGE COEFFICIENTS

7-1 INTRODUCTION

In the group-theoretical classification of states, the set of quantum numbers $l^nSM_SLM_L$ is augmented by the irreducible representations of certain groups, rather than by the nondescript symbol γ. The immediate advantage is that many states can be uniquely specified without going to the lengths of calculating the particular linear combinations of determinantal product states to which they correspond; however, if we are obliged to construct these linear combinations in order to evaluate the matrix elements of the various operators in the Hamiltonian, nothing is gained. The problem of circumventing the explicit introduction of determinantal product states has been given an elegant solution by Racah,[76] who elaborated an idea of Bacher and Goudsmit.[79]

The principle, which is very simple, is most clearly described if the sets of quantum numbers defining states of the configurations l^n, l^{n-1}, and l are abbreviated as follows:

$$\gamma SM_SLM_L \equiv \Omega$$
$$\bar{\gamma}\bar{S}\bar{M}_S\bar{L}\bar{M}_L \equiv \bar{\Omega}$$
$$m_sm_l \equiv \omega$$

It is understood that Ω refers to a state of l^n, $\bar{\Omega}$ to a state of l^{n-1}, and ω to a single-electron state. Since every determinantal product state for an n-electron system is a sum of terms, each of which involves an eigenfunction $\psi_n(slm_sm_l)$ of the nth electron, we may write

$$|\Omega) = \sum_{\bar{\Omega},\omega} (\bar{\Omega};\omega|\Omega)|\bar{\Omega})|\omega_n) \tag{7-1}$$

The subscript n to ω makes it clear that the final state refers to the nth electron. Owing to the equivalence of the electrons, the matrix element of a single-particle operator $F = \sum_i f_i$ between two states of the configuration l^n is n times the corresponding matrix element for f_n, that is

$$(\Omega|F|\Omega') = n(\Omega|f_n|\Omega')$$

Making use of Eq. (7-1), we get

$$\begin{aligned}(\Omega|F|\Omega') &= n \sum_{\bar{\Omega},\bar{\Omega}',\omega,\omega'} (\Omega|\bar{\Omega};\omega)(\bar{\Omega}|\bar{\Omega}')(\omega_n|f_n|\omega_n')(\bar{\Omega}';\omega'|\Omega') \\ &= n \sum_{\bar{\Omega},\omega,\omega'} (\Omega|\bar{\Omega};\omega)(\omega_n|f_n|\omega_n')(\bar{\Omega};\omega'|\Omega') \end{aligned} \tag{7-2}$$

The evaluation of the matrix elements of f_n for single-electron states is trivial, and the problem reduces to finding the coefficients $(\Omega|\bar{\Omega};\omega)$ and $(\bar{\Omega};\omega'|\Omega')$. We shall assume that the phases of the states of the configurations l^n, l^{n-1}, and l have been chosen to ensure that the coefficients are real; hence

$$(\bar{\Omega};\omega|\Omega) = (\Omega|\bar{\Omega};\omega)$$

The obvious way of extending the method to two-particle operators of the type $G = \sum_{i>j} g_{ij}$ is to factor out the eigenfunctions of electron $n - 1$ as well as those of electron n. The product eigenfunctions of these two electrons can be expressed as linear combinations of the states of l^2; hence we may write

$$|\Omega) = \sum_{\Pi,\pi} (\Pi;\pi|\Omega)|\Pi)|\pi_{n,n-1}) \tag{7-3}$$

where Π and π denote states of l^{n-2} and l^2, respectively. The relation

$$(\Omega|G|\Omega') = \tfrac{1}{2}n(n-1)(\Omega|g_{n,n-1}|\Omega')$$

leads at once to

$$(\Omega|G|\Omega') = \tfrac{1}{2}n(n-1) \sum_{\Pi,\pi,\pi'} (\Omega|\Pi;\pi)(\pi_{n,n-1}|g_{n,n-1}|\pi'_{n,n-1})(\Pi;\pi'|\Omega') \tag{7-4}$$

Although this result is of value, there exists a more convenient way of evaluating the matrix elements of two-particle operators. Of the $\frac{1}{2}n(n-1)$ components g_{ij} $(i > j)$ comprising G, $\frac{1}{2}(n-1)(n-2)$ of them do not involve the nth electron. Hence

$$(\Omega|G|\Omega') = [n/(n-2)](\Omega| \sum_{j<i\neq n} g_{ij}|\Omega')$$

The expansion (7-1) is used to give the equation

$$(\Omega|G|\Omega') = [n/(n-2)] \sum_{\bar{\Omega},\bar{\Omega}',\omega} (\Omega|\bar{\Omega};\omega)(\bar{\Omega}| \sum_{j<i\neq n} g_{ij}|\bar{\Omega}')(\bar{\Omega}';\omega|\Omega') \quad (7\text{-}5)$$

For many cases of interest, g_{ij} is a scalar operator, and all terms in the sum vanish unless $\bar{\Omega}$ and $\bar{\Omega}'$ share a common set of quantum numbers $\bar{S}\bar{M}_S\bar{L}\bar{M}_L$. Equation (7-5) relates the matrix elements of G for l^n to those for l^{n-1}, thus necessitating the construction of a lengthy chain calculation for configurations of large n. This disadvantage is offset by the fact that the coefficients $(\Omega|\bar{\Omega};\omega)$ are precisely the ones that occur in the calculation of the matrix elements of single-particle operators; we are therefore able to avoid the evaluation of the coefficients $(\Omega|\Pi;\pi)$ altogether.

In view of these remarks, our attention is directed to the coefficients $(\bar{\Omega};\omega|\Omega)$, which, in detail, are written

$$(l^{n-1}\bar{\gamma}\bar{S}\bar{M}_S\bar{L}\bar{M}_L;lm_sm_l|l^n\gamma SM_SLM_L) \quad (7\text{-}6)$$

For a given term γSL of l^n, the terms $\bar{\gamma}\bar{S}\bar{L}$ of l^{n-1} for which the coefficients (7-6) do not vanish are called the *parents* of γSL. Conversely, for a given term $\bar{\gamma}\bar{S}\bar{L}$ of l^{n-1}, the terms γSL for which the coefficients do not vanish are called the *daughters*, or *offspring*, of $\bar{\gamma}\bar{S}\bar{L}$. Although the calculation of these coefficients may not be easy, it has, at any rate, to be done only once. The results are then available for the construction of the matrix elements of any single-particle or two-particle operator.

7-2 THE FACTORIZATION OF THE COEFFICIENTS $(\bar{\Omega};\omega|\Omega)$

Under the operations of the group U_{4l+2}, the ${}^{4l+2}C_n$ states $|\Omega)$ transform according to the irreducible representation $\mathfrak{A}_n$ of U_{4l+2}. Similarly, the ${}^{4l+2}C_{n-1}$ states $|\bar{\Omega})$ transform according to $\mathfrak{A}_{n-1}$, and the $4l+2$ states $|\omega)$ transform according to $\mathfrak{A}_1$. Equation (7-1) can therefore be interpreted as combining the basis functions of the Kronecker product $\mathfrak{A}_{n-1} \times \mathfrak{A}_1$ so that the resulting functions form a basis for $\mathfrak{A}_n$; from this point of view the coefficients $(\bar{\Omega};\omega|\Omega)$ are examples of the quantities $(\Gamma_Q j;\Gamma_R k|\Gamma_S \beta l)$ introduced in Sec. 2-6.

To take advantage of this correspondence, we introduce a theorem of Racah.[69] It is quoted here without proof. We denote the irreducible representations of a group $\mathcal{G}$ by A and those of a subgroup $\mathcal{H}$ of $\mathcal{G}$ by B. The symbol b is used to denote a particular basis function of the irreducible representation B. It is supposed that the basis functions for every A of $\mathcal{G}$ can be taken directly as basis functions for the various irreducible representations $B, B', \ldots$ into which A decomposes, without going through the process of constructing linear combinations. We may therefore label a particular basis function for the irreducible representation A of $\mathcal{G}$ by $A\beta Bb$; the index β is required when two or more identical irreducible representations B occur in the decomposition of A and serves to make the specification complete. If we suppose that A occurs not more than once in the decomposition of $A_1 \times A_2$ and B not more than once in the decomposition of $B_1 \times B_2$, Racah's theorem can be concisely expressed by the equation

$$(A_1\beta_1B_1b_1;A_2\beta_2B_2b_2|A\beta Bb) \\ = (B_1b_1;B_2b_2|Bb)(A_1\beta_1B_1 + A_2\beta_2B_2|A\beta B) \quad (7\text{-}7)$$

The coefficients of the type appearing on the left decompose the Kronecker product $A_1 \times A_2$ into irreducible representations A of $\mathcal{G}$; similarly, the first coefficient on the right of Eq. (7-7) is a transformation coefficient connected with the decomposition of the Kronecker product $B_1 \times B_2$ into irreducible representations B of $\mathcal{H}$. The final coefficient is a quantity independent of b_1, b_2, and b.

Racah's theorem is first applied to U_{4l+2} and its subgroup $U_2 \times U_{2l+1}$. The symbols β are not required, and the second factor on the right of Eq. (7-7) becomes

$$(\mathcal{A}_{n-1}[\bar{\lambda}]\dagger[\bar{\lambda}] + \mathcal{A}_1[1][1]|\mathcal{A}_n[\lambda]\dagger[\lambda])$$

which can be simply written as

$$(l^{n-1}[\bar{\lambda}] + l|l^n[\lambda])$$

without loss of information. The symbols B in the first factor stand for representations of $U_2 \times U_{2l+1}$; hence the factor itself breaks up into a spin part

$$([\bar{\lambda}]\dagger\bar{S}\bar{M}_S;[1]sm_s|[\lambda]\dagger SM_S)$$

which is nothing else but the VC coefficient

$$(\bar{S}\bar{M}_Ssm_s|\bar{S}sSM_S)$$

and an orbital part

$$([\bar{\lambda}]\bar{W}\xi\bar{L}\bar{M}_L;[1]lm_l|[\lambda]W\xi LM_L)$$

Repeated applications of Racah's theorem give

$$\begin{aligned}([\bar{\lambda}]\bar{W}\xi\bar{L}\bar{M}_L;[1]lm_l|[\lambda]W\xi LM_L) &= (\bar{W}\xi\bar{L}\bar{M}_L;(10\cdots0)lm_l|W\xi LM_L)([\bar{\lambda}]\bar{W}+[1](10\cdots0)|[\lambda]W)\\ &= (\bar{L}\bar{M}_Llm_l|\bar{L}lLM_L)(\bar{W}\xi\bar{L}+l|W\xi L)([\bar{\lambda}]\bar{W}+[1](10\cdots0)|[\lambda]W)\end{aligned}$$

It is convenient to recombine two of the factors and write

$$(l^{n-1}[\bar{\lambda}]+l|l^n[\lambda])([\bar{\lambda}]\bar{W}+[1](10\cdots0)|[\lambda]W) = (l^{n-1}\bar{v}\bar{S}+l|\}l^nvS) \qquad (7\text{-}8)$$

No information in the description of the states is lost by this contraction. The symbol v is preferred to W because, as is shown later, the coefficients on the right of Eq. (7-8) can be expressed as simple algebraic functions of n, l, v, S, $\bar{v}$, and $\bar{S}$. The single symbol l is used here and in other coefficients to denote the relevant quantum numbers for a single electron. The successive factorizations can be summarized by the formula

$$\begin{aligned}(l^{n-1}\bar{W}\xi\bar{S}\bar{M}_S\bar{L}\bar{M}_L;lm_sm_l|l^nW\xi SM_SLM_L) &= (\bar{S}\bar{M}_Ssm_s|\bar{S}sSM_S)(\bar{L}\bar{M}_Llm|\bar{L}lLM_L)\\ &\quad\times(\bar{W}\xi\bar{L}+l|W\xi L)(l^{n-1}\bar{v}\bar{S}+l|\}l^nvS)\end{aligned} \qquad (7\text{-}9)$$

The part remaining after the two VC coefficients have been factored out defines the quantities

$$(l^{n-1}\bar{W}\xi\bar{S}\bar{L}|\}l^nW\xi SL) = (\bar{W}\xi\bar{L}+l|W\xi L)(l^{n-1}\bar{v}\bar{S}+l|\}l^nvS) \qquad (7\text{-}10)$$

which are called *fractional parentage coefficients* or *coefficients of fractional parentage*, frequently abbreviated to cfp.

For f electrons, the groups used for classifying the states are augmented by G_2, and a further factorization can be performed. In detail,

$$\begin{aligned}(\bar{W}\bar{U}\bar{\tau}\bar{L}\bar{M}_L;(100)(10)fm_l|WU\tau LM_L) &= (\bar{U}\bar{\tau}\bar{L}\bar{M}_L;(10)fm_l|U\tau LM_L)(\bar{W}\bar{U}+f|WU)\\ &= (\bar{L}\bar{M}_Lfm_l|\bar{L}fLM_L)(\bar{U}\bar{\tau}\bar{L}+f|U\tau L)(\bar{W}\bar{U}+f|WU)\end{aligned}$$

Thus

$$(f^{n-1}\bar{W}\bar{U}\bar{\tau}\bar{S}\bar{L}|\}f^nWU\tau SL) = (\bar{U}\bar{\tau}\bar{L}+f|U\tau L)(\bar{W}\bar{U}+f|WU)(f^{n-1}\bar{v}\bar{S}+f|\}f^nvS) \qquad (7\text{-}11)$$

The factorizations exhibited in Eqs. (7-10) and (7-11) substantially reduce the amount of tabulating that has to be done.

If the group Sp_{4l+2} is used to classify the many-electron states, the factorization (7-9) still results; however, the final factor arises, not through a contraction of the type represented by Eq. (7-8), but instead in virtue of

$$(l^{n-1}\bar{v} + l|l^n v)(\bar{v}\bar{S} + l|vS) = (l^{n-1}\bar{v}\bar{S} + l|\}l^n vS) \tag{7-12}$$

7-3 ORTHOGONALITY RELATIONS

In view of the orthonormality of the n-electron states, we have

$$\sum_{\bar{\Omega},\omega} (\bar{\Omega};\omega|\Omega)(\bar{\Omega};\omega|\Omega') = \delta(\Omega,\Omega')$$

In full detail,

$$\begin{aligned}\Sigma(l^{n-1}\bar{W}\bar{\xi}\bar{S}\bar{M}_S\bar{L}\bar{M}_L;lm_sm_l|l^nW\xi SM_SLM_L)\\ \times (l^{n-1}\bar{W}\bar{\xi}\bar{S}\bar{M}_S\bar{L}\bar{M}_L;lm_sm_l|l^nW'\xi'S'M'_SL'M'_L)\\ = \delta(W,W')\delta(\xi,\xi')\delta(S,S')\delta(M_S,M'_S)\delta(L,L')\delta(M_L,M'_L)\end{aligned} \tag{7-13}$$

The sum runs over $\bar{W}$, $\bar{\xi}$, $\bar{S}$, $\bar{M}_S$, $\bar{L}$, $\bar{M}_L$, m_s, and m_l. The distribution of the quantum numbers among the four coefficients of Eq. (7-9) gives each factor a special responsibility in ensuring that Eq. (7-13) is satisfied. On making the substitution (7-9), the sums over $\bar{M}_S$, m_s, $\bar{M}_L$, and m_l can be immediately carried out and give the factor

$$\delta(S,S')\delta(M_S,M'_S)\delta(L,L')\delta(M_L,M'_L)$$

The occurrence of the remaining delta functions is guaranteed if we insist that the equations

$$\sum_{\bar{L},\bar{\xi}} (\bar{W}\bar{\xi}\bar{L} + l|W\xi L)(\bar{W}\bar{\xi}\bar{L} + l|W'\xi'L) = \delta(W,W')\delta(\xi,\xi') \tag{7-14}$$

and

$$\sum_{\bar{v},\bar{S}} (l^{n-1}\bar{v}\bar{S} + l|\}l^n vS)^2 = 1 \tag{7-15}$$

are satisfied. For f electrons, Eq. (7-14) is replaced by the pair

$$\sum_{\bar{L},\bar{\tau}} (\bar{U}\bar{\tau}\bar{L} + f|U\tau L)(\bar{U}\bar{\tau}\bar{L} + f|U'\tau'L) = \delta(U,U')\delta(\tau,\tau') \tag{7-16}$$

and

$$\sum_{\bar{U}} (\bar{W}\bar{U} + f|WU)(\bar{W}\bar{U} + f|W'U) = \delta(W,W') \tag{7-17}$$

7-4 THE CFP $(p^2\bar{S}\bar{L}|\}p^3SL)$

The advantage of the factorizations indicated in Eqs. (7-10) and (7-11) is that the tables of the separate factors are much smaller and more manageable than the single table that would be required to list the complete cfp. In the preparation of these smaller tables, we naturally discard methods that give complete cfp in preference to those which determine the entries directly. However, as a first example of the evaluation of some cfp, it is convenient to introduce a method of the first kind; although extremely tedious for all but the most elementary configurations, it is much easier to describe than other more powerful and sophisticated techniques. The objection that the calculation gives complete cfp instead of their component factors can be nullified by choosing configurations of the type p^n: the groups R_{2l+1} and R_3 are identical, and the coefficient $(\bar{W}\bar{\xi}\bar{L} + l|W\xi L)$ reduces to at most a phase factor. We write

$$(p^{n-1}\bar{S}\bar{L}|\}p^nSL)$$

for the complete cfp and consider in detail the case of $n = 3$.

The starting point is Eq. (7-1). We make the substitutions $l = 1$ and $n = 3$ and factor out the VC coefficients,

$$|p^3SM_SLM_L) = \sum_{\bar{S},\bar{L}} (p^2\bar{S}\bar{L}|\}p^3SL) \sum_{m_s,m_l,\bar{M}_S,\bar{M}_L} (\bar{S}\bar{M}_S\tfrac{1}{2}m_s|\bar{S}\tfrac{1}{2}SM_S) \times (\bar{L}\bar{M}_L1m_l|\bar{L}1LM_L)|p^2\bar{S}\bar{M}_S\bar{L}\bar{M}_L)|p_3m_sm_l)$$

Suppose we insist that the expression on the right be antisymmetrical with respect to the interchange of any two electrons. It is obviously antisymmetrical with respect to the interchange of electrons 1 and 2; to test for the antisymmetry between electrons 2 and 3, we make the expansion

$$|p^2\bar{S}\bar{M}_S\bar{L}\bar{M}_L) = \sum_{m'_s,m''_s,m'_l,m''_l} (\tfrac{1}{2}m'_s\tfrac{1}{2}m''_s|\tfrac{1}{2}\tfrac{1}{2}\bar{S}\bar{M}_S)(1m'_l1m''_l|11\bar{L}\bar{M}_L) \times |p_1m'_sm'_l)|p_2m''_sm''_l)$$

and couple electrons 2 and 3 by writing

$$|p_2m''_sm''_l)|p_3m_sm_l) = \sum_{S',M'_S,L',M'_L} (\tfrac{1}{2}\tfrac{1}{2}S'M'_S|\tfrac{1}{2}m''_s\tfrac{1}{2}m_s)(11L'M'_L|1m''_l1m_l) \times |p^2S'M'_SL'M'_L)$$

The VC coefficients are converted to 3-j symbols, and Eq. (3-6) is used twice—once in the orbital space and once in the spin space. We obtain

$$|p^3SM_SLM_L) = \sum_{\bar{S},\bar{L}} (p^2\bar{S}\bar{L}|\}p^3SL) \sum_{S',M'_S,L',M'_L,m'_s,m'_l} \{[\bar{L}][L'][\bar{S}][S']\}^{\frac{1}{2}} \times (-1)^{S'+L'} \begin{Bmatrix} L & 1 & L' \\ 1 & 1 & \bar{L} \end{Bmatrix} \begin{Bmatrix} S & \frac{1}{2} & S' \\ \frac{1}{2} & \frac{1}{2} & \bar{S} \end{Bmatrix} (S'M'_S\tfrac{1}{2}m'_s|S'\tfrac{1}{2}SM_S) \times (L'M'_L1m'_l|L'1LM_L)|p^2S'M'_SL'M'_L)|p_1m'_sm'_l)$$

The antisymmetry with respect to the interchange of electrons 2 and 3 is guaranteed if $S' + L'$ is even; hence, if $S' + L'$ is odd, the equation

$$\sum_{\bar{S},\bar{L}} (p^2\bar{S}\bar{L}|\}p^3SL)\{[\bar{L}][\bar{S}]\}^{\frac{1}{2}} \begin{Bmatrix} L & 1 & L' \\ 1 & 1 & \bar{L} \end{Bmatrix} \begin{Bmatrix} S & \frac{1}{2} & S' \\ \frac{1}{2} & \frac{1}{2} & \bar{S} \end{Bmatrix} = 0 \quad (7\text{-}18)$$

must be valid. Let us choose the daughter ^2P. The forbidden terms of p^2 are ^3S, ^1P, and ^3D. Putting $S = \frac{1}{2}$, $L = 1$, $S' = 1$, and $L' = 0$ in Eq. (7-18), we get

$$\tfrac{1}{6}(p^2\ {}^1S|\}p^3\ {}^2P) - \tfrac{1}{6}(p^2\ {}^3P|\}p^3\ {}^2P) + \tfrac{1}{6}(5)^{\frac{1}{2}}(p^2\ {}^1D|\}p^3\ {}^2P) = 0 \quad (7\text{-}19)$$

Similarly, for $S' = 0$, $L' = 1$, corresponding to the forbidden term ^1P of p^2, we obtain

$$\tfrac{1}{6}(p^2\ {}^1S|\}p^3\ {}^2P) + \tfrac{1}{4}(p^2\ {}^3P|\}p^3\ {}^2P) - \tfrac{1}{12}(5)^{\frac{1}{2}}(p^2\ {}^1D|\}p^3\ {}^2P) = 0 \quad (7\text{-}20)$$

Equations (7-19) and (7-20) determine the ratios, one to another, of the three cfp; the equation

$$\sum_{\bar{S},\bar{L}} (p^2\bar{S}\bar{L}|\}p^3SL)^2 = 1$$

fixes their magnitude. The procedure can be repeated for the remaining terms of p^3, namely, ^2D and ^4S. The results are collected in Table 7-1. An arbitrary choice of phase has been made for each row, but the relative signs of the entries in a row are not at our disposal.

TABLE 7-1 THE CFP $(p^2\bar{S}\bar{L}|\}p^3SL)$

Offspring (terms of p^3)	Parents (terms of p^2)		
	^1S	^3P	^1D
^4S	0	1	0
^2P	$(\frac{2}{9})^{\frac{1}{2}}$	$-(\frac{1}{2})^{\frac{1}{2}}$	$-(\frac{5}{18})^{\frac{1}{2}}$
^2D	0	$(\frac{1}{2})^{\frac{1}{2}}$	$-(\frac{1}{2})^{\frac{1}{2}}$

By making use of the complementary nature of electrons and holes, matrix elements between states of p^{6-n} can be related to the corresponding matrix elements between states of p^n. The configurations for which $n \leq 2$ can be dealt with by elementary techniques; hence Table 7-1 is the only table of cfp required for configurations of p electrons.

7-5 THE COEFFICIENTS $(\bar{U}\bar{\tau}\bar{L} + f|U\tau L)$

The method described in the previous section can be extended to other configurations. For d^n, the tables of cfp are much larger than Table 7-1, but not impossibly so. However, there are so many terms in configurations of the type f^n that the simple approach we have adopted for finding the cfp is quite impracticable. The table for $(f^5||f^6)$, for example, has 119 rows and 73 columns. A method for constructing tables of the separate factors that go to make up the complete cfp has been described by Racah.[58] The special properties of the configurations f^n need some discussion, and we shall therefore illustrate Racah's method by selecting states from configurations of this type; at the same time it will become clear how to proceed in other cases.

The first factor of the complete cfp of Eq. (7-11) is

$$(\bar{U}\bar{\tau}\bar{L} + f|U\tau L) \tag{7-21}$$

The procedure we follow in order to evaluate these coefficients consists of a rather elaborate chain calculation, moving in the direction of increasing $D(\bar{U})$, the dimension of the irreducible representation $\bar{U}$. At the same time as the coefficients (7-21) are found, tables of the reduced matrix elements

$$(f^n W\, U\tau SL||V^{(5)}||f^n W\, U\tau' SL')$$

are constructed. As will be shown in Sec. 8-5, these matrix elements are independent of S, W, and n; we accordingly drop these descriptions and write simply

$$(U\tau L||V^{(5)}||U\tau' L')$$

For the moment, we may regard this simplification as an extension to G_2 of the familiar result that $\mathbf{V}^{(1)}$, the set of infinitesimal operators of R_3, is proportional to $\mathbf{L}$ [see Eq. (5-38)] and consequently independent of S, W, n, U, and τ.

In order to obtain equations involving the coefficients (7-21), we break $V_q^{(5)}$ into two parts. The first, $(V_q^{(5)})'$, acts only between parent states, while the second, $v_q^{(5)}$, links states of the added electron. The combination

$$\delta(m_l,m_l')(\bar{U}\bar{\tau}\bar{L}\bar{M}_L|(V_q^{(5)})'|\bar{U}'\bar{\tau}'\bar{L}'\bar{M}_L') + \delta(\bar{\Omega},\bar{\Omega}')(fm_l|v_q^{(5)}|fm_l') \\ = (\bar{U}\bar{\tau}\bar{L}\bar{M}_L;fm_l|V_q^{(5)}|\bar{U}'\bar{\tau}'\bar{L}'\bar{M}_L';fm_l')$$

is expressed as a sum over matrix elements between coupled states,

$$(\bar{U}\bar{\tau}\bar{L}\bar{M}_L;fm_l|V_q^{(5)}|\bar{U}'\bar{\tau}'\bar{L}'\bar{M}'_L;fm'_l)$$
$$= \Sigma(\bar{U}\bar{\tau}\bar{L}\bar{M}_L;fm_l|U''\tau''L''M''_L)(U''\tau''L''M''_L|V_q^{(5)}|U'\tau'L'''M'''_L)$$
$$\times (U'\tau'L'''M'''_L|\bar{U}'\bar{\tau}'\bar{L}'\bar{M}'_L;fm'_l) \quad (7\text{-}22)$$

The sum runs over U'', τ'', L'', M''_L, U', τ', L''', and M'''_L. Since $V_q^{(5)}$ is an infinitesimal operator for G_2, only those terms in the sum contribute for which $U' = U''$. The coefficients are now factorized, e.g.,

$$(\bar{U}\bar{\tau}\bar{L}\bar{M}_L;fm_l|U''\tau''L''M''_L) = (\bar{L}\bar{M}_Lfm_l|\bar{L}fL''M''_L)(\bar{U}\bar{\tau}\bar{L}+f|U''\tau''L'')$$

Next, both sides of Eq. (7-22) are multiplied by

$$(\bar{L}fLM_L|\bar{L}\bar{M}_Lfm_l)(\bar{L}'\bar{M}'_Lfm'_l|\bar{L}fL'M'_L)$$

and sums over $\bar{M}_L$, m_l, $\bar{M}'_L$, and m'_l are carried out. The quantum numbers M_L, q, and M'_L can be removed from the resulting equation by using the Wigner-Eckart theorem, and, on putting $\bar{U}' = \bar{U}$, we get

$$(\bar{U}\bar{\tau}\bar{L},f,L\|V^{(5)}\|\bar{U}\bar{\tau}'\bar{L}',f,L')$$
$$= \sum_{U'',\tau'',\tau'} (\bar{U}\bar{\tau}\bar{L}+f|U''\tau''L)(U''\tau''L\|V^{(5)}\|U''\tau'L')$$
$$\times (U''\tau'L'|\bar{U}\bar{\tau}'\bar{L}'+f)$$

A more convenient form for this equation is obtained by multiplying both sides by $(U\tau L|\bar{U}\bar{\tau}\bar{L}+f)$ and summing over $\bar{\tau}$ and $\bar{L}$. Owing to Eq. (7-16), we get the important result

$$\sum_{\bar{\tau},\bar{L}} (U\tau L|\bar{U}\bar{\tau}\bar{L}+f)(\bar{U}\bar{\tau}\bar{L},f,L\|V^{(5)}\|\bar{U}\bar{\tau}'\bar{L}',f,L')$$
$$= \sum_{\tau'} (U\tau L\|V^{(5)}\|U\tau'L')(U\tau'L'|\bar{U}\bar{\tau}'\bar{L}'+f) \quad (7\text{-}23)$$

A companion to this equation can be easily derived by using Eq. (7-16) again. It is

$$(U\tau L\|V^{(5)}\|U\tau'L')$$
$$= \sum_{\bar{L},\bar{L}',\bar{\tau},\bar{\tau}'} (U\tau L|\bar{U}\bar{\tau}\bar{L}+f)(\bar{U}\bar{\tau}\bar{L},f,L\|V^{(5)}\|\bar{U}\bar{\tau}'\bar{L}',f,L')$$
$$\times (\bar{U}\bar{\tau}'\bar{L}'+f|U\tau'L') \quad (7\text{-}24)$$

We begin the chain calculation by setting $n = 1$ in Eq. (7-1). Making use of Eqs. (7-9) and (7-10), we easily prove

$$(f^0\,{}^1S|\}f^1\,{}^2F) = 1$$

Our immediate interest lies, not in the complete cfp, but rather in the factor $((00)S + f|(10)F)$. On making the substitutions

$$\bar{U} = (00) \qquad U' = U = (10) \qquad L = 3$$

in Eq. (7-16), the sum over $\bar{L}$ and $\bar{\tau}$ reduces to a single term, corresponding to $\bar{L} = 0$. Hence

$$((00)S + f|(10)F)^2 = 1$$

We may take either the positive or the negative root, since the phases of the remaining factors are still at our disposal. Here, as elsewhere, we make Racah's choice of phase, and the table of coefficients deriving from the parent $(00)S$ comprises the single entry

$$((00)S + f|(10)F) = +1$$

From Eq. (5-13),

$$((10)F\|V^{(5)}\|(10)F) = (11)^{\frac{1}{2}} \tag{7-25}$$

The next set of coefficients to consider is that corresponding to the parent $(10)F$. The decomposition

$$(10) \times (10) = (00) + (10) + (11) + (20)$$

indicates that the offspring are $(00)S$; $(10)F$; $(11)P, H$; and $(20)D, G, I$. By putting $n = 2$ in Eq. (7-1) we get

$$(f^1\,{}^2F|\}f^2SL) = 1 \tag{7-26}$$

The factors we are interested in satisfy

$$((10)F + f|UL)^2 = 1$$

The phases of the coefficients are not entirely arbitrary; for Eq. (7-26) must be satisfied, and the remaining factors of the complete cfp, whose

TABLE 7-2 THE COEFFICIENTS $((10)F + f|UL)$

U	L	$((10)F + f\|UL)$
(00)	S	-1
(10)	F	1
(11)	P	1
	H	1
(20)	D	1
	G	1
	I	1

phases are still at our disposal, are independent of L. This means that all the coefficients $((10)F + f|UL)$ for a given U must have the same sign. The actual choices made are indicated in Table 7-2. We maintain our program for finding the reduced matrix elements of $\mathbf{V}^{(5)}$ by using Eqs. (7-24), (3-37), and (3-38) to derive the following results:

$((11)L\|V^{(5)}\|(11)L')$:

$$\begin{array}{c} \\ P \\ H \end{array}\!\!\begin{array}{c} \begin{array}{cc} P & H \end{array} \\ \begin{pmatrix} 0 & (\frac{165}{14})^{\frac{1}{2}} \\ (\frac{165}{14})^{\frac{1}{2}} & -(\frac{143}{7})^{\frac{1}{2}} \end{pmatrix} \end{array} \tag{7-27}$$

$((20)L\|V^{(5)}\|(20)L')$:

$$\begin{array}{c} \\ D \\ G \\ I \end{array}\!\!\begin{array}{c} \begin{array}{ccc} D & G & I \end{array} \\ \begin{pmatrix} 0 & -(\frac{80}{7})^{\frac{1}{2}} & (\frac{65}{6})^{\frac{1}{2}} \\ -(\frac{80}{7})^{\frac{1}{2}} & -(\frac{39}{77})^{\frac{1}{2}} & (\frac{260}{11})^{\frac{1}{2}} \\ (\frac{65}{6})^{\frac{1}{2}} & (\frac{260}{11})^{\frac{1}{2}} & (\frac{221}{33})^{\frac{1}{2}} \end{pmatrix} \end{array} \tag{7-28}$$

We pass on to the parents corresponding to $\bar{U} = (11)$. The decomposition

$$(11) \times (10) = (10) + (20) + (21)$$

shows that, if the coefficients are not to vanish, U must be (10), (20), or (21). A number of coefficients among the restricted set we are considering can be seen to be zero by noting that $\mathfrak{D}_{L'}$ occurs in the decomposition of $\mathfrak{D}_L \times \mathfrak{D}_3$ only if $|L' - L| \leq 3$. Those which remain to be calculated are indicated by asterisks in Table 7-3; according

TABLE 7-3 PATTERN OF ENTRIES FOR $((11)\bar{L} + f|UL)$

U	L	$\bar{L}$	
		P	H
(10)	F	*	*
(20)	D	*	*
	G	*	*
	I	0	1
(21)	D	*	*
	F	*	*
	G	*	*
	H	0	1
	K	0	1
	L	0	1

to Eq. (7-16), the sum of the squares of the entries in every row must be unity, and this enables us to insert four 1's beside the zeros. In doing this, four arbitrary phase factors are fixed.

To proceed further, we turn to Eq. (7-23). The symbols τ can be dropped, and only one summation, that over $\bar{L}$, remains. We now pick a set of quantum numbers UL for which the coefficients $(UL|\bar{U}\bar{L} + f)$

are known; for example, we may take $UL = (21)H$. Equation (7-23) becomes

$$((11)H,f,H\|V^{(5)}\|(11)\bar{L}',f,L') \\ = ((21)H\|V^{(5)}\|(21)L')((21)L'|(11)\bar{L}'+f) \quad (7\text{-}29)$$

The matrix element on the left can be evaluated by means of Eqs. (3-37) and (3-38),

$$((11)H,f,H\|V^{(5)}\|(11)\bar{L}',f,L') \\ = (-1)^{L'+1}\{11(2L'+1)\}^{\frac{1}{2}}\begin{Bmatrix}5 & 5 & L' \\ \bar{L}' & 3 & 5\end{Bmatrix}((11)H\|(V^{(5)})'\|(11)\bar{L}') \\ + 11\delta(\bar{L}',5)(2L'+1)^{\frac{1}{2}}\begin{Bmatrix}5 & 5 & L' \\ 3 & 5 & 3\end{Bmatrix}$$

The reduced matrix elements of $(\mathbf{V}^{(5)})'$ that are required are given by (7-27), and the results of the calculation are entered in Table 7-4.

TABLE 7-4 THE MATRIX ELEMENTS $((11)H,f,H\|V^{(5)}\|(11)\bar{L}',f,L')$

$\bar{L}'$	L'		
	D	F	G
P	$(\frac{33}{7})^{\frac{1}{2}}$	$(\frac{33}{28})^{\frac{1}{2}}$	$-(\frac{165}{28})^{\frac{1}{2}}$
H	$-(\frac{81}{14})^{\frac{1}{2}}$	$-(\frac{9}{28})^{\frac{1}{2}}$	$-(\frac{1089}{364})^{\frac{1}{2}}$

For $L' = 2$, $\bar{L}' = 1$, Eq. (7-29) becomes

$$(\tfrac{33}{7})^{\frac{1}{2}} = ((21)H\|V^{(5)}\|(21)D)((21)D|(11)P+f) \quad (7\text{-}30)$$

and for $L' = 2$, $\bar{L}' = 5$, we get

$$-(\tfrac{81}{14})^{\frac{1}{2}} = ((21)H\|V^{(5)}\|(21)D)((21)D|(11)H+f) \quad (7\text{-}31)$$

These equations determine the ratio of the two coefficients on their extreme right; since the sum of the squares of these coefficients must be unity, we have

$$((21)D|(11)P+f) = -(\tfrac{22}{49})^{\frac{1}{2}}$$

and

$$((21)D|(11)H+f) = (\tfrac{27}{49})^{\frac{1}{2}}$$

Table 7-4 can be used again to find the coefficients for $U = (21)$, $L' = 3, 4$; for each L' an arbitrary choice of phase is made. Thus, for the part of Table 7-3 corresponding to $U = (21)$, six such choices are made in all.

The six entries of Table 7-3 still outstanding can be found by using the orthogonality properties of rows with the same L. However, the phases of $((11)\bar{L}+f|(20)L)$ are not at our disposal. We have already decided to take $((11)H+f|(20)I)$ to be $+1$, and in using the equation

$$((20)I|(11)H+f)((11)H,f,I\|V^{(5)}\|(11)\bar{L}',f,L') = ((20)I\|V^{(5)}\|(20)L')((20)L'|(11)\bar{L}'+f)$$

to derive the coefficients on the extreme right, we must make use of (7-28) for the reduced matrix elements of $\mathbf{V}^{(5)}$ between states for which $U = (20)$.

By methods such as these, the chain calculation can be extended indefinitely. In fixing arbitrary phase factors, care must be taken to ensure that the freedom to make such a choice really exists. If two distinct states $U\tau L$ and $U\tau' L$ occur [as is the case for $U = (31)$ and (40)], we may make the entries in the rows $U\tau L$ and $U\tau' L$ in an infinite number of ways and still satisfy all the necessary equations. Once a particular choice has been made, the states are determined and the symbols τ and τ' have a well-defined significance. The completed form of Table 7-3 is to be found in Appendix 2, where all the necessary factors of the cfp are given for calculations involving terms of the highest and next-to-highest multiplicities of f^n. These terms are the ones of greatest interest experimentally. The original calculation was carried out by Racah.[69]

7-6 RECIPROCITY

The method just described for finding the coefficients $(\bar{U}\bar{\tau}\bar{L}+f|U\tau L)$ has made no use of any symmetry properties these quantities might possess. The presentation of the method is thereby simplified: the calculations can be performed in a systematic way, by use of as few equations as possible. However, from a practical point of view, any additional relation that can be established between coefficients is of value, if only because of the opportunity it gives for independent checks. An equation that provides a very useful supplement to the techniques of Sec. 7-5 is

$$(W_1\xi_1L_1+W_2\xi_2L_2|W_3\xi_3L_3) = (-1)^{L_3-L_2-L_1+x}\{[L_1]D(W_3)/[L_3]D(W_1)\}^{\frac{1}{2}} \times (W_3\xi_3L_3+W_2\xi_2L_2|W_1\xi_1L_1) \quad (7\text{-}32)$$

where x is independent of ξ_1, ξ_2, ξ_3, L_1, L_2, and L_3 and depends solely on W_1, W_2, and W_3. For the proof of this result, the reader is referred to a paper of Racah.[69] We shall merely make it plausible by combining it with

$$(L_1M_1L_2M_2|L_1L_2L_3 - M_3) = (-1)^{L_2+M_2}\{[L_3]/[L_1]\}^{\frac{1}{2}}(L_3M_3L_2M_2|L_3L_2L_1 - M_1) \quad (7\text{-}33)$$

to give the equation

$$(W_1\xi_1L_1M_1;W_2\xi_2L_2M_2|W_3\xi_3L_3 - M_3) = (-1)^{L_3-L_1+M_2+x}\{D(W_3)/D(W_1)\}^{\frac{1}{2}} \times (W_3\xi_3L_3M_3;W_2\xi_2L_2M_2|W_1\xi_1L_1 - M_1) \quad (7\text{-}34)$$

Since $2L + 1$ is the dimension of the representation $\mathfrak{D}_L$ of R_3, Eq. (7-34) can be regarded as the generalization to R_{2l+1} of Eq. (7-33).

In order to use Eq. (7-32), the set of quantum numbers $W_2\xi_2L_2$ is replaced by those corresponding to a single electron. If a number of arbitrary choices of phase have already been made in the construction of the tables of $(\bar{W}\bar{\xi}\bar{L} + l|W\xi L)$, then x is likely to be sometimes even, sometimes odd; however, x may equally well be picked in advance, thereby reducing the number of disposable phase factors. This has been done by Racah,[69] who put $x = l$ for all W. Equation (7-32) reduces to

$$(\bar{W}\bar{\xi}\bar{L} + l|W\xi L) = (-1)^{L-\bar{L}}\{[\bar{L}]D(W)/[L]D(\bar{W})\}^{\frac{1}{2}}(W\xi L + l|\bar{W}\bar{\xi}\bar{L}) \quad (7\text{-}35)$$

and the reciprocal nature of the sets of quantum numbers $W\xi L$ and $\bar{W}\bar{\xi}\bar{L}$ is clearly demonstrated.

For f electrons, Eq. (7-35) breaks up into

$$(\bar{W}\bar{U} + f|WU) = \{D(\bar{U})D(W)/D(U)D(\bar{W})\}^{\frac{1}{2}}(WU + f|\bar{W}\bar{U}) \quad (7\text{-}36)$$

and

$$(\bar{U}\bar{\tau}\bar{L} + f|U\tau L) = (-1)^{L-\bar{L}}\{[\bar{L}]D(U)/[L]D(\bar{U})\}^{\frac{1}{2}}(U\tau L + f|\bar{U}\bar{\tau}\bar{L}) \quad (7\text{-}37)$$

The latter equation eliminates a disposable phase factor when the coefficients $(\bar{U}\bar{\tau}\bar{L} + f|U\tau L)$ are being evaluated for quantum numbers $U\tau L$ that have previously served in a parental capacity. The coefficient $((10)F + f|(00)S)$ is the first example of this, and the reason for the somewhat bizarre entry of -1 at the head of the column in Table 7-2 is now apparent. All pairs of reciprocal entries in the tables of Appendix 2 satisfy Eq. (7-37).

7-7 THE COEFFICIENTS $(\bar{W}\bar{U} + f|WU)$

The second factor of the complete cfp for f electrons may be calculated by an extension of the techniques of Sec. 7-5. Equation (7-23) is generalized to

$$\sum_{\bar{U},\bar{\tau},\bar{L}} (WU\tau L|\bar{W}\bar{U}\bar{\tau}\bar{L} + f)(\bar{W}\bar{U}\bar{\tau}\bar{L},f,L\|V^{(3)}\|\bar{W}\bar{U}'\bar{\tau}'\bar{L}',f,L')$$
$$= \sum_{U',\tau'} (WU\tau L\|V^{(3)}\|WU'\tau'L')(WU'\tau'L'|\bar{W}\bar{U}'\bar{\tau}'\bar{L}' + f) \quad (7\text{-}38)$$

The components of $\mathbf{V}^{(3)}$ are infinitesimal operators of R_7 but not of G_2; this is the natural extension of the relation that the components of $\mathbf{V}^{(5)}$ bear to the groups G_2 and R_3. Factorizations of the type

$$(WU\tau L|\bar{W}\bar{U}\bar{\tau}\bar{L} + f) = (WU|\bar{W}\bar{U} + f)(U\tau L|\bar{U}\bar{\tau}\bar{L} + f)$$

allow the coefficients $(WU|\bar{W}\bar{U} + f)$ to be studied. The sums over U' and τ' in Eq. (7-38) can be eliminated by introducing $(\bar{U}'\bar{\tau}'\bar{L}' + f|U''\tau''L')$ on both sides of Eq. (7-38) and summing over $\bar{\tau}'$ and $\bar{L}'$.

The result is

$$\sum_{\bar{U},\bar{\tau},\bar{L},\bar{\tau}',\bar{L}'} (WU|\bar{W}\bar{U} + f)(U\tau L|\bar{U}\bar{\tau}\bar{L} + f)$$
$$\times (\bar{W}\bar{U}\bar{\tau}\bar{L},f,L\|V^{(3)}\|\bar{W}\bar{U}'\bar{\tau}'\bar{L}',f,L')(\bar{U}'\bar{\tau}'\bar{L}' + f|U''\tau''L')$$
$$= (WU\tau L\|V^{(3)}\|WU''\tau''L')(WU''|\bar{W}\bar{U}' + f) \quad (7\text{-}39)$$

This equation can be used as a basis for a chain calculation involving the coefficients $(WU|\bar{W}\bar{U} + f)$ in much the same way as Eq. (7-23) is used to solve for the coefficients $(U\tau L|\bar{U}\bar{\tau}\bar{L} + f)$. Specific representations W, U, and $\bar{W}$ are selected, and the sums over $\bar{U}$, $\bar{\tau}$, $\bar{L}$, $\bar{\tau}'$, and $\bar{L}'$ are carried out for the various representations $\bar{U}'$ into which $\bar{W}$ decomposes. The quantum numbers L and L' are chosen to make this procedure as simple as possible.

As an example, we consider the tabulation of the coefficients $((110)\bar{U} + f|WU)$. The labels for the columns and rows are easily obtained from the decomposition

$$(110) \times (100) = (100) + (111) + (210)$$

and from the branching rules for the reduction $R_7 \to G_2$. The structure of the table is given in Table 7-5; three zeros have been inserted in those positions $U\bar{U}$ for which U does not occur in the decomposition of $\bar{U} \times (10)$. Equation (7-17) permits the rows in which these zeros occur to be completed. In order to find the row (210)(20), we set $WUL = (210)(21)H$ in Eq. (7-39), with $\bar{W}$ equal to (110), of course. At once $\bar{U}\bar{L}$ is limited to $(11)H$, and the sum runs over $\bar{L}'$ only. Moreover, if we set $L' = 6$, then, for both $\bar{U}' = (10)$ and $\bar{U}' = (11)$, the

TABLE 7-5 PRELIMINARY TABLE FOR THE COEFFICIENTS $((110)\bar{U}+f|WU)$

W	U	$\bar{U}$ (10)	$\bar{U}$ (11)
(100)	(10)	*	*
(111)	(00)	1	0
	(10)	*	*
	(20)	*	*
(210)	(11)	1	0
	(20)	*	*
	(21)	0	1

sum over $\bar{L}'$ reduces to a single term. The matrix elements

$$((110)(11)H,f,H\|V^{(3)}\|(110)\bar{U}'\bar{L}',f,I)$$

are evaluated in a similar way to the corresponding elements of $\mathbf{V}^{(5)}$, and we obtain

$$\begin{aligned}((110)(11)H,f,H\|V^{(3)}\|(110)(10)F,f,I) &= -(\tfrac{5}{18})^{\frac{1}{2}} \\ &= ((210)(21)H\|V^{(3)}\|(210)(20)I)((210)(20)|(110)(10)+f)\end{aligned}$$

and

$$\begin{aligned}((110)(11)H,f,H\|V^{(3)}\|(110)(11)H,f,I) &= -(\tfrac{5}{63})^{\frac{1}{2}} \\ &= ((210)(21)H\|V^{(3)}\|(210)(20)I)((210)(20)|(110)(11)+f)\end{aligned}$$

Owing to the normalization condition on the coefficients, we get, with an arbitrary choice of phase,

$$((210)(20)|(110)(10)+f) = (\tfrac{7}{9})^{\frac{1}{2}}$$

and

$$((210)(20)|(110)(11)+f) = (\tfrac{2}{9})^{\frac{1}{2}}$$

This completes that part of Table 7-5 for which $W = (210)$; the remainder can be found by similar methods or by making use of the orthogonality between rows and the reciprocity relation (7-36). The entire table is given, with others, in Appendix 2.

7-8 THE COEFFICIENTS $(l^{n-1}\bar{v}\bar{S}+l|\}l^n vS)$

The remaining factor of the complete cfp, namely,

$$(l^{n-1}\bar{v}\bar{S}+l|\}l^n vS) \tag{7-40}$$

differs from its associated factors in that it lends itself to an explicit algebraic construction. No advantage is gained by restricting the treatment to f electrons, and this specialization is therefore dropped. The basic idea of the method is to calculate the matrix elements of operators whose eigenvalues are already known, thereby providing equations involving the cfp. In order to avoid the coefficients $(\bar{W}\xi\bar{L} + l|W\xi L)$, we choose operators whose eigenvalues do not depend on L; the undesired coefficients can then be eliminated by means of Eq. (7-14). That this approach is feasible is largely due to the fact that for a given v and S there are at most four coefficients (7-40), corresponding to $\bar{S} = S \pm \frac{1}{2}$ and $\bar{v} = v \pm 1$. The condition on $\bar{S}$ follows at once from

$$\mathfrak{D}_{\bar{S}} \times \mathfrak{D}_{\frac{1}{2}} = \mathfrak{D}_{\bar{S}+\frac{1}{2}} + \mathfrak{D}_{\bar{S}-\frac{1}{2}}$$

that on $\bar{v}$ from

$$(\underbrace{11 \cdots 1}_{\bar{v}}0 \cdots 0) \times (10 \cdots 0) = (\underbrace{11 \cdots 1}_{\bar{v}-1}0 \cdots 0) + (\underbrace{11 \cdots 1}_{\bar{v}+1}0 \cdots 0) + (2\underbrace{11 \cdots 1}_{\bar{v}-1}0 \cdots 0)$$

The second representation on the right occurs only if $\bar{v} < 2l + 1$; the third representation does not arise in the decomposition of $\mathfrak{A}_n$ and is of no interest to us.

To begin with, we set

$$G = 2 \sum_{i>j} \mathbf{s}_i \cdot \mathbf{s}_j$$

in Eq. (7-5), with $\Omega = \Omega' = v\xi SM_SLM_L$. The VC coefficients are factored out from $(\Omega|\bar{\Omega};\omega)$ and $(\bar{\Omega}';\omega|\Omega')$. Since G is a scalar, we have $\bar{\Omega}' = \bar{\Omega}$ and the sums over the M quantum numbers can be carried out. Equation (7-14) is used to dispose of the coefficients $(\bar{W}\xi\bar{L} + l|W\xi L)$, and we obtain

$$S(S+1) - \tfrac{3}{4}n = \frac{n}{n-2} \sum_{\bar{v},\bar{S}} (l^{n-1}\bar{v}\bar{S} + l|\}l^n vS)^2[\bar{S}(\bar{S}+1) - \tfrac{3}{4}(n-1)]$$

We are able to use this equation in conjunction with the normalization condition (7-15) to derive the pair of equations

$$(l^{n-1} \quad v-1 \quad S-\tfrac{1}{2} \quad +l|\}l^n vS)^2 + (l^{n-1} \quad v+1 \quad S-\tfrac{1}{2} \quad +l|\}l^n vS)^2 = \frac{S(n+2S+2)}{n(2S+1)} \qquad (7\text{-}41)$$

$$(l^{n-1} \quad v-1 \quad S+\tfrac{1}{2} \quad +l|\}l^n vS)^2 + (l^{n-1} \quad v+1 \quad S+\tfrac{1}{2} \quad +l|\}l^n vS)^2 = \frac{(S+1)(n-2S)}{n(2S+1)} \qquad (7\text{-}42)$$

For the special case $n = v$, these become

$$(l^{v-1}\quad v-1\quad S-\tfrac{1}{2}\quad +l|\}l^v vS)^2 = \frac{S(v+2S+2)}{v(2S+1)} \tag{7-43}$$

$$(l^{v-1}\quad v-1\quad S+\tfrac{1}{2}\quad +l|\}l^v vS)^2 = \frac{(S+1)(v-2S)}{v(2S+1)} \tag{7-44}$$

We now follow a similar procedure with the operator

$$G' = \Xi_2 + \sum_{i>j} \mathbf{s}_i \cdot \mathbf{s}_j$$

where Ξ_2 is defined as in Sec. 6-4. The eigenvalues of G' are

$$\tfrac{1}{8}v(4+4l-v) - \tfrac{3}{8}n - \tfrac{1}{2}nl$$

and the analogues of Eqs. (7-41) and (7-42) are

$$\begin{aligned}(l^{n-1}\quad v-1\quad S-\tfrac{1}{2}\quad +l|\}l^n vS)^2 + (l^{n-1}\quad v-1\quad S+\tfrac{1}{2}\quad +l|\}l^n vS)^2 \\ = \frac{v(4l+4-n-v)}{2n(2+2l-v)}\end{aligned} \tag{7-45}$$

$$\begin{aligned}(l^{n-1}\quad v+1\quad S-\tfrac{1}{2}\quad +l|\}l^n vS)^2 + (l^{n-1}\quad v+1\quad S+\tfrac{1}{2}\quad +l|\}l^n vS)^2 \\ = \frac{(n-v)(4l+4-v)}{2n(2+2l-v)}\end{aligned} \tag{7-46}$$

The factorization indicated in Eq. (7-12) now plays a vital role in furthering the calculation. For

$$\frac{(l^{n-1}\quad v-1\quad S-\frac{1}{2}\quad +l|\}l^n vS)^2}{(l^{n-1}\quad v-1\quad S+\frac{1}{2}\quad +l|\}l^n vS)^2} = \frac{(v-1\quad S-\frac{1}{2}\quad +l|vS)^2}{(v-1\quad S+\frac{1}{2}\quad +l|vS)^2}$$

and the ratio is therefore independent of n. In particular, it is equal to

$$\frac{(l^{v-1}\quad v-1\quad S-\frac{1}{2}\quad +l|\}l^v vS)^2}{(l^{v-1}\quad v-1\quad S+\frac{1}{2}\quad +l|\}l^v vS)^2} = \frac{S(v+2S+2)}{(S+1)(v-2S)}$$

from Eqs. (7-43) and (7-44). We have now sufficient equations to solve for all the coefficients. Thus, on putting

$$\begin{aligned}(l^{n-1}\quad v-1\quad S+\tfrac{1}{2}\quad +l|\}l^n vS)^2 \\ = \left[\frac{(S+1)(v-2S)}{S(v+2S+2)}\right](l^{n-1}\quad v-1\quad S-\tfrac{1}{2}\quad +l|\}l^n vS)^2\end{aligned}$$

in Eq. (7-45), we get

$$(l^{n-1} \quad v-1 \quad S-\tfrac{1}{2} \quad +l|\}l^n vS)^2 = \frac{(4l+4-n-v)(v+2S+2)S}{2n(2l+2-v)(2S+1)} \tag{7-47}$$

The remaining coefficients are

$$(l^{n-1} \quad v-1 \quad S+\tfrac{1}{2} \quad +l|\}l^n vS)^2 = \frac{(4l+4-n-v)(v-2S)(S+1)}{2n(2l+2-v)(2S+1)} \tag{7-48}$$

$$(l^{n-1} \quad v+1 \quad S-\tfrac{1}{2} \quad +l|\}l^n vS)^2 = \frac{(n-v)(4l+6-v+2S)S}{2n(2l+2-v)(2S+1)} \tag{7-49}$$

$$(l^{n-1} \quad v+1 \quad S+\tfrac{1}{2} \quad +l|\}l^n vS)^2 = \frac{(n-v)(4l+4-v-2S)(S+1)}{2n(2l+2-v)(2S+1)} \tag{7-50}$$

There remains the question of phase when the square roots of the expressions in Eqs. (7-47) to (7-50) are taken. If we have already fixed the phases of $(\bar{W}\bar{\xi}\bar{L}+l|W\xi L)$, the present choice is severely limited, since there is only one disposable phase for every row of the complete table of cfp. If we wish to have an explicit formula for the phases of the coefficients $(l^{n-1}\bar{v}\bar{S}+l|\}l^n vS)$, then the phases of $(\bar{W}\bar{\xi}\bar{L}+l|W\xi L)$ must be adjusted accordingly. We follow Racah in introducing the following phase factors for the coefficients $(f^{n-1}\bar{v}\bar{S}+f|f^n vS)$:

$$\begin{array}{ll} (-1)^{\bar{S}} & \text{if } v \text{ is odd} \\ (-1)^{\bar{S}+\frac{1}{2}(\bar{v}-v)} & \text{if } v \text{ is even} \end{array}$$

The phases of the remaining factors that are tabulated in Appendix 2 are consistent with this choice.

7-9 MATRIX ELEMENTS

In Eq. (7-2), a matrix element of a single-particle operator is expressed as a sum over matrix elements of the nth electron. The factorization of the coefficients $(\bar{\Omega};\omega|\Omega)$ into cfp and VC coefficients usually permits the equation to be simplified. For example, suppose that matrix elements of the tensor operator $\mathbf{V}^{(k)}$ are required. We make the abbreviations

$$\begin{aligned} W\xi SL &\equiv \theta \\ W'\xi'S'L' &\equiv \theta' \\ \bar{W}\bar{\xi}\bar{S}\bar{L} &\equiv \bar{\theta} \end{aligned}$$

with the understanding that θ and θ' define terms of l^n, while $\bar{\theta}$ defines a term of l^{n-1}. Equation (7-2) gives

$$(l^n W\xi SM_S LM_L|V_q^{(k)}|l^n W'\xi'S'M_S'LM_L') = n\sum_{\bar{\theta}} (\theta\{|\bar{\theta})(\theta'\{|\bar{\theta})$$
$$\times \sum_{m_l,\bar{M}_L,m_l'} (\bar{L}\bar{M}_L l m_l|\bar{L}lLM_L)(\bar{L}\bar{M}_L l m_l'|\bar{L}lL'M_L')(lm_l|v_q^{(k)}|lm_l')$$
$$\times \sum_{m_s,\bar{M}_S,m_s'} (\bar{S}\bar{M}_S s m_s|\bar{S}sSM_S)(\bar{S}\bar{M}_S s m_s'|\bar{S}sS'M_S')\delta(m_s,m_s') \quad (7\text{-}51)$$

The sum over m_s, $\bar{M}_S$, m_s' yields $\delta(S,S')\delta(M_S,M_S')$. Since $\mathbf{V}^{(k)}$ is independent of spin, these delta functions are expected. Another necessary result is that the sum over m_l, $\bar{M}_L$, and m_l' should have the same dependence on M_L, q, and M_L' as

$$(-1)^{L-M_L}\begin{pmatrix} L & k & L' \\ -M_L & q & M_L' \end{pmatrix}$$

for otherwise the Wigner-Eckart theorem would be violated. It is easy to see that Eq. (3-6) leads to this result and at the same time introduces a 6-j symbol. In fact, the calculations can be performed without explicitly considering the z components of the angular momenta. Instead of Eq. (7-51), we write

$$(l^n W\xi SL\|V^{(k)}\|l^n W'\xi'S'L')$$
$$= n\delta(S,S')\sum_{\bar{\theta}} (\theta\{|\bar{\theta})(\theta'\{|\bar{\theta})(\bar{L},l_n,L\|(v^{(k)})_n\|\bar{L},l_n,L')$$
$$= n\delta(S,S')\{[L][k][L']\}^{\frac{1}{2}}$$
$$\times \sum_{\bar{\theta}} (\theta\{|\bar{\theta})(\theta'\{|\bar{\theta})(-1)^{\bar{L}+l+L+k}\begin{Bmatrix} L & k & L' \\ l & \bar{L} & l \end{Bmatrix} \quad (7\text{-}52)$$

The last step follows from Eqs. (3-38) and (5-13).

This procedure can be extended to deal with more complicated cases. Consider, for example, the operator

$$\mathbf{X}^{(K)} = \sum_i \{\mathbf{z}_i^{(\kappa)}\mathbf{y}_i^{(k)}\}^{(K)}$$

where $\mathbf{z}_i^{(\kappa)}$ and $\mathbf{y}_i^{(k)}$ are tensor operators acting on the spin and on the orbit, respectively, of electron i. We calculate a reduced matrix element in the scheme $l^n W\xi SLJM_J$. Making use of Eq. (3-35), we get, with some manipulation,

$$(l^nW\xi SLJ\|X^{(K)}\|l^nW'\xi'S'L'J') = \{[J][K][J']\}^{\frac{1}{2}}\{[k][\kappa]\}^{-\frac{1}{2}}\begin{Bmatrix} S & S' & \kappa \\ L & L' & k \\ J & J' & K \end{Bmatrix}$$
$$\times\ (s\|z^{(\kappa)}\|s)(l\|y^{(k)}\|l)(l^nW\xi SL\|W^{(\kappa k)}\|l^nW'\xi'S'L') \quad (7\text{-}53)$$

where, in analogy to Eq. (7-52)

$$(l^nW\xi SL\|W^{(\kappa k)}\|l^nW'\xi'S'L')$$
$$= n\{[S][\kappa][S'][L][k][L']\}^{\frac{1}{2}}\sum_{\bar\theta}(\theta\{|\bar\theta)(\theta'\{|\bar\theta)(-1)^{\bar S+s+S+\kappa+\bar L+l+L+k}$$
$$\times\begin{Bmatrix} S & \kappa & S' \\ s & \bar S & s \end{Bmatrix}\begin{Bmatrix} L & k & L' \\ l & \bar L & l \end{Bmatrix} \quad (7\text{-}54)$$

Since $\sum_i \mathbf{s}_i\cdot\mathbf{l}_i = -(3)^{\frac{1}{2}}\sum_i(\mathbf{s}^{(1)}\mathbf{l}^{(1)})_i{}^{(0)}$,

$$(l^nW\xi SLJM_J|\textstyle\sum \mathbf{s}\cdot\mathbf{l}|l^nW'\xi'S'L'J'M'_J)$$
$$= \delta(J,J')\delta(M_J,M'_J)(-1)^{S'+L+J}\left[\frac{l(l+1)(2l+1)}{6}\right]^{\frac{1}{2}}$$
$$\times\begin{Bmatrix} S & S' & 1 \\ L' & L & J \end{Bmatrix}(l^nW\xi SL\|W^{(11)}\|l^nW'\xi'S'L') \quad (7\text{-}55)$$

where $(l^nW\xi SL\|W^{(11)}\|l^nW'\xi'S'L') = 3n\{[S][S'][L][L']\}^{\frac{1}{2}}(-1)^{\frac{1}{2}+l+S+L}$

$$\times\sum_{\bar\theta}(\theta\{|\bar\theta)(\theta'\{|\bar\theta)(-1)^{\bar S+\bar L}\begin{Bmatrix} S & 1 & S' \\ \frac{1}{2} & \bar S & \frac{1}{2} \end{Bmatrix}\begin{Bmatrix} L & 1 & L' \\ l & \bar L & l \end{Bmatrix} \quad (7\text{-}56)$$

As an example, consider

$$(f^7\ {}^8S_{\frac{7}{2}}|\zeta\Sigma\mathbf{s}\cdot\mathbf{l}|f^7\ {}^6P_{\frac{7}{2}})$$

There is only one parent of the term ^8S of f^7, namely, ^7F of f^6. The WU descriptions of the terms can be obtained from Tables 5-1 to 5-3, the seniorities from Table 6-1. We find

$$\theta \equiv (000)(00)\ {}^8S \qquad v = 7$$
$$\theta' \equiv (110)(11)\ {}^6P \qquad v = 5$$
$$\bar\theta \equiv (100)(10)\ {}^7F \qquad v = 6$$

Hence, from Appendix 2,

$$
\begin{aligned}
(\theta\{|\bar{\theta}) &= ((10)F + f|(00)S)((100)(10) + f|(000)(00)) \\
&\qquad \times (f^6, \bar{v} = 6, \bar{S} = 3||\}f^7, v = 7, S = \tfrac{7}{2}) \\
&= (-1)(+1)(-1) = 1
\end{aligned}
$$

Again, $$
\begin{aligned}
(\theta'\{|\bar{\theta}) &= ((10)F + f|(11)P)((100)(10) + f|(110)(11)) \\
&\qquad \times (f^6, \bar{v} = 6, \bar{S} = 3||\}f^7, v = 5, S = \tfrac{5}{2}) \\
&= (+1)(+1)[-(6)^{-\frac{1}{2}}] = -(6)^{-\frac{1}{2}}
\end{aligned}
$$

Thus $$
\begin{aligned}
(f^7\, {}^8S_{\frac{7}{2}}|\zeta\Sigma \mathbf{s}\cdot\mathbf{l}|f^7\, {}^6P_{\frac{7}{2}}) &= 84\zeta(21)^{\frac{1}{2}} \begin{Bmatrix} \frac{7}{2} & \frac{5}{2} & 1 \\ 1 & 0 & \frac{7}{2} \end{Bmatrix} \begin{Bmatrix} \frac{7}{2} & 1 & \frac{5}{2} \\ \frac{1}{2} & 3 & \frac{1}{2} \end{Bmatrix} \begin{Bmatrix} 0 & 1 & 1 \\ 3 & 3 & 3 \end{Bmatrix} \\
&= -\zeta(14)^{\frac{1}{2}}
\end{aligned}
$$

agreeing to within a phase factor with the result quoted in Prob. 1-11.

The sums over $\bar{\theta}$ in equations such as (7-52) and (7-56) can often be simplified with a little ingenuity. The device of separating the parents $\bar{\theta}$ into two classes, corresponding to $\bar{S} = S + \frac{1}{2}$ and $\bar{S} = S - \frac{1}{2}$, is sometimes useful; for example, the sum

$$
\sum_{\bar{\theta}} (\theta\{|\bar{\theta})(\theta'\{|\bar{\theta})(-1)^{\bar{S}+\bar{L}} \begin{Bmatrix} S & 1 & S \\ \frac{1}{2} & \bar{S} & \frac{1}{2} \end{Bmatrix} \begin{Bmatrix} L & 1 & L' \\ l & \bar{L} & l \end{Bmatrix}
$$

which occurs in Eq. (7-56) if $S' = S$, can be written as

$$
\begin{aligned}
&(-1)^{S-\frac{1}{2}} \begin{Bmatrix} S & 1 & S \\ \frac{1}{2} & S-\frac{1}{2} & \frac{1}{2} \end{Bmatrix} \sum_{\bar{\theta}} (\theta\{|\bar{\theta})(\theta'\{|\bar{\theta})(-1)^{\bar{L}} \begin{Bmatrix} L & 1 & L' \\ l & \bar{L} & l \end{Bmatrix} \\
&\qquad + (-1)^{S+\frac{1}{2}} \left[\begin{Bmatrix} S & 1 & S \\ \frac{1}{2} & S+\frac{1}{2} & \frac{1}{2} \end{Bmatrix} + \begin{Bmatrix} S & 1 & S \\ \frac{1}{2} & S-\frac{1}{2} & \frac{1}{2} \end{Bmatrix} \right] \\
&\qquad\qquad \times {\sum_{\bar{\theta}}}' (\theta\{|\bar{\theta})(\theta'\{|\bar{\theta})(-1)^{\bar{L}} \begin{Bmatrix} L & 1 & L' \\ l & \bar{L} & l \end{Bmatrix} \qquad (7\text{-}57)
\end{aligned}
$$

where the prime on the sigma indicates that the sum over $\bar{\theta}$ is restricted to those parents for which $\bar{S} = S + \frac{1}{2}$. Now the sum

$$
\sum_{\bar{\theta}} (\theta\{|\bar{\theta})(\theta'\{|\bar{\theta})(-1)^{\bar{L}} \begin{Bmatrix} L & 1 & L' \\ l & \bar{L} & l \end{Bmatrix}
$$

can be explicitly evaluated by setting $S' = S$, $k = 1$ in Eq. (7-52) and combining the result with the equation

$$(l^nW\xi SL\|V^{(1)}\|l^nW'\xi'SL') = \delta(\theta,\theta')\left[\frac{3L(L+1)(2L+1)}{l(l+1)(2l+1)}\right]^{\frac{1}{2}} \quad (7\text{-}58)$$

which follows from Eq. (5-38). The restricted sum over parents for which $\bar{S} = S + \frac{1}{2}$ is usually much less extensive than the original sum over all parents and is therefore more easily evaluated. Further simplifications for special cases are given in Prob. 7-7.

7-10 EXPLICIT FORMULAS FOR THE CFP

Only one factor of the three that go to make up the complete cfp for f electrons has been expressed as an explicit function of its arguments, and it is natural to ask whether or not it might be possible to derive formulas for the other factors. These factors form parts of coefficients that decompose Kronecker products, as exemplified by the equation

$$(\bar{U}\bar{\tau}\bar{L}\bar{M}_L;(10)fm_l|U\tau LM_L) = (\bar{U}\bar{\tau}\bar{L} + f|U\tau L)(\bar{L}\bar{M}_Lfm_l|\bar{L}fLM_L).$$

The coefficient on the left can be regarded as a generalization to G_2 of the VC coefficients, which decompose the Kronecker products of representations of R_3. However, the only higher analogues of the explicit expressions for the VC coefficients that have so far been found are for the group R_4, a result obtained by exploiting the isomorphism between R_4 and the direct product $R_3 \times R_3$ (see Prob. 5-9). The properties of R_3 lend themselves to an algebraic description: those of the other groups that we are considering rarely do. The fact that the branching rules for the reduction $G_2 \to R_3$ are given in a table rather than a formula points to the difficulty of treating the representations U algebraically. In addition, some irreducible representations of G_2 occur more than once in the decomposition of the Kronecker products $U \times U'$, whereas for R_3 each allowed representation $\mathfrak{D}_{L''}$ occurs just once in the decomposition of $\mathfrak{D}_L \times \mathfrak{D}_{L'}$. The method of Sec. 7-5 for finding the coefficients $(\bar{U}\bar{\tau}\bar{L} + f|U\tau L)$ is essentially a generalization of the shift-operator technique that is used in Sec. 1-5 to find the VC coefficients, but the formulation of the process in algebraic terms presents difficulties that are not encountered in the derivation of an explicit expression for the VC coefficients.

Redmond[80] has approached the problem of obtaining explicit formulas for the cfp from another direction. By expressing in operator form the demand that nuclear eigenfunctions be antisymmetric with respect to the interchange of any two like nucleons, he derived a formula that for electrons reduces to

$$N_{\theta\tilde{\theta}}(\theta\{|\bar{\theta}) = \delta(\bar{\theta},\tilde{\theta}) + (n-1)\sum_{\bar{\bar{\theta}}} (\bar{\theta}\{|\bar{\bar{\theta}})(\tilde{\theta}\{|\bar{\bar{\theta}})(-1)^{\bar{S}+\tilde{S}+\bar{L}+\tilde{L}}$$

$$\times \{[\bar{S}][\bar{L}][\tilde{S}][\tilde{L}]\}^{\frac{1}{2}} \begin{Bmatrix} \frac{1}{2} & \bar{\bar{S}} & \bar{S} \\ \frac{1}{2} & S & \tilde{S} \end{Bmatrix} \begin{Bmatrix} l & \bar{\bar{L}} & \bar{L} \\ l & L & \tilde{L} \end{Bmatrix} \qquad (7\text{-}59)$$

The symbols θ, $\bar{\theta}$, $\tilde{\theta}$, and $\bar{\bar{\theta}}$ stand for states of l^n, l^{n-1}, l^{n-1}, and l^{n-2}, respectively. To use Eq. (7-59), a particular state $\tilde{\theta}$—the *godparent*—of l^{n-1} is selected. The various cfp are evaluated for a given $\theta = \gamma SL$ with the aid of Eq. (7-59), and $N_{\theta\tilde{\theta}}$ is chosen to normalize them. A second godparent is chosen, and a similar calculation is made. If only one term with that particular S and L occurs in l^n, then the second set of cfp will simply be a multiple of the first. In general, if x terms of a kind exist, then x godparents must be chosen; the x sets of cfp will usually be linearly independent. Needless to say, the various sets of cfp will seldom reproduce those obtained from the group-theoretical classification of states, and for this reason we shall not make any use of Redmond's formula.

PROBLEMS

7-1.‡ The states of l, l^2, l^{n-3}, l^{n-2}, l^{n-1}, and l^n are denoted by ω, π, $\bar{\Pi}$, Π, $\bar{\Omega}$. and Ω, respectively. Derive the equation

$$\sum_{\Pi} (\bar{\Pi};\omega|\Pi)(\Pi;\pi|\Omega) = \sum_{\bar{\Omega}} (\bar{\Pi};\pi|\bar{\Omega})(\bar{\Omega};\omega|\Omega)$$

7-2. Prove that the eigenvalues of

$$q_{ij} = -\frac{1}{2} - 2\mathbf{s}_i \cdot \mathbf{s}_j - 2\sum_{\text{odd } k} (\mathbf{v}_i^{(k)} \cdot \mathbf{v}_j^{(k)})$$

are zero for all terms of l^2 except ${}^1S$. Make the substitution

$$G = \sum_{i>j} q_{ij}$$

in Eq. (7-4), and deduce that

$$(l^{n-2}v\xi SL + l^2({}^1S)|\}l^n v\xi SL)^2 = \frac{(n-v)(4+4l-n-v)}{2n(n-1)(2l+1)}$$

7-3. By setting $\pi \equiv {}^1S$ in Prob. 7-1, obtain the equation

$$(l^{n-3}\bar{v}\bar{\xi}\bar{S}\bar{L}|\}l^{n-2}v\xi SL)^2 = \frac{n(n-\bar{v}-1)(5+4l-n-\bar{v})}{(n-2)(n-v)(4+4l-n-v)}(l^{n-1}\bar{v}\bar{\xi}\bar{S}\bar{L}|\}l^n v\xi SL)^2$$

‡ The equations in this problem and the two following ones are due to Racah.[76]

Show that this is consistent with Eqs. (7-47) to (7-50).

7-4. Write down the orthogonality relation for the coefficients

$$([\bar{\lambda}]\bar{W} + [1](10 \cdots 0)|[\lambda]W)$$

Prove that

$$(l^{n-1}[\bar{\lambda}] + l|l^n[\lambda])^2 = \frac{S(n + 2S + 2)}{n(2S + 1)}$$

if the spins associated with $[\lambda]$ and $[\bar{\lambda}]$ are S and $S - \frac{1}{2}$, respectively. Derive the companion to this equation, corresponding to spins S and $S + \frac{1}{2}$.

Prove that the number of standard tableaux that can be constructed from the partition $[ab]$ is

$$D[ab] = \frac{(a + b)!(a - b + 1)}{b!(a + 1)!}$$

and show that

$$(l^{n-1}[\bar{\lambda}] + l|l^n[\lambda])^2 = \frac{D[\bar{\lambda}]}{D[\lambda]}$$

(This result, which is equally valid for shapes $[\lambda]$ possessing more than two columns, is due to Racah.[58])

7-5. For d electrons, the cfp factorize according to the equation

$$(d^{n-1}\bar{W}\bar{S}\bar{L}|\}d^nWSL) = (\bar{W}\bar{L} + d|WL)(d^{n-1}\bar{v}\bar{S} + d|\}d^nvS)$$

Construct tables for the two factors on the right, and check your results against Racah's tables[76] for the complete cfp. [Jahn[81] has tabulated the required values of $(\bar{W}\bar{L} + d|WL)$.]

7-6. Prove

$$(l^nW\xi SLJ\|V^{(k)}\|l^nW'\xi'S'L'J') = n\delta(S,S')(-1)^{S+J+L+L'+l}\{[J][L][k][L'][J']\}^{\frac{1}{2}} \times \begin{Bmatrix} L & k & L' \\ J' & S & J \end{Bmatrix} \sum_{\bar{\theta}} (-1)^{\bar{L}}(\theta\{|\bar{\theta})(\theta'\{|\bar{\theta}) \begin{Bmatrix} L & k & L' \\ l & \bar{L} & l \end{Bmatrix}$$

7-7. The states θ and θ' of l^n possess the same spin S. Prove

$$(\theta JM_J|\zeta\Sigma\mathbf{s}\cdot\mathbf{l}|\theta'JM_J) = \left(\frac{\zeta}{4S}\right)\delta(\theta,\theta')[J(J+1) - L(L+1) - S(S+1)]$$
$$+ \tfrac{1}{2}n\zeta(-1)^{J+S+l}\begin{Bmatrix} L & 1 & L' \\ S & J & S \end{Bmatrix}\left[\frac{l(l+1)(2l+1)(2L+1)(2L'+1)(2S+1)^2}{S(S+1)}\right]^{\frac{1}{2}} \times \sum_{\bar{\theta}}{}' (\theta\{|\bar{\theta})(\theta'\{|\bar{\theta})(-1)^{\bar{L}}\begin{Bmatrix} L & 1 & L' \\ l & \bar{L} & l \end{Bmatrix}$$

Show that when $L' = L$ this equation reduces to

$$(\theta JM_J|\zeta\Sigma\mathbf{s}\cdot\mathbf{l}|\theta'JM_J) = \zeta X\,\frac{J(J+1) - L(L+1) - S(S+1)}{8SL(L+1)}$$

where

$$X = \delta(\theta,\theta')[(2S + 2 - n)L(L + 1) - (n - 2S)l(l + 1)] \\ + n\frac{2S+1}{S+1}\sum_{\bar{\theta}}{}'(\theta\{|\bar{\theta})(\theta'\{|\bar{\theta})\bar{L}(\bar{L} + 1)$$

Verify that the factor λ in the equivalent operator $\lambda\mathbf{S}\cdot\mathbf{L}$ is given by

$$\lambda = \pm 1/2S$$

for terms of maximum multiplicity, and by

$$\lambda = \pm\frac{-l(l+1)\delta(\theta,\theta') + (n-1)\sum_{\bar{\theta}}{}'(\theta\{|\bar{\theta})(\theta'\{|\bar{\theta})\bar{L}(\bar{L}+1)}{(n-2)L(L+1)}$$

for terms of the next highest multiplicity. The plus sign applies in the first half of the shell, the minus sign in the second half. (These formulas are given by Elliott et al.[82])

7-8. Plot the array of weights (m_1,m_2) for the representation $(u_1u_2) = (20)$ of G_2. Use the equations

$$[4(\tfrac{1}{7})^{\frac{1}{2}}V_0^{(1)} + 6(\tfrac{1}{21})^{\frac{1}{2}}V_0^{(5)}]|(20)I, M_L = 6) = 2|(20)I,6)$$
$$[(\tfrac{1}{7})^{\frac{1}{2}}V_0^{(1)} - 9(\tfrac{1}{21})^{\frac{1}{2}}V_0^{(5)}]|(20)I,6) = 0$$

to derive $((20)I\|V^{(5)}\|(20)I)$, and extend the method to obtain the matrix (7-28).

8

CONFIGURATIONS OF MORE THAN TWO EQUIVALENT ELECTRONS

8-1 INTRODUCTION

The coefficients of fractional parentage allow the matrix elements of an operator to be evaluated for any configuration l^n. In spite of its power, this method does not expose many simple properties that the matrix elements possess; on the contrary, it frequently obscures them. For example, the reduced matrix elements of $\mathbf{V}^{(1)}$, being proportional to those of $\mathbf{L}$, are given by a very simple closed expression [see Eq. (7-58)]; yet the general equation for the reduced matrix elements of $\mathbf{V}^{(k)}$ [namely, Eq. (7-52)] does not immediately give this expression if we set $k = 1$. In fact, the lack of explicit formulas for the cfp makes it appear virtually impossible to obtain the required expression in this way. With this difficulty, it is natural to turn to group theory in order to investigate the general properties of the matrix elements, that is, properties they possess in virtue of the particular representations of groups such as Sp_{4l+2}, R_{2l+1}, or G_2 that label the states. By so doing, we might hope to augment the results of the straightforward cfp approach. But if we are to proceed along these lines, we need to know how the various contributions to the Hamiltonian, which appear as operators in the matrix elements, stand with respect to the higher groups. The situation is analogous to that at the beginning of Chap. 4; there, we express quantities such as H_1 and H_2 in tensor-operator form and apply the theory to f^2. Our present objective is to express these operators as the components of generalized tensors and to illustrate the techniques by applying the theory to a configuration of the type l^n for which $n > 2$. We choose the configuration f^6 for this role: it has been observed as the ground configuration in SmI,[83] EuIV,[84] PuI,[85,86] and AmIV.[87]‡

8-2 GENERALIZED TENSORS

In Sec. 2-7, it is shown that, if the operator $\mathbf{T}^{(k)}$ is to transform like the spherical harmonics Y_{kq}, then its components have to satisfy commutation relations of the type

$$\begin{aligned} [J_z,T_q^{(k)}] &= qT_q^{(k)} \\ \text{and} \qquad [J_\pm,T_q^{(k)}] &= [k(k+1) - q(q \pm 1)]^{\frac{1}{2}}\, T_{q\pm1}^{(k)} \end{aligned} \tag{8-1}$$

It is no accident that an eigenstate $|J = k, M = q)$ satisfies the very similar pair of equations

$$\begin{aligned} J_z|k,q) &= q|k,q) \\ \text{and} \qquad J_\pm|k,q) &= [k(k+1) - q(q \pm 1)]^{\frac{1}{2}}|k, q \pm 1) \end{aligned} \tag{8-2}$$

since the transformation properties of both $T_q^{(k)}$ and $|k,q)$ are directly related to those of Y_{kq}. Indeed, given Eqs. (8-1) and (8-2), we could immediately deduce that the operators $T_q^{(k)}$ transform according to the representation of R_3 for which the eigenstates $|k,q)$ serve as a basis, since J_z, J_+, and J_- form a complete set of infinitesimal operators for R_3.

This principle is easily extended to other groups. Consider, for example, the state of l^2 defined by

$$S = \kappa_2 \qquad M_S = \pi_2 \qquad L = k_2 \qquad M_L = q_2$$

It is straightforward to show

$$\begin{aligned} W_{\pi_1 q_1}^{(\kappa_1 k_1)}|l^2\kappa_2\pi_2k_2q_2) &= \sum_{\kappa_3,\pi_3,k_3,q_3} (-1)^{2s+2l-\pi_3-q_3}[(-1)^{\kappa_1+\kappa_2+\kappa_3+k_1+k_2+k_3} + (-1)^{\kappa_1+k_1}] \\ &\times \begin{Bmatrix} \kappa_1 & \kappa_2 & \kappa_3 \\ s & s & s \end{Bmatrix} \begin{Bmatrix} k_1 & k_2 & k_3 \\ l & l & l \end{Bmatrix} \begin{pmatrix} \kappa_1 & \kappa_2 & \kappa_3 \\ \pi_1 & \pi_2 & -\pi_3 \end{pmatrix} \begin{pmatrix} k_1 & k_2 & k_3 \\ q_1 & q_2 & -q_3 \end{pmatrix} \\ &\times \{[\kappa_1][\kappa_2][\kappa_3][k_1][k_2][k_3]\}^{\frac{1}{2}}|l^2\kappa_3\pi_3k_3q_3) \end{aligned} \tag{8-3}$$

‡ The possibility of using Sm^{++} ions in CaF_2 as an optical maser has stimulated interest in the spectrum of Sm III, the ground configuration of which is also of the type f^6. [See G. H. Dieke and R. Sarup, *J. Chem. Phys.*, **36**, 371 (1962), and D. L. Wood and W. Kaiser, *Phys. Rev.*, **126**, 2079 (1962).]

On comparing this equation with Eq. (6-9), we see that the coefficient of $|l^2\kappa_3\pi_3k_3q_3)$ in the development of

$$W^{(\kappa_1k_1)}_{\pi_1q_1}|l^2\kappa_2\pi_2k_2q_2)$$

is identical to the coefficient of $W^{(\kappa_3k_3)}_{\pi_3q_3}$ in the expansion of

$$[W^{(\kappa_1k_1)}_{\pi_1q_1},W^{(\kappa_2k_2)}_{\pi_2q_2}]$$

provided that $\kappa_1 + k_1$ is odd. This establishes the correspondence

$$W_{\pi q}{}^{(\kappa k)} \leftrightarrow |l^2\kappa\pi kq)$$

which is the generalization of

$$T_q{}^{(k)} \leftrightarrow |k,q)$$

The condition that $\kappa_1 + k_1$ be odd is satisfied by the infinitesimal operators of Sp_{4l+2}, R_{2l+1}, and G_2; we can therefore characterize sets of tensor operators by the representations of these groups to which the corresponding states of l^2 belong. Thus, since the terms ^3P, ^3F, . . . and ^1D, ^1G, . . . together form the basis functions for the irreducible representation $(110 \cdots 0)$ of Sp_{4l+2}, the tensors

$$\mathbf{W}^{(11)}, \mathbf{W}^{(02)}, \mathbf{W}^{(13)}, \mathbf{W}^{(04)}, \ldots, \mathbf{W}^{(0\ 2l)}$$

together can be regarded as the $2l(4l + 3)$ components of a *single* tensor that transforms according to the irreducible representation $(110 \cdots 0)$ of Sp_{4l+2}. The transformation properties of the tensors $\mathbf{W}^{(\kappa k)}$ for which $\kappa + k$ is odd can be found by investigating the forbidden terms of l^2. Being totally symmetric with respect to the interchange of the two electrons, the latter form a basis for the representation [2] of U_{4l+2}. The branching rules[75] for the reduction $U_{4l+2} \rightarrow Sp_{4l+2}$ give simply

$$[2] \rightarrow (20 \cdots 0)$$

Accordingly, the tensors

$$\mathbf{W}^{(10)}, \mathbf{W}^{(01)}, \mathbf{W}^{(12)}, \mathbf{W}^{(03)}, \ldots, \mathbf{W}^{(1\ 2l)}$$

form the $(2l + 1)(4l + 3)$ components of a single tensor that transforms according to $(20 \cdots 0)$.

We can proceed in a similar manner for the group R_{2l+1}. The terms D, G, . . . of l^2 form the basis for the irreducible representation $(20 \cdots 0)$ of R_{2l+1}; therefore the tensors

$$\mathbf{W}^{(\kappa 2)}, \mathbf{W}^{(\kappa 4)}, \ldots, \mathbf{W}^{(\kappa\ 2l)}$$

for any given projection π of κ form the $l(2l + 3)$ components of a single tensor that transforms according to $(20 \cdots 0)$ of R_{2l+1}. Similarly, for a given π, the tensors

$$\mathbf{W}^{(\kappa 1)}, \mathbf{W}^{(\kappa 3)}, \ldots, \mathbf{W}^{(\kappa\, 2l-1)}$$

form the $l(2l+1)$ components of a single tensor that transforms according to $(110 \cdots 0)$ of R_{2l+1}, since the terms $P, F, \ldots$ of l^2 form a basis for this representation. Finally, the three sets of tensors

$$\mathbf{W}^{(\kappa 2)}, \mathbf{W}^{(\kappa 4)}, \mathbf{W}^{(\kappa 6)}$$
$$\mathbf{W}^{(\kappa 1)}, \mathbf{W}^{(\kappa 5)}$$

and

$$\mathbf{W}^{(\kappa 3)}$$

for a given π form the components of three tensors that transform according to the representations (20), (11), and (10) of G_2, respectively.

8-3 APPLICATIONS OF THE WIGNER-ECKART THEOREM

The knowledge of the transformation properties of operators with respect to groups such as G_2, R_{2l+1} or Sp_{4l+2} is of great value in the calculation of their matrix elements. Consider, for example, the matrix elements

$$(f^n W\, U\tau S M_S L M_L | V_q^{(k)} | f^n W'\, U'\tau' S M_S L' M_L') \qquad (8\text{-}4)$$

for various L, k, L', M_L, q, and M_L'. Since the tensors $\mathbf{V}^{(k)}$ are proportional to $\mathbf{W}^{(0k)}$, the components $V_q^{(k)}$ are restricted to the components of a single generalized tensor that transforms according to, say, U'' of G_2, if k is restricted to one of the sequences (2,4,6), (1,5), or (3). Suppose that the representation U is contained once in the decomposition of $U'' \times U'$. Then the matrix elements (8-4) are proportional to

$$(U'\tau' L' M_L'; U'' kq | U\tau L M_L)$$

It is convenient at this point to reverse the positions of $U'\tau' L' M_L'$ and $U''kq$ in this coefficient; the dependence on L', k, L, M_L', q, and M_L is unaltered. The factorization

$$(U''kq; U'\tau' L' M_L' | U\tau L M_L) = (kqL'M_L' | kL'LM_L)(U''k + U'\tau' L' | U\tau L)$$

gives at once the VC coefficient to be expected from the usual application of the Wigner-Eckart theorem; in addition, it follows that the dependence of the reduced matrix element

$$(f^n W\, U\tau S L \| V^{(k)} \| f^n W'\, U'\tau' S L')$$

on L, L', and the restricted set of values of k is represented by

$$(-1)^{L+k-L'}[L]^{\frac{1}{2}}(U''k + U'\tau' L' | U\tau L)$$

The factor of proportionality can be found by using Eq. (7-52) once. For example, we find

$$(f^6(210)(21)^5L\|V^{(k)}\|f^6(111)(10)^5F)$$
$$= (-1)^{L+1}(2)^{-\frac{1}{2}}[L]^{\frac{1}{2}}((20)k + (10)f|(21)L) \quad (8\text{-}5)$$

where $k = 2, 4$, or 6 and $L = 2, 3, 4, 5, 7$, or 8. The coupling coefficients on the right can be found from Appendix 2, and hence at a single stroke we obtain $3 \times 6 = 18$ matrix elements, 14 of which are nonzero.

Since only a few tables of the coefficients $(U''k + U'\tau'L'|U\tau L)$ are available, equations like Eq. (8-5) are frequently of no immediate value in an actual calculation. The second form of the Wigner-Eckart theorem, represented by Eq. (2-24), is more useful; for example, a constant A independent of L and k must exist such that

$$(f^6(210)(21)^5L\|V^{(k)}\|f^6(111)(10)^5F)$$
$$= A(f^3(210)(21)^2L\|V^{(k)}\|f^3(100)(10)^2F) \quad (8\text{-}6)$$

for $k = 2, 4, 6$ and $L = 2, 3, 4, 5, 7, 8$. In this way we can often relate matrix elements that are tedious to evaluate to corresponding ones in a less complex configuration.

The simplicity of Eqs. (8-5) and (8-6) is due to the fact that the representation (21) of G_2 occurs once only in the reduction of $(20) \times (10)$. In the notation of Sec. 5-11, $c((20)(10)(21)) = 1$. Instances where $c(UU'U'') > 1$ or $c(WW'W'') > 1$ occur frequently. For example, $c((20)(20)(20)) = 2$; hence we may write

$$(f^6(210)(20)^5L\|V^{(k)}\|f^6(210)(20)^5L')$$
$$= A_1(f^3(210)(20)^2L\|V^{(k)}\|f^3(210)(20)^2L')$$
$$+ A_2(f^3(111)(20)^4L\|V^{(k)}\|f^3(111)(20)^4L') \quad (8\text{-}7)$$

for k, L, $L' = 2, 4, 6$ (in any combination), provided that the two sets of matrix elements on the right are linearly independent. The reduced matrix elements of $\mathbf{U}^{(k)}$, defined by $\mathbf{U}^{(k)} = \mathbf{V}^{(k)}[k]^{-\frac{1}{2}}$, have been tabulated[88] for f^3, so we have only to use Eq. (7-52) twice to find A_1 and A_2, and the remaining matrix elements follow at once.

The proportionality factors in equations such as (8-6) and (8-7) are often very simple; for example,

$$A = (\tfrac{1}{2})^{\frac{1}{2}} \qquad A_1 = A_2 = \tfrac{2}{3}$$

In some cases they can be obtained without resorting to detailed calculations. For example, it can be shown for the group Sp_{4l+2} that, if $v > 0$,

$$c((v)(v)(200\cdots0)) = 1$$

where (v) stands for the symbol $(11 \cdots 10 \cdots 0)$ in which v ones appear. Hence a constant A' must exist such that

$$(l^n v\xi SL\|W^{(\kappa k)}\|l^n v\xi' S'L') = A'(l^v v\xi SL\|W^{(\kappa k)}\|l^v \xi' S'L') \tag{8-8}$$

where $\kappa + k$ is odd. The factor A' is immediately seen to be $+1$ by setting $\kappa = 0$, $k = 1$, $\xi = \xi'$, $S = S'$, $L = L'$ and using Eq. (7-58). Equation (8-8) was first obtained by Racah[76] by different methods.

8-4 THE SPIN-ORBIT INTERACTION

The contribution H_2 to the Hamiltonian is easier to treat by the methods of the previous section than the Coulomb interaction H_1, and for this reason we prefer to examine it first. From Eq. (7-55), we see that the matrix elements of H_2 depend on the reduced matrix elements of $\mathbf{W}^{(11)}$. For f electrons, this double tensor forms part of a generalized tensor whose classification according to the groups G_2, R_7,

TABLE 8-1 THE NUMBERS $c(UU'(11))$

U	U'								
	(00)	(10)	(11)	(20)	(21)	(30)	(22)	(31)	(40)
(00)	—	—	1	—	—	—	—	—	—
(10)	—	1	—	1	1	—	—	—	—
(11)	1	—	1	1	—	1	1	—	—
(20)	—	1	1	1	1	1	—	1	—
(21)	—	1	—	1	2	1	—	1	1
(30)	—	—	1	1	1	1	1	1	1
(22)	—	—	1	—	—	1	1	1	—
(31)	—	—	—	1	1	1	1	2	1
(40)	—	—	—	—	1	1	—	1	1

and Sp_{14} is (11), (110), and (1100000), respectively. The numbers $c(UU'(11))$, $c(WW'(110))$, and $c((v)(v')(1100000))$ are given in Tables 8-1 to 8-3; the first two tables have been derived by McLellan.[89]

The selection rule $\Delta L = 0, \pm 1$ on H_2 obtained in Sec. 4-4 can be restated by saying that all matrix elements vanish if $c(\mathfrak{D}_L \mathfrak{D}_{L'} \mathfrak{D}_1) = 0$. Every zero in Tables 8-1 to 8-3 implies the vanishing of a collection of matrix elements; for example, $c((11)(21)(11)) = 0$, and therefore

$$(f^n W(21)SL\|W^{(11)}\|f^n W'(11)S'L') = 0 \tag{8-9}$$

TABLE 8-2 THE NUMBERS $c(WW'(110))$

W	W' (000)	(100)	(110)	(200)	(111)	(210)	(211)	(220)	(221)	(222)
(000)	—	—	1	—	—	—	—	—	—	—
(100)	—	1	—	—	1	1	—	—	—	—
(110)	1	—	1	1	1	—	1	1	—	—
(200)	—	—	1	1	—	—	1	—	—	—
(111)	—	1	1	—	1	1	1	—	1	—
(210)	—	1	—	—	1	2	1	—	1	—
(211)	—	—	1	1	1	1	2	1	1	1
(220)	—	—	1	—	—	—	1	1	1	—
(221)	—	—	—	—	1	1	1	1	2	1
(222)	—	—	—	—	—	—	1	—	1	1

TABLE 8-3 THE NUMBERS $c((v)(v')(1100000))$

(v)	(v') (0)	(1)	(2)	(3)	(4)	(5)	(6)	(7)
(0)	—	—	1	—	—	—	—	—
(1)	—	1	—	1	—	—	—	—
(2)	1	—	1	—	1	—	—	—
(3)	—	1	—	1	—	1	—	—
(4)	—	—	1	—	1	—	1	—
(5)	—	—	—	1	—	1	—	1
(6)	—	—	—	—	1	—	1	—
(7)	—	—	—	—	—	1	—	—

for all n, W, W', S, S', L, and L'. In the tables[90] of the matrix elements of H_2 for f^3, 32 zeros occur that are not covered by the selection rules ΔL, $\Delta S = 0, \pm 1$; of these, 8 are examples of Eq. (8-9). The zeros in Table 8-2 can be interpreted in a similar way, but the analogues of Eq. (8-9) appear to be less useful. (For example, none of the remaining 24 zeros for f^3 are accounted for by these equations.) The entries in Table 8-3 form a regular pattern, and we can deduce that all matrix elements of H_2 vanish unless $\Delta v = 0, \pm 2$.

When $c(UU'(11)) = 1$, parameters A (independent of L, L', τ, and τ') can always be found such that

$$(f^n W U \tau S L \| W^{(11)} \| f^n W' U' \tau' S' L')$$
$$= A(f^{n'} W'' U \tau S'' L \| W^{(11)} \| f^{n'} W''' U' \tau' S''' L')$$

for all L, L', τ, and τ'. It is interesting to observe at this point that $\mathbf{V}^{(1)}$, like $\mathbf{W}^{(11)}$, is part of a generalized tensor whose classification according to the groups R_7 and G_2 is (110)(11); hence, when $c(UU'(11)) = 1$, all matrix elements of $\mathbf{W}^{(11)}$ are proportional to those of $\mathbf{V}^{(1)}$. Unless $U = U'$, the proportionality factor is ∞; so the only useful result is

$$\begin{aligned}(f^nW\,U\tau SL\|W^{(11)}\|f^nW'\,U\tau'S'L')\\ &= A_1(f^nW''\,U\tau S''L\|V^{(1)}\|f^nW''\,U\tau'S''L')\\ &= A_2\delta(\tau,\tau')\delta(L,L')[L(L+1)(2L+1)]^{\frac{1}{2}} \qquad (8\text{-}10)\end{aligned}$$

from Eq. (7-58). The delta functions give the selection rules $\Delta\tau = 0$, $\Delta L = 0$ when $c(UU(11)) = 1$. The vanishing of

$$(f^6(211)(30)^4L\|W^{(11)}\|f^6(211)(30)^4L') \qquad (8\text{-}11)$$

when $L \neq L'$ illustrates this.

Similar remarks can be made when $c(WW'(110)) = 1$. For example, if $W = W'$,

$$\begin{aligned}(f^nW\,U\tau SL\|W^{(11)}\|f^nW\,U'\tau'S'L')\\ &= A_3(f^nW\,U\tau S''L\|V^{(1)}\|f^nW\,U'\tau'S''L')\\ &= A_4\delta(U,U')\delta(\tau,\tau')\delta(L,L')[L(L+1)(2L+1)]^{\frac{1}{2}} \qquad (8\text{-}12)\end{aligned}$$

Thus when $c(WW(110)) = 1$, spin-orbit matrix elements are zero unless diagonal with respect to U, τ, and L. The vanishing of

$$(f^3(111)(20)^4D\|W^{(11)}\|f^3(111)(10)^4F)$$

is an example of this kind. [The vanishing of the matrix element (8-11) is not an example of this kind because $c((211)(211)(110)) = 2$.] The terms of maximum multiplicity of a particular configuration f^n all belong to a single representation W_n of R_7. Moreover,

$$c(W_nW_n(110)) = 1$$

for all n [except in the trivial case $W_n = (000)$]. We find from Eqs. (7-55) and (8-12) that the diagonal matrix elements of H_2 have a dependence on J and L given by

$$J(J+1) - L(L+1) - S(S+1)$$

which accounts for the familiar fact that the multiplet splittings of terms of maximum multiplicity can be matched, so that all the levels of any J coincide, by displacing the centers of gravity of the multiplets. This property, which is illustrated in Fig. 8-1 for the quartet terms

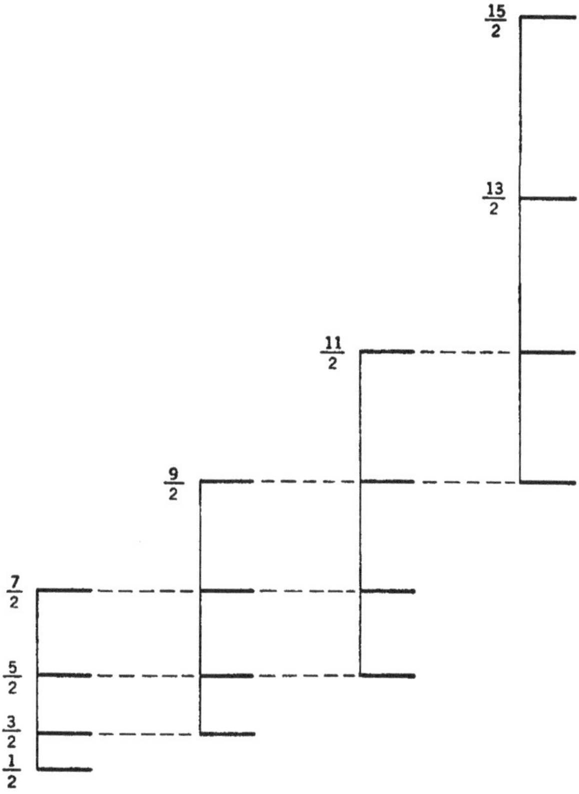

FIG. 8-1 In the limit of LS coupling, the multiplets of f^3 with $S = \frac{3}{2}$ can be displaced so that the energies of all levels with the same J coincide. A similar property holds for any set of terms of l^n belonging to an irreducible representation W of R_{2l+1} for which $c(WW(110 \cdots 0)) = 1$.

of f^3, can also be derived from the equation $\lambda = \pm 1/2S$ given in Prob. 7-7.

The generalization of Table 8-3 to electrons of azimuthal quantum number l is straightforward, and we find $c((v)(v)(110 \cdots 0)) \leq 1$. We can therefore write

$$(l^n v \xi SL \| W^{(\kappa k)} \| l^n v \xi' S'L') = A_5 (l^v v \xi SL \| W^{(\kappa k)} \| l^v v \xi' S'L') \quad (8\text{-}13)$$

for $\kappa + k$ even. The explicit expression

$$A_5 = \frac{2l + 1 - n}{2l + 1 - v} \quad (8\text{-}14)$$

has been given by Racah[76] (see Prob. 8-3).

When $c(UU'(11))$ or $c(WW'(110))$ is 2, every set of matrix elements between states labeled by U and U' or W and W' can be expressed as a linear combination of two independent sets, in analogy with Eq. (8-7). From Tables 8-1 and 8-2, it can be seen that this occurs only when $U = U'$ or $W = W'$; hence one of the sets can be taken to be the matrix elements of $\mathbf{V}^{(1)}$. To ensure that the other set is not a simple multiple of this set, it is necessary only that a matrix element off diagonal with respect to L should be nonzero. Thus, since $c((21)(21)(11)) = 2$,

$$\begin{aligned}(f^nW(21)SL\|W^{(11)}\|f^nW'(21)S'L')\\ = A_6(f^3(210)(21)^2L\|W^{(11)}\|f^3(210)(21)^2L')\\ + A_7(f^3(210)(21)^2L\|V^{(1)}\|f^3(210)(21)^2L') \quad (8\text{-}15)\end{aligned}$$

The matrix element of $\mathbf{W}^{(11)}$ between the states $(210)(21)^2G$ and $(210)(21)^2F$ of f^3 can be shown to vanish, apparently accidentally, by using Eq. (7-56). Since the matrix of $\mathbf{V}^{(1)}$ is diagonal with respect to L, it follows from Eq. (8-15) that all matrix elements of H_2 between states of the type $(21)G$ and $(21)F$ in any configuration f^n are zero. Powerful though techniques of this kind are, they do not account for all the zeros that appear in tables of H_2 (see Probs. 8-4 and 8-5).

TABLE 8-4 ORDER OF CONSTRUCTING MATRIX ELEMENTS OF H_2 FOR SOME LEVELS OF f^6 FOR WHICH $J = 1$

	(100) (10) $^7F_1$	(210) (20) $^5D_1$	(210) (21) $^5D_1$	(111) (20) $^5D_1$	(210) (21) $^5F_1$	(111) (10) $^5F_1$
$(100)(10)^7F_1$	6	1	1	5	1	5
$(210)(20)^5D_1$	1	2	2	3	0	3
$(210)(21)^5D_1$	1	2	2	3	2	3
$(111)(20)^5D_1$	5	3	3	4	0	0
$(210)(21)^5F_1$	1	0	2	0	2	3
$(111)(10)^5F_1$	5	3	3	0	3	4

To conclude this section, we indicate how advantage can be taken of the existing[90] spin-orbit matrices for f^3 in order to minimize the labor in constructing some matrix elements of H_2 for f^6. We select the levels of f^6 for which $J = 1$, $S \geq 2$, $L \geq 2$; by choosing these special cases, the ground is prepared for the detailed illustration of the hyperfine structure of PuI given in Sec. 8-9. Table 8-4 indicates the order of constructing the matrix elements. Apart from the zeros, which correspond to vanishing matrix elements, the numbers in the table show which matrix elements derive from a common equation;

for example, the first entries are made by performing the summation (7-56) once and then using

$$(f^6(100)(10)^7F\|W^{(11)}\|f^6(210)U^5L) \\ = A_8(f^3(100)(10)^2F\|W^{(11)}\|f^3(210)U^2L)$$

where A_8 is independent of U and L. The results are given in Table 8-5.

TABLE 8-5 THE MATRIX ELEMENTS $(f^6WUSLJ|H_2|f^6W'U'S'L'J)$ FOR WHICH $S \geq 2$, $L \geq 2$, $J = 1$
(All entries must be multiplied by ζ.)

	(100) (10) $^7F_1$	(210) (20) $^5D_1$	(210) (21) $^5D_1$	(111) (20) $^5D_1$	(210) (21) $^5F_1$	(111) (10) $^5F_1$
$(100)(10)^7F_1$	$-\frac{11}{6}$	$-(\frac{16}{21})^{\frac{1}{2}}$	$-(\frac{176}{63})^{\frac{1}{2}}$	$(\frac{32}{3})^{\frac{1}{2}}$	$-(\frac{11}{18})^{\frac{1}{2}}$	$-(\frac{2}{9})^{\frac{1}{2}}$
$(210)(20)^5D_1$	$-(\frac{16}{21})^{\frac{1}{2}}$	$-\frac{5}{42}$	$5(33)^{\frac{1}{2}}/21$	$-5(14)^{\frac{1}{2}}/84$	0	$-(\frac{6}{7})^{\frac{1}{2}}$
$(210)(21)^5D_1$	$-(\frac{176}{63})^{\frac{1}{2}}$	$5(33)^{\frac{1}{2}}/21$	$-\frac{95}{84}$	$5(462)^{\frac{1}{2}}/42$	$(\frac{14}{36})^{\frac{1}{2}}$	$(\frac{11}{126})^{\frac{1}{2}}$
$(111)(20)^5D_1$	$(\frac{32}{3})^{\frac{1}{2}}$	$-5(14)^{\frac{1}{2}}/84$	$5(462)^{\frac{1}{2}}/42$	$-\frac{5}{12}$	0	0
$(210)(21)^5F_1$	$-(\frac{11}{18})^{\frac{1}{2}}$	0	$(\frac{14}{36})^{\frac{1}{2}}$	0	$\frac{2}{3}$	$-(\frac{11}{9})^{\frac{1}{2}}$
$(111)(10)^5F_1$	$-(\frac{2}{9})^{\frac{1}{2}}$	$-(\frac{6}{7})^{\frac{1}{2}}$	$(\frac{11}{126})^{\frac{1}{2}}$	0	$-(\frac{11}{9})^{\frac{1}{2}}$	$-\frac{2}{3}$

8-5 MATRIX ELEMENTS OF $V^{(5)}$

In Sec. 7-5 it is stated that reduced matrix elements of $\mathbf{V}^{(5)}$ can be defined unambiguously by using the limited set of quantum numbers U, τ, and L to label the states. To justify this remark, we first observe that, since the components of $\mathbf{V}^{(5)}$ are infinitesimal operators for R_7 and G_2, we need concern ourselves only with matrix elements diagonal with respect to W and U. The fact that $\mathbf{V}^{(1)}$ and $\mathbf{V}^{(5)}$ together form the components of a tensor that transforms according to (11) of G_2 implies that, when $c(UU(11)) = 1$, there exists a constant A_9 such that

$$(f^nWU\tau SL\|V^{(k)}\|f^nWU\tau'SL') \\ = A_9(f^{n'}\,W'U\tau S'L\|V^{(k)}\|f^{n'}\,W'U\tau'S'L') \quad (8\text{-}16)$$

for $k = 1$ and 5 and all permitted values of L, L', τ, and τ'. If we set $k = 1$, $L = L'$, and $\tau = \tau'$, then an application of Eq. (7-58) yields at once $A_9 = 1$. This proves that the set of symbols f^nWS can be omitted from the matrix elements when $c(UU(11)) = 1$.

From Table 8-1 it can be seen that $c(UU(11)) = 2$ when $U = (21)$ or (31). We may take

$$
\begin{aligned}
(f^n W\, U_\tau SL \| V^{(k)} \| f^n W\, U_\tau' SL') \\
&= A_{10}(f^{n'} W'\, U_\tau S'L \| V^{(k)} \| f^{n'} W'\, U_\tau' S'L') \\
&\qquad + A_{11}(f^{n'} W'\, U_\tau S'L \| W^{(1k)} \| f^{n'} W'\, U_\tau' S'L')
\end{aligned}
$$

as the analogue of Eq. (8-16). If we set $k = 1$, $L \neq L'$, then the matrix elements of $\mathbf{V}^{(k)}$ vanish. But it can be proved that, under these conditions, some matrix elements of $\mathbf{W}^{(11)}$ are nonzero; indeed,

$$(f^3(210)(21)\,^2L_{13/2}|H_2|f^3(210)(21)\,^2K_{13/2}) = (\tfrac{255}{64})^{\frac{1}{2}}\zeta$$

and

$$(f^5(221)(31)\,^2O_{21/2}|H_2|f^5(221)(31)\,^2N_{21/2}) = (\tfrac{680}{121})^{\frac{1}{2}}\zeta$$

Therefore $A_{11} = 0$. Similar arguments to those which led to $A_9 = 1$ now yield $A_{10} = 1$, thus completing the proof that, for the matrix elements of $\mathbf{V}^{(5)}$, the set of symbols f^nWS can be omitted from the description of the states.

8-6 THE COULOMB INTERACTION H_1

It has been seen that the tensor $\mathbf{W}^{(11)}$, whose scalar component is proportional to the spin-orbit interaction, can itself be interpreted as a component of a generalized tensor. The situation for two-particle operators is more complex. If we express H_1 in terms of the equivalent operator

$$\sum_{i>j} [7F_0(\mathbf{v}_i^{(0)} \cdot \mathbf{v}_j^{(0)}) + 84F_2(\mathbf{v}_i^{(2)} \cdot \mathbf{v}_j^{(2)}) + 154F_4(\mathbf{v}_i^{(4)} \cdot \mathbf{v}_j^{(4)}) + 924F_6(\mathbf{v}_i^{(6)} \cdot \mathbf{v}_j^{(6)})] \quad (8\text{-}17)$$

then we are confronted with the difficulty that the scalar products $(\mathbf{v}_i^{(k)} \cdot \mathbf{v}_j^{(k)})$, which transform, of course, according to $\mathfrak{D}_0$ of R_3, do not separately have simple transformation properties with respect to the groups G_2, R_7, and Sp_{14}. The first objective is therefore to find those linear combinations

$$\sum_k a_k(\mathbf{v}_i^{(k)} \cdot \mathbf{v}_j^{(k)}) \quad (8\text{-}18)$$

which are the components of generalized tensors with well-defined transformation properties. The problem of constructing these components and that of evaluating their matrix elements were solved by Racah in his paper of 1949.

Since the tensors $\mathbf{v}^{(2)}$, $\mathbf{v}^{(4)}$, and $\mathbf{v}^{(6)}$ form the components of a generalized tensor transforming according to (200)(20), the quantities

$$\sum_{k,q,k',q'} (v_q^{(k)})_i (v_{q'}^{(k')})_j ((200)(20)kq;(200)(20)k'q| WUKQ) \qquad (8\text{-}19)$$

possess the same transformation properties as the states $|WUKQ)$. In order to evaluate the coefficients, we perform a recoupling:

$$\begin{aligned}
&((200)(20)kq;(200)(20)k'q'|WUKQ)\\
&\quad = \sum_{m,m',W',U',K',Q'} ((200)(20)k'q'|(100)(10)Fm;(100)(10)Fm')\\
&\qquad \times ((200)(20)kq;(100)(10)Fm|W'U'K'Q')\\
&\qquad \times (W'U'K'Q';(100)(10)Fm'|WUKQ)\\
&\qquad \times (((200)(100))W',(100),W|(200),((100)(100))(200),W)\\
&\quad = \sum_{W',U',K'} ((200)(20)k'|(100)(10)F+f)\\
&\qquad \times ((200)(20)k+f|W'U'K')(W'U'K'+f|WUK)\\
&\qquad \times (((200)(100))W',(100),W|(200),((100)(100))(200),W)\\
&\qquad \times (-1)^K\{[K'][k']\}^{\frac{1}{2}}(kqk'q'|kk'KQ)\begin{Bmatrix} k & k' & K\\ 3 & K' & 3\end{Bmatrix} \qquad (8\text{-}20)
\end{aligned}$$

The coupling coefficient that involves only representations of R_7 is a generalization of the coupling coefficient of Eq. (3-4); its properties may be investigated by methods similar to those of Sec. 3-1. For the present purposes such a digression is unnecessary, however.‡ On setting $K = Q = 0$ and noting which representations W and U occur in the decompositions of the Kronecker products (200) × (200) and (20) × (20) and which, at the same time, give rise to the representation $\mathfrak{D}_0$ under the reductions $R_7 \to R_3$ and $G_2 \to R_3$, we may easily show that the only permitted values of WUK are

$$(000)(00)0 \qquad (400)(40)0 \qquad (220)(22)0 \qquad (8\text{-}21)$$

An examination of the Kronecker products of those representations involved in the coupling coefficients of Eq. (8-20) reveals that, for each of the values (8-21) of WUK, the summation of that equation reduces to a single term. The sole values of $W'U'K'$ are, in order,

$$(100)(10)F \qquad (300)(30)F \qquad (210)(21)F$$

‡ Analogues of the n-j symbols can be defined in terms of the appropriate coupling coefficients for any group of interest to us. J. S. Griffith [*Molecular Phys.*, **3**, 285 (1960)] has studied the 6-Γ and 9-Γ coefficients of the octahedral group O; E. Mauza and J. Batarunas [*Trudy Akad. Nauk. Litovsk S.S.R.*, B**3**(**26**), 27 (1961)] have tabulated 6-Γ coefficients for the double octahedral group. See also Y. Tanabe and H. Kamimura [*J. Phys. Soc. Japan*, **13**, 394 (1958)].

Two of the three coefficients $(W_1U_1L_1 + f|W_2U_2L_2)$ on the right side of Eq. (8-20) become unity, and the dependence of

$$((200)(20)kq;(200)(20)k,-q|WU00)$$

on k and q is given by

$$(-1)^q[k]^{-\frac{1}{2}}((20)k + f|U'F)$$

where $U' = (10)$, (30), or (21) according as $WU = (000)(00)$, $(400)(40)$, or $(220)(22)$, respectively. The coefficients may be found from

TABLE 8-6 THE PRODUCTS $[k]^{-\frac{1}{2}}((20)k + f|U'F)$

U'	Factor common to all elements in row	k		
		2	4	6
(10)	$-(108)^{-\frac{1}{2}}$	2	2	2
(30)	$(1081080)^{-\frac{1}{2}}$	286	−260	70
(21)	$-(5544)^{-\frac{1}{2}}$	22	8	−14

Appendix 2; they are collected in Table 8-6 with the associated factors $[k]^{-\frac{1}{2}}$. This table enables the linear combinations (8-19) to be constructed. Following Racah,[69] we define the operators

$$e_0 = 7 \sum_{i>j} (\mathbf{v}_i^{(0)} \cdot \mathbf{v}_j^{(0)})$$

$$e_1 = \sum_{i>j} [9(\mathbf{v}_i^{(0)} \cdot \mathbf{v}_j^{(0)}) + 2(\mathbf{v}_i^{(2)} \cdot \mathbf{v}_j^{(2)}) + 2(\mathbf{v}_i^{(4)} \cdot \mathbf{v}_j^{(4)}) + 2(\mathbf{v}_i^{(6)} \cdot \mathbf{v}_j^{(6)})] \quad (8\text{-}22)$$

$$e_2 = \sum_{i>j} [286(\mathbf{v}_i^{(2)} \cdot \mathbf{v}_j^{(2)}) - 260(\mathbf{v}_i^{(4)} \cdot \mathbf{v}_j^{(4)}) + 70(\mathbf{v}_i^{(6)} \cdot \mathbf{v}_j^{(6)})]$$

$$e_3 = \sum_{i>j} [22(\mathbf{v}_i^{(2)} \cdot \mathbf{v}_j^{(2)}) + 8(\mathbf{v}_i^{(4)} \cdot \mathbf{v}_j^{(4)}) - 14(\mathbf{v}_i^{(6)} \cdot \mathbf{v}_j^{(6)})]$$

which correspond to $WUL = (000)(00)0$, $(000)(00)0$, $(400)(40)0$, and $(220)(22)0$, respectively. It is now a simple matter to show that the expression (8-17) can be replaced by

$$e_0E^0 + e_1E^1 + e_2E^2 + e_3E^3 \quad (8\text{-}23)$$

where

$$E^0 = F_0 - 10F_2 - 33F_4 - 286F_6$$

$$E^1 = \frac{70F_2 + 231F_4 + 2002F_6}{9}$$

$$E^2 = \frac{F_2 - 3F_4 + 7F_6}{9} \quad (8\text{-}24)$$

and

$$E^3 = \frac{5F_2 + 6F_4 - 91F_6}{3}$$

The linear combination (8-23) expresses the Coulomb interaction in terms of the components of generalized tensors and completes the first part of the problem.

The eigenvalues of the operators e_0 and e_1 are easy to obtain. Those of e_0 are simply $\frac{1}{2}n(n-1)$ for all terms of the configuration f^n and can be ignored if we are interested solely in the relative energies of the terms. The eigenvalues of e_1 can be immediately derived by adding $9n(n-1)/14$ to the eigenvalues of 2Ξ, where Ξ is defined as in Sec. 6-4. Using (6-23), we get

$$(f^n v U \tau SL|e_1|f^n v U \tau SL) = \frac{9(n-v)}{2} + \tfrac{1}{4}v(v+2) - S(S+1)$$

Since e_1 is a scalar with respect to R_3, G_2, and R_7, all nondiagonal matrix elements vanish.

The operator e_2 corresponds to $WUL = (400)(40)0$. It is not difficult to show that, for the irreducible representations W and W' used in classifying the states of f^n, $c(WW'(400)) = 0$ unless $W = W'$ and $w_1 = 2$. Thus the matrix elements of e_2 are diagonal with respect to seniority and are zero for all terms of maximum multiplicity in f^n. Racah[69] has shown that it is permissible to write

$$(f^n v W U \tau SL|e_2|f^n v W U' \tau' SL) = \pm(W U \tau L|e_2|W U' \tau' L) \quad (8\text{-}25)$$

where the upper or lower sign is taken according to whether $v + 2S$ is less than or greater than 7. From the Wigner-Eckart theorem, the matrix element on the right of Eq. (8-25) may be factorized by writing

$$(W U \tau L|e_2|W U' \tau' L) = \sum_{\gamma} x_\gamma(W,UU')(U\tau|\chi_\gamma(L)|U'\tau') \quad (8\text{-}26)$$

The coefficients x are the analogues of the symbols A_j of Secs. 8-4 and 8-5. The number of terms in the sum over γ is determined by $c(UU'(40))$; but although this sometimes becomes quite large [for example, $c((31)(31)(40)) = 5$], Racah found that most of the linear combinations (8-26) are proportional to one another, and it is possible to express the results through one table of matrix elements

$$(U\tau|\chi_\gamma(L)|U'\tau') \quad (8\text{-}27)$$

for every couple UU', with the sole exception of $U = U' = (21)$, in which case two terms ($\gamma = 1$ and 2) are required. Racah has tabulated the quantities (8-27) and $x_\gamma(W,UU')$ for all terms of f^n.

The numbers $c(UU'(22))$ and $c(WW'(220))$ very often exceed unity, and this leads us to expect that a large number of tables analogous to those for x_γ and χ_γ will be required. However, Racah achieved a striking simplification by considering the eigenvalues, not just of e_3, but of $e_3 + \Omega$, where

$$\Omega = 11(\mathbf{V}^{(1)})^2 - 3(\mathbf{V}^{(5)})^2$$

This operator can also be written as

$$\Omega = -462^{\frac{1}{2}} \sum_{k,k',q,q'} V_q^{(k)} V_{q'}^{(k')} \times ((110)(11)kq;(110)(11)k'q'|(220)(22)00) \qquad (8\text{-}28)$$

and therefore has the same tensorial properties as e_3. It turns out that the analogous sum to (8-26) reduces to a single term,

$$(f^n v U \tau SL|e_3 + \Omega|f^n v' U'\tau' SL) = y(f^n, vSU, v'SU')(U\tau|\varphi(L)|U'\tau') \quad (8\text{-}29)$$

Both factors have been tabulated by Racah. From Eq. (5-49), it is clear that Ω can be written as $\frac{1}{2}\mathbf{L}^2 - 12G(G_2)$; Eq. (5-54) shows that the eigenvalues of this operator are

$$\tfrac{1}{2}L(L + 1) - (u_1^2 + u_2^2 + u_1u_2 + 5u_1 + 4u_2) \qquad (8\text{-}30)$$

and hence we may easily subtract from Eq. (8-29) the contribution from Ω.

For the terms of maximum multiplicity of f^n, a remarkable analysis can be carried out. First, we observe that, owing to Eq. (6-19), the eigenvalues of Ω are the same as those of the operator

$$X = \sum_{i>j} [22(\mathbf{v}_i^{(1)} \cdot \mathbf{v}_j^{(1)}) - 6(\mathbf{v}_i^{(5)} \cdot \mathbf{v}_j^{(5)})]$$

Second, it can be shown that $c((110)(110)(220)) = 1$; hence a constant A exists such that

$$(f^2(110)U^3L|e_3|f^2(110)U^3L) = A(f^2(110)U^3L|X|f^2(110)U^3L) \quad (8\text{-}31)$$

An actual calculation gives $A = -3$. Third, we notice that X, like e_3, is an operator of the type G of Sec. 7-1, and therefore either X or e_3 can be substituted into Eq. (7-4). If the states $(\Omega|$ and $|\Omega')$ of this equation correspond to the maximum spin for f^n, then the states $(\pi|$ and $|\pi')$ correspond to the maximum spin for f^2, provided that $n \leq 7$. But Eq. (8-31) indicates that it is immaterial whether we take G to be e_3 or $-3X$; hence

$$\begin{aligned}(f^n\, {}^{n+1}L|e_3|f^n\, {}^{n+1}L) &= -3(f^n\, {}^{n+1}L|X|f^n\, {}^{n+1}L) \\ &= 3(u_1^2 + u_2^2 + u_1u_2 + 5u_1 + 4u_2) - \frac{3L(L+1)}{2}\end{aligned} \qquad (8\text{-}32)$$

where $(u_1u_2) = U$ is the representation of G_2 labeling the term ${}^{n+1}L$ of f^n. Since e_2 does not give contributions to the terms of maximum multiplicity of f^n, while the contribution of e_1 is the same for all such

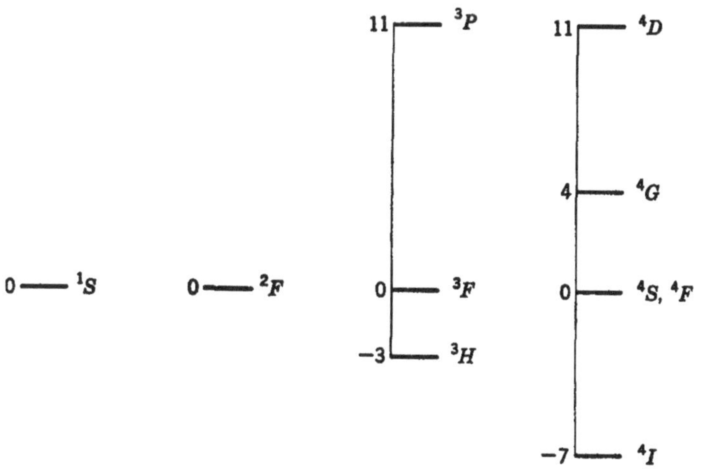

Fig. 8-2 From left to right, the four schemes above give the energies of the terms of maximum multiplicity in the configurations f^0, f^1, f^2, and f^3 as multiples of $3E^3$. Terms of maximum multiplicity for f^n, f^{7-n}, f^{7+n}, and f^{14-n} (where $0 \leq n \leq 7$) belong to the same irreducible representation W of R_7, and a single energy-level pattern is common to all. (For example, the relative positions of the quintets of f^4, expressed as multiples of $3E^3$, are identical to the corresponding quantities for the quartets of f^3.)

terms, we have the result that the energy spacings of terms of maximum multiplicity are simple multiples of the single parameter E^3, given by

$$E^3 = \frac{5F_2 + 6F_4 - 91F_6}{3}$$

Equation (8-32) has been used to plot out in Fig. 8-2 the positions of the terms of maximum multiplicity of f^n, where $0 \leq n \leq 3$, on the assumption that E^3 is the same for all configurations. In making the first term analysis of f^3, Satten[91] successfully used the known positions of the triplets of f^2 as a guide to predict the positions of the quartets of f^3. The surprising coincidence of the terms ${}^4S$ and ${}^4F$, which, when the spin-orbit coupling H_2 is included, corresponds to the coincidence of ${}^4S_{\frac{3}{2}}$ and ${}^4F_{\frac{3}{2}}$ near the LS limit, has been observed to be almost perfectly fulfilled in several neodymium crystals.[92,93]

The explicit construction of the matrices of H_1 for all configurations of the type f^n has been completed. Those for f^3 were first derived by Racah,[18] those for f^4 by Reilly.[94] Wybourne[95] has used Racah's tables to construct the complete matrices for f^5 and f^6. The half-filled shell, f^7, has been treated by Runciman.[96] The matrices of H_1 for f^{14-n} are identical to those for f^n, owing to the correspondence between electrons and holes; hence there seems little point in reproducing Racah's tables of x_γ, $\chi_\gamma(L)$, y, and $\varphi(L)$ here.

8-7 THE CONFIGURATION f^6

The group-theoretical classification of the 119 terms of f^6 can be obtained from Tables 5-1 to 5-3. There is only one term in the configuration for which $S = 3$, namely, ^7F, and this is the lowest in energy. Of the 16 quintets, 5 have already been encountered in the construction of Tables 8-4 and 8-5; they are $(210)(20)^5D$, $(210)(21)^5D$, $(111)(20)^5D$, $(210)(21)^5F$, and $(111)(10)^5F$. Of course, W and U are not good quantum numbers, and the eigenfunctions of a term that results from diagonalizing H_1 are expected to be of the type

$$\sum_{W,U,\tau} a(WU\tau)|f^nWU\tau SM_SLM_L)$$

The spin-orbit interaction further complicates matters by mixing the levels that derive from different terms, leaving only J a good quantum number. The problem of simultaneously diagonalizing the interactions H_1 and H_2 for f^6 is much more difficult than for f^2; in the latter case no equations more complicated than cubics arise, whereas for f^6 a matrix with as many as 46 rows and columns occurs. To spare ourselves such complexities, we consider first H_1 and then treat H_2 by perturbation theory.

The only levels that have as yet been established with certainty in configurations of the type nf^6 are those deriving from the ground multiplet ^7F and the first excited multiplet ^5D. It is therefore natural to select the three ^5D terms for special study, since the septet can be treated very simply by the conventional methods of Condon and Shortley.[1] The matrix of H_1 is derived by means of Racah's tables; ordering the terms by the sequence

$$(210)(20)^5D,\ (210)(21)^5D,\ (111)(20)^5D$$

we find that the matrix takes the form

$$\begin{pmatrix} 15E^0 + 6E^1 + 858E^2/7 + 11E^3 & 468(33)^{\frac{1}{2}}E^2/7 & 22(14)^{\frac{1}{2}}E^3/7 \\ 468(33)^{\frac{1}{2}}E^2/7 & 15E^0 + 6E^1 - 1131E^2/7 + 18E^3 & 12(462)^{\frac{1}{2}}E^3/7 \\ 22(14)^{\frac{1}{2}}E^3/7 & 12(462)^{\frac{1}{2}}E^3/7 & 15E^0 + 9E^1 - 11E^3 \end{pmatrix} \tag{8-33}$$

The seniority of the ^5D terms labeled by (210) is 6; that of (111)(20)^5D is 4. If $E^3 = 0$, the coupling between the terms of different seniority vanishes and v becomes a good quantum number. It is easy to see from Eqs. (8-24) that this is satisfied for the hypothetical F_k ratios of Eqs. (6-17). If we take the ratios of Eqs. (4-10), corresponding to a $4f$ hydrogenic eigenfunction, the matrix (8-33) can be diagonalized numerically, and we find the eigenfunctions of the three ^5D terms that we might hope realistically reproduce the eigenfunctions that actually occur in SmI or EuIV. The lowest of the three ^5D terms is found to correspond to the mixture

$$-0.196|(210)(20)^5D) + 0.770|(210)(21)^5D) - 0.607|(111)(20)^5D) \tag{8-34}$$

clearly demonstrating that neither W nor U is a good quantum number.

This procedure can be repeated for other sets of terms that share common values of S and L. Having used the F_k ratios of Eqs. (4-10) in the calculations, we can express all eigenvalues of H_1 as a constant plus a multiple of F_2. Gruber and Conway[97] have in this manner found the relative energies of the 119 terms of f^6; for the calculations, they used three sets of ratios for the integrals F_k, including those defined by Eqs. (4-10) and (4-11). The positions of the 10 lowest terms are given in Fig. 8-3 for ratios corresponding to a $4f$ hydrogenic eigenfunction. The calculations predict that the lowest of the 118 excited multiplets is of the type ^5D, and this result agrees with experiment.[98,99] A detailed investigation of the variation of the parameter F_2 along the rare-earth series[100] suggests a value of 360 cm^{-1} for EuIV. This may be compared with 310 cm^{-1} for PrIV, which can be deduced from the data for Fig. 4-1. With $F_2 = 360$ cm^{-1}, we find that ^5D lies 18610 cm^{-1} above ^7F. A detailed comparison between experiment and theory can be made only after the effect of the spin-orbit coupling has been allowed for; however, it is clear from Fig. 8-4 that the centers of the multiplets ^5D and ^7F are separated by approximately this amount.

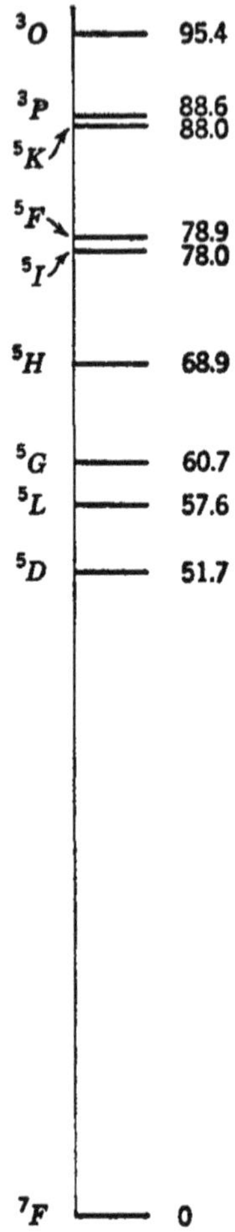

FIG. 8-3 The 10 lowest terms of f^6, calculated on the basis of a $4f$ hydrogenic eigenfunction (after Gruber and Conway[97]). The energies of the terms are given in units of F_2 relative to ^7F. On the basis of a generalization of Hund's rule for terms of maximum S, we might expect the term ^5L to be the lowest of the quintets, ^3O the lowest of the triplets, and ^1Q the lowest of the singlets; for these three terms possess the maximum L for a given S. Curiously, all three are not lowest, but next to lowest, being ousted from the former position by terms of small L, namely, ^5D, ^3P and ^1S, respectively. (^1Q and ^1S are too high to be included in the figure.)

Near the LS limit, H_2 can be replaced by $\lambda\mathbf{S}\cdot\mathbf{L}$(see Sec. 4-4); with the aid of Table 8-5, we find that $\lambda(^7F) = \zeta/6$, while, for the lowest of the three ^5D terms, the linear combination (8-34) leads to the exceptionally large value $\lambda(^5D) = 0.737\zeta$. This implies that, if LS coupling is a good approximation, the energies $E(^{2S+1}L_J)$ of the

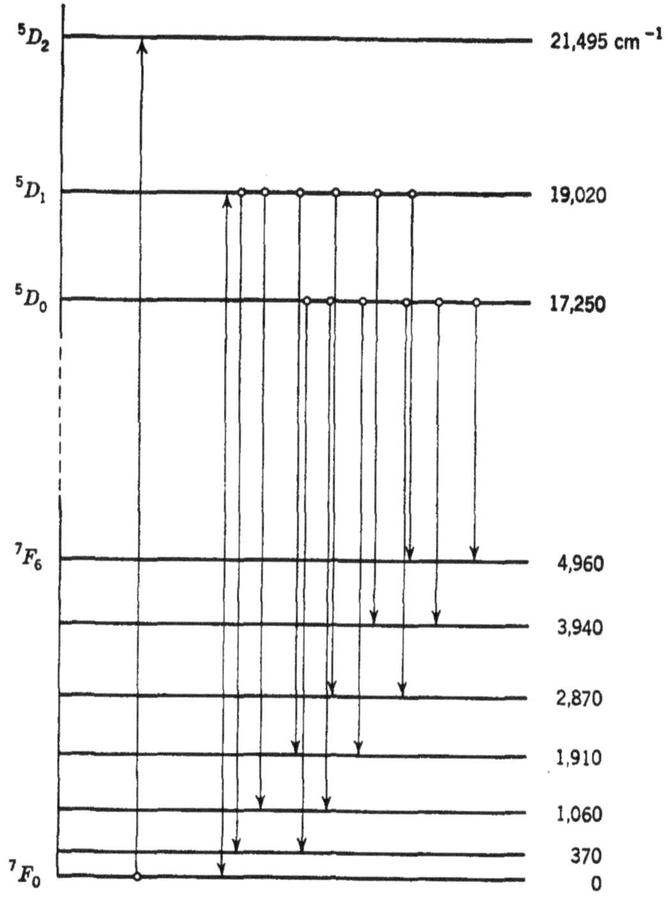

FIG. 8-4 Observed transitions in $Eu(C_2H_5SO_4)_3{\cdot}9H_2O$.

levels $^{2S+1}L_J$ should satisfy

$$\frac{E(^5D_J) - E(^5D_0)}{E(^7F_J) - E(^7F_0)} = \frac{0.737}{0.167} = 4.41 \tag{8-35}$$

The observed positions, taken from Sayre and Freed's analysis[84] of the absorption and fluorescence spectra of $Eu(C_2H_5SO_4)_3{\cdot}9H_2O$, are indicated in Fig. 8-4; they satisfy the equations

$$\frac{19020 - 17250}{370} = 4.78,$$

and

$$\frac{21495 - 17250}{1060} = 4.00 \tag{8-36}$$

for $J = 1$ and 2, respectively. The energies are given in cm^{-1} relative to $^7F_0$. Although the ratios (8-36) are quite close to 4.41, it is to be noticed that the Landé interval rule is not exactly obeyed for the multiplet ^7F_J. Presumably the second-order spin-orbit effects that produce the distortion also perturb the multiplet ^5D_J in such a way that the good agreement between experiment and theory is maintained for some distance away from the *LS*-coupling extreme. Deviations from *LS* coupling are much more severe in AmIV.[87]

8-8 CRYSTAL FIELD EFFECTS

Lines corresponding to transitions between the multiplets ^5D and ^7F have been observed in several europium salts. Most of the lines show a fine structure, varying markedly from salt to salt, and the splittings of the levels can thereby be deduced. For the purpose of illustrating the theory, we fix our attention on the two levels $^5D_1$ and $^7F_1$. The splittings of these levels are given in Fig. 8-5 for several europium compounds.

The contribution H_3 to the Hamiltonian arising from crystal field effects is expanded in spherical harmonics in Eq. (2-1). If we intend to calculate the splittings of levels for which $J = 1$, it is apparent from the discussion at the end of Sec. 2-7 that we may disregard all terms in the expansion except those for which $k = 2$. In setting up the matrix of H_3, it is convenient to label the rows and columns by $M_J = 1, 0$, and -1. The matrices for any two $J = 1$ levels are very similar, since the same crystal field parameters and 3-*j* symbols enter in corresponding positions in both matrices; in fact, the only respect in which they differ is through the reduced matrix element of a second-rank tensor, which multiplies every element of a given matrix. Hence the crystal field splitting of the level $^5D_1$ should be similar to the splitting of $^7F_1$, but contracted (or expanded) by the factor

$$R = ({}^5D_1\|V^{(2)}\|{}^5D_1)/({}^7F_1\|V^{(2)}\|{}^7F_1)$$

This can be regarded as the ratio between the two operator equivalent factors α for the two levels. From Eq. (3-38),

$$R = \frac{\begin{Bmatrix} 1 & 2 & 1 \\ 2 & 2 & 2 \end{Bmatrix} ({}^5D\|V^{(2)}\|{}^5D)}{\begin{Bmatrix} 1 & 2 & 1 \\ 3 & 3 & 3 \end{Bmatrix} ({}^7F\|V^{(2)}\|{}^7F)}$$

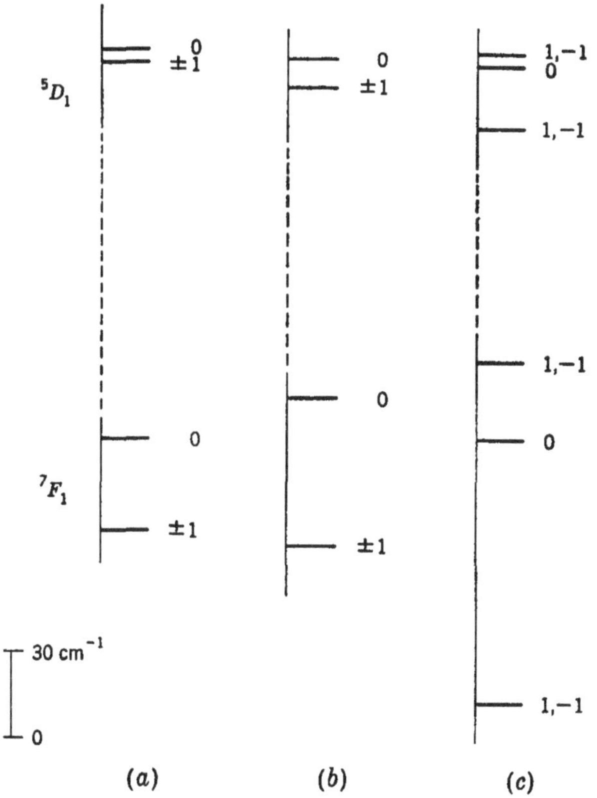

FIG. 8-5 The crystal field splittings of $^5D_1$ and $^7F_1$ are given for (a) $Eu(C_2H_5SO_4)_3\cdot 9H_2O$, from the data of Hellwege et al,[101] (b) $Eu(BrO_3)_3\cdot 9H_2O$, from the data of Hellwege and Kahle,[99] and (c) $EuCl_3\cdot 6H_2O$, also from the data of Hellwege and Kahle.[98] The numbers opposite the levels give values of M_J; the symbol ± 1 denotes a doublet whose two components can be taken to correspond to $M_J = \pm 1$. In the case of the chloride, the symbols 1, -1 label a singlet corresponding to admixtures from both $M_J = +1$ and $M_J = -1$. According to the theory, the splitting of $^5D_1$ for any crystal should be similar to that for $^7F_1$, but contracted by the factor 0.298.

It is not difficult to show that

$$(f^6\,{}^7F\|V^{(2)}\|f^6\,{}^7F) = -(5)^{1/2}$$

(see Prob. 8-6), and we are left with the reduced matrix element of ^5D to contend with. Although the techniques of Sec. (8-3) can be used to shorten the calculations, it seems preferable to use Eq. (7-52) for all the reduced matrix elements of $\mathbf{V}^{(2)}$; equations such as (8-7) then provide excellent checks. Ordering the three ^5D terms in the same manner as for the matrix (8-33), we find the following values for $(f^6WU^5D\|V^{(2)}\|f^6W'U'\,^5D)$:

$$\begin{pmatrix} \dfrac{52(30)^{\frac{1}{2}}}{441} & \dfrac{5(110)^{\frac{1}{2}}}{49} & \dfrac{67(105)^{\frac{1}{2}}}{441} \\ \dfrac{5(110)^{\frac{1}{2}}}{49} & \dfrac{11(30)^{\frac{1}{2}}}{98} & \dfrac{2(385)^{\frac{1}{2}}}{49} \\ \dfrac{67(105)^{\frac{1}{2}}}{441} & \dfrac{2(385)^{\frac{1}{2}}}{49} & \dfrac{-17(30)^{\frac{1}{2}}}{126} \end{pmatrix}$$

Taking the mixture (8-34) for the lower ^5D term, we find

$$R = 0.298$$

The contraction factors for the three cases illustrated in Fig. 8-5 are given in Table 8-7. The crystal splittings for the chloride and the bromate have been measured by Hellwege and Kahle;[98,99] the papers of Sayre and Freed[84] and Hellwege et al.[101] provide data for the ethyl sulfate. The observed values of R are all slightly smaller than the theoretical figure, a result due, no doubt, to admixtures of other levels for which $J = 1$ in both $^7F_1$ and $^5D_1$. The variation of R from salt to salt shows that J is not a good quantum number in all cases, and this is to be ascribed to second-order crystal field effects.

TABLE 8-7 THE CONTRACTION FACTOR R

Salt	Experimental R	Theoretical R
$Eu(C_2H_5SO_4)_3 \cdot 9H_2O$	0.119	
$Eu(BrO_3)_3 \cdot 9H_2O$	0.170	0.298
$EuCl_3 \cdot 6H_2O$	0.217	

8-9 HYPERFINE STRUCTURE IN PuI

The magnetic moment μ_N in nuclear magnetons of ^{239}Pu has been measured in two ways. Paramagnetic-resonance experiments by Bleaney et al.[102] on crystals of $UO_2Rb(NO_3)_3$ containing small amounts of Pu substituted for U give $|\mu_N| = 0.5 \pm 0.2$. Hubbs et al.[85] have obtained the much smaller value of 0.021 from their atomic-beam experiments on free plutonium atoms.‡ The electronic configuration

‡ A later measurement has been made by L. A. Korostyleva [*Optika i Spectroscopiya*, **12**, 671 (1962)]. By analyzing the hyperfine structure of PuII $5f^67s$, she finds $\mu_N = +0.15 \pm 0.05$. Like the value reported by Bleaney et al., it is an order of magnitude larger than that obtained from the atomic-beam experiments.

surrounding the nucleus of the plutonium atom is quite different in the two cases, and in order to reconcile the results both systems should no doubt be studied. However, for the purpose of illustrating the theory, we confine our attention to the free plutonium atom.

As measured by Bovey and Gerstenkorn,[86] $^7F_1$ lies 2203.6 cm^{-1} above $^7F_0$. In interpreting the hyperfine splitting of the former level, Hubbs et al. assumed that it could be treated as a pure level for which $S = L = 3$ and $J = 1$. At first sight this seems plausible, since the Landé g value is 1.497, very close to the theoretical figure of 1.5. Normally, deviations from perfect LS coupling produce significant changes in g. However, the g value of $^5D_1$ is also 1.5, so a serious admixing of the states corresponding to this pure level into those corresponding to $^7F_1$ could take place without affecting g. In order to estimate this admixture, we do not need F_2 and ζ separately, but only their ratio, ζ/F_2. The reason for this will become evident as soon as the actual calculation is started. Lämmermann and Conway have obtained an excellent fit between the observed and calculated levels of PuIV $5f^5$ with the respective values 245 and 2290 cm^{-1} for these quantities.[103] On adding electrons to the plutonium ion, the $5f$ orbitals expand; we therefore expect that the corresponding parameters for PuI $5f^67s^2$ are slightly smaller than these values. For our purposes, it is unnecessary to go into these details, and we shall simply assume $\zeta/F_2 = 10$ for the problem in hand. The uncertainty in this ratio scarcely makes it worthwhile to repeat the analysis of Sec. 8-7 for a $5f$ hydrogenic eigenfunction, and we shall assume that the linear combination (8-34) can again be taken to represent the lowest ^5D term.

The strong perturbing influence of the lowest $^5D_1$ level on $^7F_1$ is readily established. For

$$\begin{aligned}(^5D_1|H_1 + H_2|^5D_1) - (^7F_1|H_1 + H_2|^7F_1) \\ &= 51.7F_2 - 0.737 \times 5\zeta + \frac{11\zeta}{6} \\ &= 3.318\zeta\end{aligned}$$

while from the expansion (8-34) and Table 8-5 it is easy to show

$$(^5D_1|H_2|^7F_1) = -3.10\zeta$$

Neglecting all other levels for which $J = 1$, we obtain

$$|A) = 0.858|^7F_1) + 0.514|^5D_1)$$

for the eigenfunction of the lower level.

We now examine the matrix elements of the hyperfine operator. The program consists in evaluating $(A\|N\|A)$, where

$$\mathbf{N} = \sum_i \mathbf{N}_i = \mathbf{L} - (10)^{\frac{1}{2}} \sum_i (\mathbf{s}\mathbf{C}^{(2)})_i^{(1)}$$

and then comparing it with $({}^7F_1\|N\|{}^7F_1)$. The ratio

$$P = (A\|N\|A)/({}^7F_1\|N\|{}^7F_1) \tag{8-37}$$

is the factor that the magnetic moment of ^{239}Pu, as calculated by Hubbs et al., ought to be divided by to allow for the hybrid nature of the level. Using Eq. (7-53), we find

$$\begin{aligned}({}^7F_1\|N\|{}^7F_1) &= 3\begin{Bmatrix}1 & 1 & 1\\ 3 & 3 & 3\end{Bmatrix}({}^7F\|L\|{}^7F) - (10)^{\frac{1}{2}}({}^7F_1\|\sum_i(\mathbf{s}\mathbf{C}^{(2)})_i^{(1)}\|{}^7F_1)\\ &= (\tfrac{3}{2})^{\frac{1}{2}} - (\tfrac{1}{6})^{\frac{1}{2}} = (\tfrac{2}{3})^{\frac{1}{2}} = 0.817\end{aligned}$$

The 9-j symbol

$$\begin{Bmatrix}3 & 3 & 1\\ 3 & 3 & 2\\ 1 & 1 & 1\end{Bmatrix} = \frac{1}{84}$$

is required to obtain this result. Since $({}^7F_1\|J\|{}^7F_1) = (6)^{\frac{1}{2}}$, we see $\mathbf{N} = \mathbf{J}/3$ for ${}^7F_1$.

In taking account of the effect of ${}^5D_1$, our efforts are directed to that part of $\mathbf{N}$ involving the tensor $(\mathbf{s}\mathbf{C}^{(2)})_i^{(1)}$, since the matrix elements of $\mathbf{L}$ are easy to obtain. Apart from a numerical factor, $\sum_i (\mathbf{s}\mathbf{C}^{(2)})_i^{(1)}$ is the vector part of the double tensor $\mathbf{W}^{(12)}$. From Sec. 8-2, we see that this in turn is part of a generalized tensor transforming according to (200)(20). We also note that it is an infinitesimal operator for Sp_{14}, and is therefore diagonal with respect to seniority. The required values of

$$(f^nWUSL\|W^{(12)}\|f^nW'U'S'L')$$

are given in Table 8-8. In constructing this table, we can augment

TABLE 8-8 SOME MATRIX ELEMENTS $(f^6WUSL\|W^{(12)}\|f^6W'U'S'L')$

	$(100)(10)^7F$	$(210)(20)^5D$	$(210)(21)^5D$	$(111)(20)^5D$
$(100)(10)^7F$	$-(\frac{70}{3})^{\frac{1}{2}}$	$(\frac{120}{7})^{\frac{1}{2}}$	$-3(\frac{55}{14})^{\frac{1}{2}}$	0
$(210)(20)^5D$	$(\frac{120}{7})^{\frac{1}{2}}$	$335(2)^{\frac{1}{2}}/196$	$15(66)^{\frac{1}{2}}/98$	0
$(210)(21)^5D$	$-3(\frac{55}{14})^{\frac{1}{2}}$	$15(66)^{\frac{1}{2}}/98$	$-405(2)^{\frac{1}{2}}/196$	0
$(111)(20)^5D$	0	0	0	$-85(2)^{\frac{1}{2}}/28$

Eq. (7-54) with equations of the type introduced in Sec. 8-3. For example, $c((100)(200)(210)) = 1$, and we find

$$(f^6(100)(10)\,{}^7F\|W^{(12)}\|f^2(210)U\,{}^5L) = -(\tfrac{21}{2})^{\frac{1}{2}}(f^3(100)(10)\,{}^2F\|V^{(2)}\|f^3(210)U\,{}^2L)$$

Since $c((210)(200)(210)) = 2$, we know that constants A_1 and A_2 independent of L and L' can be found such that

$$\begin{aligned}(f^6(210)U\,{}^5L\|W^{(12)}\|f^6(210)U'\,{}^5L') &= A_1(f^6(210)U\,{}^5L\|V^{(2)}\|f^6(210)U'\,{}^5L')\\ &\quad + A_2(f^3(210)U\,{}^2L\|V^{(2)}\|f^3(210)U'\,{}^2L')\end{aligned}$$

since the two sets of matrix elements on the right are linearly independent. It turns out, rather surprisingly, that $A_1 = 0$. Moreover, the entry in Table 8-8 involving the state $(111)(20)\,{}^5D$ can also be included in the general formula

$$(f^6WU\,{}^5L\|W^{(12)}\|f^6W'U'\,{}^5L') = -(\tfrac{15}{4})^{\frac{1}{2}}(f^3WUL\|V^{(2)}\|f^3W'U'L') \qquad (8\text{-}38)$$

Some understanding of simplifications of this kind can be obtained by studying the symmetry properties of the corresponding determinantal product states;[104] a group-theoretical interpretation is yet to be given. Equation (8-38) embraces all the quintets of f^6 and all the terms of f^3 except $(100)(10)\,{}^2F$. The vanishing of $\mathbf{W}^{(12)}$ between states of different seniority in f^6 corresponds to the vanishing of $\mathbf{V}^{(2)}$ between states of different spin in f^3. Since all the matrix elements of $\mathbf{V}^{(2)}$ are known in f^3,[88] it is a trivial matter to write down the matrix elements of $\mathbf{W}^{(12)}$ within the quintets of f^6. Matrix elements of this type other than those given in Table 8-8 would be needed in a study of the hyperfine structure of ${}^7F_2$.

To complete the calculation, we use

$$\begin{aligned}-(10)^{\frac{1}{2}}(f^nW\,USLJ\|\sum_i(\mathbf{sC}^{(2)})_i^{(1)}\|f^nW'U'S'L'J) &= -[J](2)^{\frac{1}{2}}(s\|s\|s)(f\|C^{(2)}\|f)\\ &\quad\times(f^nWUSL\|W^{(12)}\|f^nW'U'S'L')\begin{Bmatrix}S & S' & 1\\ L & L' & 2\\ J & J & 1\end{Bmatrix}\end{aligned}$$

The following 9-j symbols are required:

$$\begin{Bmatrix}2 & 2 & 1\\ 2 & 2 & 2\\ 1 & 1 & 1\end{Bmatrix} = \frac{1}{30}\left(\frac{7}{15}\right)^{\frac{1}{2}} \qquad \begin{Bmatrix}3 & 2 & 1\\ 3 & 2 & 2\\ 1 & 1 & 1\end{Bmatrix} = \frac{1}{15}\left(\frac{1}{21}\right)^{\frac{1}{2}}$$

We find

$$(A\|N\|A) = 0.280$$

and hence P, defined in Eq. (8-37), is given by

$$P = \frac{0.280}{0.817} = 0.343$$

Thus the nuclear moment of 0.021 nuclear magneton, which was estimated on the assumption of a pure $^7F_1$ level, should be corrected to 0.061 nuclear magneton. The simple treatment given here has made allowance for only one excited level; and although its effect is certainly the most important, the small value of $(A\|N\|A)$ makes this quantity very sensitive to the influence of other levels. If, for example, these additional perturbations reduce $(^7F_1\|N\|^7F_1)$ by only a third of the amount that those due to $^5D_1$ do, which is by no means unreasonable, then the figure of 0.061 nuclear magneton must be more than doubled. It is probable that the remaining discrepancy between the experimental results of Hubbs et al. and Bleaney et al. could be resolved along these lines.

PROBLEMS

8-1. Justify the existence of a constant A, independent of U, L, U', and L', in the equation

$$(f^4(111)U\,^5L\|V^{(k)}\|f^4(111)U'\,^5L') = A(f^3(111)U\,^4L\|V^{(k)}\|f^3(111)U'\,^4L')$$

where $k = 2$, 4, or 6. Prove $A = -1$, and generalize the equation to relate matrix elements of $\mathbf{V}^{(k)}$ between states of maximum S in the configurations l^n, l^{2l+1-n}, l^{2l+1+n}, and l^{4l+2-n}.

8-2. Prove that for irreducible representations $(w_1w_2w_3)$ of R_7,

$$\chi(w_1w_2w_3)\chi(200) = 2\chi(w_1w_2w_3) + \Sigma\chi(w_1 + x,\ w_2 + y,\ w_3 + z)$$

where the sum runs over all integral values of x, y, and z satisfying

$$|x| + |y| + |z| \leq 2$$

Construct a table for $c(WW'(200))$, and prove

$$(f^6(110)U\,^3L|H_3|f^6(221)U'\tau\,^3L') = 0$$

for all U, L, U', τ, and L'.

8-3. Prove that, when $\kappa + k$ is even,

$$(l^nv\xi SL\|W^{(\kappa k)}\|l^nv\xi'S'L') = -(l^{4l+2-n}v\xi SL\|W^{(\kappa k)}\|l^{4l+2-n}v\xi'S'L')$$

Use this equation in conjunction with the factorization (7-12) to derive Eq. (8-14). (See Racah.[76])

8-4. Use Eq. (6-9) to prove

$$(f^3(111)(20)^4DM_SM_L\|[W_{\pi q}^{(11)},V_{q'}^{(5)}]\|f^3(210)(21)^2LM'_SM'_L) = 0$$

for all M_S, M_L, M'_S, M'_L, π, q, and q', where ^2L denotes the term of f^3 for which $S = \frac{1}{2}$, $L = 8$. Hence show that

$$(f^3(111)(20)^4D\|W^{(11)}\|f^3(210)(21)^2F) = 0$$

Deduce that

$$(f^nW(20)SD\|W^{(11)}\|f^nW'(21)S'F) = 0$$

for all n, W, W', S, and S'.

8-5. By considering the matrix elements of the commutator

$$[W_{\pi q}^{(11)},V_{q'}^{(3)}]$$

between the states $(110)(11)^6P$ and $(211)U^4H$ of f^5, prove that

$$(f^n(110)(11)SL\|W^{(11)}\|f^n(211)(20)S'L') = 0$$

for all n, S, S', L, and L'.

8-6. Prove that

$$(l^nSL\|V^{(2)}\|l^nSL) = \frac{2l-2n+1}{2L-1}\left[\frac{5(2L+3)!(2l-2)!}{(2l+3)!(2L-2)!}\right]^{\frac{1}{2}}$$

if $n \le 2l+1$, $S = \frac{1}{2}n$, and $L = \frac{1}{2}n(2l-n+1)$. Check that this equation is consistent with the results of Prob. 8-1.

8-7. Obtain the following analogue to Eq. (8-38):

$$(l^{2l+1}SL\|W^{(12)}\|l^{2l+1}SL') = -\left[\frac{l(2l+1)}{2l-1}\right]^{\frac{1}{2}}(l^2L\|V^{(2)}\|l^2L')$$

where $S = \frac{1}{2}(2l-1)$.

8-8. The terms of the half-filled shell l^{2l+1} are separated into two classes; for all members of the first, $\frac{1}{2}(v+1)$ is odd, while for those of the second, it is even. Prove that there are no nonvanishing matrix elements of $\mathbf{W}^{(\kappa k)}$ between terms of different classes if $\kappa + k$ is odd and none between terms of the same class if $\kappa + k$ is even.

8-9. Verify the formula

$$(f^7\ ^6P_{\frac{7}{2}}|H_2|f^7\ ^6D_{\frac{7}{2}}) = \frac{9\zeta(5)^{\frac{1}{2}}}{10}$$

The lowest configuration of EuI is f^7, and the ground level is $^8S_{\frac{7}{2}}$. Indicate how contributions to the term $A\mathbf{I}\cdot\mathbf{J}$ in the effective Hamiltonian for the hyperfine interaction can arise through admixtures of the levels $^6P_{\frac{7}{2}}$ and $^6D_{\frac{7}{2}}$ in the ground level, and prove that, near the LS limit, the contribution to A is

$$\frac{2\beta\beta_N\mu_N}{I}\left[\frac{16\zeta^2}{5E_P{}^2}+\frac{12\zeta^2}{5E_PE_D}\right]\langle r^{-3}\rangle$$

where E_P and E_D are the energies of $^6P_{\frac{7}{2}}$ and $^6D_{\frac{7}{2}}$ above $^8S_{\frac{7}{2}}$, respectively.

8-10. For d electrons, the parameters F_k of Condon and Shortley[1] are related to the Slater integrals $F^{(k)}$ (defined in Eq. (4-9)) by the equations

$$F_0 = F^{(0)} \qquad F_2 = \frac{F^{(2)}}{49} \qquad F_4 = \frac{F^{(4)}}{441}$$

Prove that the Coulomb interaction within a shell of equivalent d electrons can be replaced by the operator

$$e_0E^0 + e_1E^1 + e_2E^2$$

where
$$e_0 = 5 \sum_{i>j} (\mathbf{v}_i^{(0)} \cdot \mathbf{v}_j^{(0)})$$

$$e_1 = \sum_{i>j} [7(\mathbf{v}_i^{(0)} \cdot \mathbf{v}_j^{(0)}) + 2(\mathbf{v}_i^{(2)} \cdot \mathbf{v}_j^{(2)}) + 2(\mathbf{v}_i^{(4)} \cdot \mathbf{v}_j^{(4)})]$$

and
$$e_2 = \sum_{i>j} [18(\mathbf{v}_i^{(2)} \cdot \mathbf{v}_j^{(2)}) - 10(\mathbf{v}_i^{(4)} \cdot \mathbf{v}_j^{(4)})]$$

with
$$E^0 = F_0 - \frac{7F_2}{2} - \frac{63F_4}{2}$$

$$E^1 = \frac{5(F_2 + 9F_4)}{2}$$

and
$$E^2 = \frac{(F_2 - 5F_4)}{2}$$

Show that e_0, e_1, and e_2 are the components of generalized tensors and may be labeled by WL = (00)0, (00)0, and (22)0, respectively, where W denotes an irreducible representation of R_5. Prove that both e_0 and e_1 are diagonal in a scheme in which the states are taken to be of the type $|d^n vSM_SLM_L)$ and that their matrix elements are given by $\frac{1}{2}n(n-1)$ and

$$\frac{7(n-v)}{2} + \frac{v(v+2)}{4} - S(S+1)$$

respectively. Obtain the values of $c(WW'(22))$ given in Table 8-9. Show that the operator

$$\Omega' = 7(\mathbf{V}^{(1)})^2 - 3(\mathbf{V}^{(3)})^2$$

is the scalar component of a tensor transforming according to (22) of R_5, and deduce that, with the possible exception of W = (21), parameters A independent of L exist such that

$$(d^nWSL|e_2|d^nWSL) = A[3w_1(w_1+3)/2 + 3w_2(w_2+1)/2 - L(L+1)]$$

TABLE 8-9 THE NUMBERS $c(WW'(22))$

W	W'					
	(00)	(10)	(11)	(20)	(21)	(22)
(00)	—	—	—	—	—	1
(10)	—	—	—	—	1	1
(11)	—	—	1	—	1	1
(20)	—	—	—	1	1	1
(21)	—	1	1	1	2	1
(22)	1	1	1	1	1	1

where $W = (w_1w_2)$. Prove that, for all representations W and W' [including the case of $W = W' = (21)$], the factorization

$$(d^nWSL|e_2 + \Omega'|d^nW'SL) = B(d^n,WW'S)(W|\mu(L)|W')$$

can be carried out.

8-11. Extend the techniques of Sec. 8-6 to double tensors, and construct two operators whose descriptions in terms of irreducible representations of Sp_{14}, R_7, G_2, and R_3 are

$$(1111000)(220)(22)S$$

and

$$(2200000)(220)(22)S$$

respectively. Show that the eigenvalues of these operators for the seven terms of f^2 are proportional to those of the operators $e_3 + \Omega$ and Ω, respectively, in the notation of Sec. 8-6. Construct the table of numbers $c((v)(v')(4))$, and prove that the selection rules on $e_3 + \Omega$ are given by

$$\Delta v = 0, \ \pm 2, \ \pm 4$$

Prove also that all matrix elements of $e_3 + \Omega$ within the manifolds of states for which $v = 6$ or 7 are zero, and show that the dependence of the matrix elements

$$(f^nv\,U\tau SL|e_3 + \Omega|f^nv'\,U'\tau'SL)$$

on U, U', τ, τ', S, and L is the same for all values of n.

Use the orthogonality property of states characterized by the labels (1111000) and (2200000) to derive the equation

$$\Sigma[S][L](f^nvW\,U\tau SL|e_3 + \Omega|f^nvW\,U\tau SL)(UL|\Omega|UL) = 0$$

where the sum runs over W, U, τ, S, and L.

8-12. Use the formula of Prob. 6-2 to prove that nonvanishing matrix elements of the spin-spin interaction are diagonal with respect to seniority.

APPENDIX 1

RADIAL INTEGRALS FOR HYDROGENIC EIGENFUNCTIONS

Integrals of the type

$$\int_0^\infty \int_0^\infty \frac{r_<^k}{r_>^k}\,[R_{nl}(r_i)R_{nl}(r_j)]^2\,dr_i\,dr_j \tag{A-1}$$

often arise in the evaluation of two-particle operators for configurations of equivalent nl electrons. In order to evaluate them, we write (A-1) as

$$2\int_0^\infty R_{nl}^2(r_i)r_i^{-k}\int_0^{r_i} r_j^k R_{nl}^2(r_j)\,dr_j\,dr_i$$

If R_{nl} is hydrogenic, every integral becomes a linear combination of integrals of the form

$$J(a,b) = \int_0^\infty \rho^a e^{-\rho}\int_0^\rho \sigma^b e^{-\sigma}\,d\sigma\,d\rho$$

It is straightforward to show that

$$\int_0^\rho \sigma^b e^{-\sigma}\,d\sigma = b! - e^{-\rho}\sum_{r=0}^{b} b!\rho^r/r!$$

and

$$J(a,b) = a!b! - b!\sum_{r=0}^{b}\frac{(a+r)!}{r!2^{a+r+1}}$$

But

$$\begin{aligned} a!b! &= 2^{-a-1}a!b!(1-\tfrac{1}{2})^{-a-1} \\ &= 2^{-a-1}b!\sum_{r=0}^{\infty}\frac{(a+r)!}{r!2^r} \end{aligned}$$

and hence

$$\begin{aligned} J(a,b) &= \sum_{r=b+1}^{\infty} b!(a+r)!/(r!2^{a+r+1}) \\ &= (a+b+1)!F(a+b+2,1;b+2;\tfrac{1}{2})/[(b+1)2^{a+b+2}] \end{aligned}$$

where the hypergeometric function is defined in the usual way as

$$F(\alpha,\beta;\gamma;z) = 1 + \frac{\alpha\beta}{\gamma 1!}z + \frac{\alpha(\alpha+1)\beta(\beta+1)}{\gamma(\gamma+1)2!}z^2 + \cdots$$

All integrals can be calculated from

$$F(u,1;u;\tfrac{1}{2}) = 2$$

by making use of the recurrence relations

$$xF(x+1,\,1;\,u;\,\tfrac{1}{2}) = 2(u-1) + 2(x-u+1)F(x,1;u;\tfrac{1}{2})$$

and $$yF(u,1;y;\tfrac{1}{2}) = 2y + (u-y)F(u,\,1;\,y+1;\,\tfrac{1}{2}) \qquad \text{(A-2)}$$

which enable either the first or the third argument of the hypergeometric function to be changed by integral amounts.‡

Consider, for example, the Slater integrals $F^{(k)}$ for a 4f hydrogenic eigenfunction. Since $R_{4f}(r) \sim \rho^4 e^{-\rho}$, where $\rho = \frac{1}{4}r$,

$$F^{(k)} \sim J(7-k,\,8+k)$$

Therefore

$$\frac{F^{(k)}}{F^{(2)}} = \frac{J(7-k,\,8+k)}{J(5,10)} = \frac{11F(17,\,1;\,k+10;\,\frac{1}{2})}{(k+9)F(17,\,1;\,12;\,\frac{1}{2})}$$

The sequence

$$\begin{aligned}
F(17,1;17;\tfrac{1}{2}) &= 2\\
F(17,1;16;\tfrac{1}{2}) &= 2 + (\tfrac{1}{16})(2) = \tfrac{17}{8}\\
F(17,1;15;\tfrac{1}{2}) &= 2 + (\tfrac{2}{15})(\tfrac{17}{8}) = \tfrac{137}{60}\\
F(17,1;14;\tfrac{1}{2}) &= 2 + (\tfrac{3}{14})(\tfrac{137}{60}) = \tfrac{697}{280}\\
F(17,1;13;\tfrac{1}{2}) &= 2 + (\tfrac{4}{13})(\tfrac{697}{280}) = \tfrac{2517}{910}\\
F(17,1;12;\tfrac{1}{2}) &= 2 + (\tfrac{5}{12})(\tfrac{2517}{910}) = \tfrac{2295}{728}
\end{aligned}$$

is easily derived from Eq. (A-2). Hence

$$\frac{F^{(4)}}{F^{(2)}} = \frac{451}{675} \qquad \frac{F^{(6)}}{F^{(2)}} = \frac{1001}{2025}$$

These equations are equivalent to Eqs. (4-10). The ratios for a 5f hydrogenic eigenfunction, which can be obtained in a similar manner, are given in Eqs. (4-11). The author has taken the opportunity to remove an arithmetical error in his original calculation.[82]

‡ These techniques can be extended to treat integrals for which the exponentials $e^{-\rho}$ and $e^{-\sigma}$ in the definition of $J(a,b)$ are replaced by $e^{-x\rho}$ and $e^{-y\sigma}$, respectively. Generalized integrals of this kind have been computed over wide ranges of the parameters a, b, x, and y by V. Vanagas, J. Glembockij, and K. Uspalis (*Tables of Radial Integrals of Atomic Spectra Theory*, Academy of Sciences Computing Center, Moscow, 1960). The specialized form $J(a,b)$ is related to the V function that these authors tabulate by the equation $V(ba;1) = \log J(a,b)$. The integrals $J(a,b)$ can also be considered as special cases of the S functions of A. Tubis [*Phys. Rev.*, **102**, 1049 (1956)]. In his notation, for example,

$$J(7-k,\,8+k) = \tfrac{1}{2}S_1{}^1(5,\,5,\,k+1)$$

APPENDIX 2

THE COEFFICIENTS $(UL|U'L' + f)$ AND $(WU|W'U' + f)$

The tables below give values of the coupling coefficients $(UL|U'L' + f)$ and $(WU|W'U' + f)$ as determined by Racah.[69]‡ The arguments W, U, L, W', U', and L' are restricted to those values occurring in the group-theoretical classification of the terms of the two highest multiplicities in every configuration of the type f^n. To make the tabulation as concise as possible, the denominator D common to all the entries in a row is given separately. In addition, all square-root signs are removed. In order to find a particular coefficient $(U_1L_1|U_2L_2 + f)$, proceed as follows:

1. Find the table $(UL|U_2L' + f)$.
2. Read off the number in the row U_1L_1 and column L_2.
3. Divide by the denominator D for the row U_1L_1.
4. Extract the positive square root unless the entry found in (2) is followed by an asterisk, in which case the negative root must be taken.

For example,

$$((21)F|(20)G + f) = -(144/1386)^{\frac{1}{2}} = -(8/77)^{\frac{1}{2}}$$

Similar rules apply to the tables for $(WU|W'U' + f)$.

‡ Several tables obtained by Racah but not included in Ref. 69 have been reproduced by B. G. Wybourne [*J. Chem. Phys.*, **36**, 2295 (1962)]. In a private communication, C. W. Nielson reports that he has derived all of the cfp for f^n in both factored and composite form. See C. W. Nielson and P. B. Nutter [*Bull. Am. Phys. Soc.*, **7**, 80 (1962)].

TABLE A1 $(UL|(00)L' + f)$

U	L	D	L'
			S
(10)	F	1	1

TABLE A2 $(UL|(10)L' + f)$

U	L	D	L'
			F
(00)	S	1	1*
(10)	F	1	1
(11)	P	1	1
	H	1	1
(20)	D	1	1
	G	1	1
	I	1	1

TABLE A3 $(UL|(11)L' + f)$

U	L	D	L'	
			P	H
(10)	F	14	3	11
(20)	D	49	27*	22*
	G	98	33	65*
	I	1	0	1
(21)	D	49	22*	27
	F	14	11*	3
	G	98	65	33
	H	1	0	1
	K	1	0	1
	L	1	0	1

TABLE A4 $(UL|(20)L' + f)$

U	L	D	L′		
			D	G	I
(10)	F	27	5*	9*	13*
(11)	P	21	10	11*	0
	H	2079	220	585	1274*
(20)	D	49	16*	33	0
	G	1617	605	375*	637
	I	11	0	3	8
(21)	D	49	33	16	0
	F	1386	605*	144*	637
	G	882	13	624	245
	H	297	143	144*	10*
	K	33	0	16	17*
	L	1	0	0	1
(30)	P	21	11	10	0
	F	4158	1573	2340*	245
	G	198	121	12	65*
	H	693	286	162	245
	I	11	0	8	3*
	K	33	0	17	16
	M	1	0	0	1

TABLE A5 $(UL|(21)L' + f)$

U	L	D	L′					
			D	F	G	H	K	L
(11)	P	1344	220	539*	585*	0	0	0
	H	4928	270*	147	297*	1078	1470	1666*
(20)	D	31360	8910	8085	351	14014*	0	0
	G	4312	330	147	1287	1078	1470*	0
	I	18304	0	1911*	1485	220	4590	10098
(21)	D	5390	375	1960	1144*	1911	0	0
	F	154	40*	49	65*	0	0	0
	G	630630	74360*	207025	226941	51744*	70560	0
	H	15730	2535*	0	1056	3179*	7260	1700
	K	2860	0	0	192*	968	85*	1615
	L	572	0	0	0	40*	285*	247
(30)	P	2688	1156	245*	1287	0	0	0
	F	112	39*	0	24	49	0	0
	G	7920	1375	2450	858*	1617	1620	0
	H	640640	42250*	207025*	3971	261954	490	124950
	I	9152	0	1274	2750	1320	2125	1683*
	K	1040	0	0	204*	136*	605	95
	M	64	0	0	0	0	15*	49*

TABLE A6 $(UL|(30)L' + f)$

U	L	D	L' P	F	G	H	I	K	M
(20)	D	490	54*	91*	189	156*	0	0	0
	G	5929	330*	910	126	594*	2184	1785*	0
	I	22022	0	245*	1755*	2310*	2106*	4320*	11286*
(21)	D	2695	578*	1092	700	325	0	0	0
	F	1694	55*	0	560*	715*	364*	0	0
	G	630630	83655*	87360*	56784*	3971*	227500	171360	0
	H	55055	0	12740	7644*	18711	7800*	8160*	0
	K	165165	0	0	16848*	77	27625*	79860	40755*
	L	17017	0	0	0	1785*	1989*	1140*	12103
(30)	P	16	0	13	3	0	0	0	0
	F	1232	429	77*	273	33*	420*	0	0
	G	55440	3465*	9555*	7203	27797	1300	6120*	0
	H	11440	0	195*	4693*	1232	2600	2720	0
	I	32032	0	5880	520	6160*	1911*	15680	1881*
	K	510510	0	0	33813	89012	216580*	8085	163020
	M	3808	0	0	0	0	153	960	2695

TABLE A7 $(WU|(000)U' + f)$

W	U	D	U' (00)
(100)	(10)	1	1

TABLE A8 $(WU|(100)U' + f)$

W	U	D	U' (10)
(000)	(00)	1	1
(110)	(10)	1	1
	(11)	1	1
(200)	(20)	1	1

TABLE A9 $(WU|(110)U'+f)$

W	U	D	U' (10)	U' (11)
(100)	(10)	3	1	2
(111)	(00)	1	1	0
	(10)	3	2	1*
	(20)	9	2	7*
(210)	(11)	1	1	0
	(20)	9	7	2
	(21)	1	0	1

TABLE A10 $(WU|(200)U'+f)$

W	U	D	U' (20)
(100)	(10)	1	1
(210)	(11)	1	1
	(20)	1	1
	(21)	1	1

TABLE A11 $(WU|(111)U'+f)$

W	U	D	U' (00)	U' (10)	U' (20)
(110)	(10)	35	3	14	18
	(11)	10	0	1*	9*
(111)	(00)	1	0	1*	0
	(10)	56	8*	21*	27
	(20)	8	0	1	7*
(211)	(10)	280	216	63*	1
	(11)	10	0	9	1*
	(20)	8	0	7	1
	(21)	1	0	0	1
	(30)	1	0	0	1

TABLE A12 $(WU|(210)U' + f)$

W	U	D	U' (11)	U' (20)	U' (21)
(110)	(10)	5	2	3	0
	(11)	35	0	3	32
(200)	(20)	105	14	27	64
(211)	(10)	5	3*	2	0
	(11)	35	0	32	3*
	(20)	21	7	6	8*
	(21)	112	21	25*	66
	(30)	7	0	1	6
(220)	(20)	105	56	48*	1
	(21)	56	7*	27	22
	(22)	1	0	0	1

TABLE A13 $(WU|(211)U' + f)$

W	U	D	U' (10)	U' (11)	U' (20)	U' (21)	U' (30)
(111)	(00)	1	1	0	0	0	0
	(10)	24	1*	8	15	0	0
	(20)	5832	1	56*	135	2560	3080
(210)	(11)	42	7*	0	15	20	0
	(20)	1701	98	448	270	500*	385
	(21)	672	0	7*	60*	220	385
(211)	(10)	72	5*	40	27*	0	0
	(11)	126	35	0	27*	64	0
	(20)	2520	245*	280*	867*	512*	616
	(21)	315	0	35	27*	176*	77
	(30)	315	0	0	27	64	224*

REFERENCES

1. E. U. Condon and G. H. Shortley, "The Theory of Atomic Spectra," Cambridge University Press, New York (1935). A general account of the subject, incorporating many of the later developments of the theory, has been given by J. C. Slater, "Quantum Theory of Atomic Structure," McGraw-Hill Book Company, Inc., New York (1960).
2. H. A. Bethe and E. E. Salpeter, "Quantum Mechanics of One- and Two-electron Atoms," Springer-Verlag OHG, Berlin (1957).
3. P. A. M. Dirac, "The Principles of Quantum Mechanics," Oxford University Press, New York (1947).
4. A. R. Edmonds, "Angular Momentum in Quantum Mechanics,"Princeton University Press, Princeton, N. J. (1960).
5. T. Regge, *Nuovo cimento*, **10**, 544 (1958).
6. Rotenberg, Bivins, Metropolis, and Wooten, "The 3-j and 6-j Symbols," Technology Press, Massachusetts Institute of Technology, Cambridge, Mass. (1959).
7. C. E. Moore, Atomic Energy Levels, *Natl. Bur. Standards Circ.* 467 (1958).
8. R. T. Sharp, *Am. J. Phys.*, **28**, 116 (1960).
9. B. R. Judd, *Proc. Roy. Soc.* (*London*), **A228**, 120 (1955).
10. K. W. H. Stevens, *Proc. Phys. Soc.* (*London*), **A65**, 209 (1952).
11. H. A. Bethe, *Ann. Physik*, **3**, 133 (1929).
12. B. Bleaney and K. W. H. Stevens, *Repts. Progr. Phys.*, **16**, 108 (1953). A more extensive review of crystal field theory, particularly for ions with unfilled d shells, has been given by J. S. Griffith, "The Theory of Transition-metal Ions," Cambridge University Press, New York (1961).
13. J. S. Lomont, "Applications of Finite Groups," Academic Press Inc., New York (1959).
14. E. P. Wigner, "Gruppentheorie," F. Vieweg und Sohn, Brunswick, Germany (1931). An expanded English translation has been prepared by J. J. Griffin, "Group Theory," Academic Press Inc., New York (1959). A formal mathematical account of the elementary theory of finite groups has been given by A. Speiser, "Die Theorie der Gruppen von Endlicher Ordnung," Springer-Verlag OHG, Berlin (1927).
15. H. Weyl, "Gruppentheorie und Quantenmechanik, S. Hirzel Verlag, Leipzig (1931); translated by H. P. Robertson, "The Theory of Groups and Quantum Mechanics," reprinted by Dover Publications, New York.
16. E. Feenberg and G. E. Pake, "Notes on the Quantum Theory of Angular Momentum," Addison-Wesley Publishing Company, Reading, Mass. (1953).
17. C. Eckart, *Revs. Modern Phys.*, **2**, 305 (1930).

18. G. Racah, *Phys. Rev.*, **62**, 438 (1942). For a description of the theory of tensor operators, see also U. Fano and G. Racah, "Irreducible Tensorial Sets," Academic Press Inc., New York (1959); M. E. Rose, "Elementary Theory of Angular Momentum," John Wiley & Sons, Inc., New York (1957); A. R. Edmonds, Ref. 4; and J. Schwinger, "On Angular Momentum," U.S. Atomic Energy Commission, NYO-3071 (1952).
19. N. F. Ramsey, "Molecular Beams," Oxford University Press, New York (1956).
20. G. F. Koster and H. Statz, *Phys. Rev.*, **113**, 445 (1959).
21. B. Bleaney, *Proc. Phys. Soc.* (*London*), **73**, 937, 939 (1959).
22. Baker, Bleaney, and Hayes, *Proc. Roy. Soc.* (*London*), **A247**, 141 (1958).
23. G. L. Goodman and M. Fred, *J. Chem. Phys.*, **30**, 849 (1959).
24. J. C. Eisenstein and M. H. L. Pryce, *Proc. Roy. Soc.* (*London*), **A255**, 181 (1960); C. A. Hutchison and B. Weinstock, *J. Chem. Phys.*, **32**, 56 (1960); Hutchison, Tsang, and Weinstock, *J. Chem. Phys.*, **37**, 555 (1962).
25. J. D. Axe, The Electronic Structure of Octahedrally Co-ordinated Protactinium (IV), *Univ. Calif. Radiation Lab. Rept.* UCRL-9293 (1960); see also Axe, Stapleton, and Jeffries, *Phys. Rev.*, **121**, 1630 (1961).
26. M. Dvir and W. Low, *Proc. Phys. Soc.* (*London*), **75**, 136 (1960).
27. Bleaney, Llewellyn, Pryce, and Hall, *Phil. Mag.*, **45**, 991, 992 (1954).
28. J. C. Eisenstein and M. H. L. Pryce, *Proc. Roy. Soc.* (*London*), **A229**, 20 (1955).
29. T. Regge, *Nuovo cimento*, **11**, 116 (1959).
30. H. A. Jahn and K. M. Howell, *Proc. Cambridge Phil. Soc.*, **55**, 338 (1959).
31. D. König, "Theorie der Endlichen und Unendlichen Graphen," reprinted, Chelsea Publications, New York (1950). See also A. Sainte-Laguë, Les Réseaux, *Mém. sci. math.* (*Paris*), **18** (1926).
32. J. P. Elliott and A. M. Lane, *Handbuch der Physik*, **39**, 393 (1957).
33. L. C. Biedenharn, *J. Math. and Phys.*, **31**, 287 (1953).
34. J. P. Elliott, *Proc. Roy. Soc.* (*London*), **A218**, 345 (1953).
35. H. A. Jahn and J. Hope, *Phys. Rev.*, **93**, 318 (1954).
36. R. J. Ord-Smith, *Phys. Rev.*, **94**, 1227 (1954).
37. J. P. Elliott and B. H. Flowers, *Proc. Roy. Soc.* (*London*), **A229**, 545 (1955).
38. F. R. Innes and C. W. Ufford, *Phys. Rev.*, **111**, 194 (1958).
39. Arima, Horie, and Tanabe, *Progr. Theoret. Phys. Japan*, **11**, 143 (1954).
40. F. H. Spedding, *Phys. Rev.*, **58**, 255 (1940).
41. Sayre, Sancier, and Freed, *J. Chem. Phys.*, **23**, 2060 (1955); G. H. Dieke and R. Sarup, *J. Chem. Phys.*, **29**, 741 (1958); J. S. Margolis, *J. Chem. Phys.*, **35**, 1367 (1961).
42. A. M. Hellwege and K. H. Hellwege, *Z. Physik*, **130**, 549 (1951); *ibid.*, **135**, 92 (1953).
43. P. F. A. Klinkenberg, *Physica*, **16**, 618 (1950).
44. G. Racah, *Physica*, **16**, 651 (1950).
45. J. G. Conway, *J. Chem. Phys.*, **31**, 1002 (1959).
46. Satten, Young, and Gruen, *J. Chem. Phys.*, **33**, 1140 (1960).
47. J. C. Eisenstein and M. H. L. Pryce, *Proc. Roy. Soc.* (*London*), **A238**, 31 (1956).
48. R. E. Trees, *Phys. Rev.*, **92**, 308 (1953).
49. C. Schwartz, *Phys. Rev.*, **97**, 380 (1955).
50. J. A. Gaunt, *Trans. Roy. Soc.* (*London*), **A228**, 151 (1929).
51. W. A. Runciman and B. G. Wybourne, *J. Chem. Phys.*, **31**, 1149 (1959).

52. E. Fermi, *Z. Physik*, **60**, 320 (1930).
53. F. R. Innes, *Phys. Rev.*, **91**, 31 (1953). The magnetic orbit-orbit interaction has been thrown into tensor-operator form by S. Yanagawa, *Progr. Theoret. Phys. Japan*, **13**, 559 (1955).
54. E. C. Ridley, *Proc. Cambridge Phil. Soc.*, **56**, 41 (1960).
55. G. H. Shortley and B. Fried, *Phys. Rev.*, **54**, 739 (1938).
56. Coles, Orton, and Owen, *Phys. Rev. Letters*, **4**, 116 (1960).
57. H. H. Marvin. *Phys. Rev.*, **71**, 102 (1947). The tables of matrix elements given in this reference for configurations of inequivalent electrons are in error. Corrected and expanded tables have been given by Jucys, Dagys, Vizbaraite, and Zvironaite, *Trudy Akad. Nauk. Litovsk S. S. R.*, B3(**26**), 53 (1961).
58. G. Racah, Group Theory and Spectroscopy, mimeographed notes, Princeton (1951). These notes are available as a CERN (Geneva) reprint.
59. S. Lie and G. Scheffers, "Vorlesungen über continuierliche Gruppen," Teubner Verlagsgesellschaft, Leipzig (1893). For more recent expositions, see L. P. Eisenhart, "Continuous Groups of Transformations," reprinted by Dover Publications, New York (1961); H. Boerner, "Darstellungen von Gruppen," Springer-Verlag OHG, Berlin (1955); M. Hamermesh, "Group Theory and its Application to Physical Problems," Addison-Wesley Publishing Company, Reading, Mass. (1962). An account of the theory of continuous groups, with reference to possible symmetries of the strong interactions between elementary particles, has been given by Behrends, Dreitlein, Fronsdal, and Lee, *Revs. Modern Phys.*, **34**, 1 (1962).
60. H. Goldstein, "Classical Mechanics," Addison-Wesley Publishing Company, Reading, Mass. (1959).
61. H. Weyl, "The Classical Groups," Princeton University Press, Princeton, N. J. (1946).
62. H. Weyl, *Math. Z.*, **23**, 271 (1925); *ibid.*, **24**, 328, 377 (1925).
63. E. Cartan, Sur la structure des groupes de transformations finis et continus, thesis, Nony, Paris (1894) (reprinted 1933).
64. J. A. Schouten, Vorlesung über die Theorie des Halbeinfachen Kontinuierlichen Gruppen, Leiden (1926–1927); mimeographed notes.
65. B. L. van der Waerden, *Math. Z.*, **37**, 446 (1933).
66. D. E. Rutherford, "Substitutional Analysis," Edinburgh University Press, Edinburgh (1948).
67. T. Yamanouchi, *Proc. Phys.-Math. Soc. Japan*, 3d ser., **19**, 436 (1937).
68. H. Casimir, *Proc. Koninkl Akad. Amsterdam*, **34**, 844 (1931).
69. G. Racah, *Phys. Rev.*, **76**, 1352 (1949).
70. D. E. Littlewood, "The Theory of Group Characters," Oxford University Press, New York (1950).
71. F. D. Murnaghan, "The Theory of Group Representations," Johns Hopkins Press, Baltimore (1938).
72. R. F. Curl and J. E. Kilpatrick, *Am. J. Phys.*, **28**, 357 (1960).
73. H. A. Jahn, *Proc. Roy. Soc. (London)*, **A201**, 516 (1950).
74. J. P. Elliott, *Proc. Roy. Soc. (London)*, **A245**, 128 (1958).
75. B. H. Flowers, *Proc. Roy. Soc. (London)*, **A212**, 248 (1952).
76. G. Racah, *Phys. Rev.*, **63**, 367 (1943).
77. O. Laporte and J. R. Platt, *Phys. Rev.*, **61**, 305 (1942).
78. E. W. Hobson, "The Theory of Spherical and Ellipsoidal Harmonics," Cambridge University Press, New York (1931).
79. R. F. Bacher and S. Goudsmit, *Phys. Rev.*, **46**, 948 (1934).

80. P. J. Redmond, *Proc. Roy. Soc.* (*London*), **A222,** 84 (1954).
81. H. A. Jahn, *Proc. Roy. Soc.* (*London*), **A205,** 192 (1950).
82. Elliott, Judd, and Runciman, *Proc. Roy. Soc.* (*London*), **A240,** 509 (1957).
83. W. Albertson, *Phys. Rev.*, **52,** 644 (1937).
84. E. V. Sayre and S. Freed, *J. Chem. Phys.*, **24,** 1213 (1956).
85. Hubbs, Marrus, Nierenberg, and Worcester, *Phys. Rev.*, **109,** 390 (1958).
86. L. Bovey and S. Gerstenkorn, *J. Opt. Soc. Am.*, **51,** 522 (1961).
87. J. B. Gruber, *J. Chem. Phys.*, **35,** 2186 (1961).
88. B. R. Judd, *Proc. Roy. Soc.* (*London*), **A250,** 562 (1959).
89. A. G. McLellan, *Proc. Phys. Soc.* (*London*), **76,** 419 (1960).
90. B. R. Judd and R. Loudon, *Proc. Roy. Soc.* (*London*), **A251,** 127 (1959).
91. R. A. Satten, *J. Chem. Phys.*, **21,** 637 (1953).
92. G. H. Dieke and L. Heroux, *Phys. Rev.*, **103,** 1227 (1956).
93. B. G. Wybourne, *J. Chem. Phys.*, **32,** 639 (1960).
94. E. F. Reilly, *Phys. Rev.*, **91,** 876 (1953).
95. B. G. Wybourne, *J. Chem. Phys.*, **35,** 340 (1961); *ibid.*, **37,** 450 (1962).
96. W. A. Runciman, *J. Chem. Phys.*, **36,** 1481 (1962).
97. J. B. Gruber and J. G. Conway, *J. Chem. Phys.*, **34,** 632 (1961).
98. K. H. Hellwege and H. G. Kahle, *Z. Physik.*, **129,** 62 (1951).
99. K. H. Hellwege and H. G. Kahle, *Z. Physik.*, **129,** 85 (1951).
100. B. R. Judd, *Proc. Phys. Soc.* (*London*), **A69,** 157 (1956).
101. Hellwege, Johnsen, Kahle, and Schaack, *Z. Physik.*, **148,** 112 (1957).
102. Bleaney, Llewellyn, Pryce, and Hall, *Phil. Mag.*, **45,** 773 (1954).
103. H. Lämmermann and J. G. Conway, The Absorption Spectrum of Pu^{3+} in Lanthanum Trichloride and Lanthanum Ethylsulfate, *Univ. Calif. Radiation Lab. Rept.* UCRL-10257 (1962).
104. B. R. Judd, *J. Math. Phys.*, **3,** 557 (1962).

INDEX

Milton Keynes UK
Ingram Content Group UK Ltd.
UKHW021808260124
436770UK00006B/507

9 780691 604275